D0942231

ALLE ZEIT WACH
1842

Flora Brasiliensis, tabula IX. *'Arbores ante Christum enatae'.* In October 1819 Von Martius (see p. 15), with his draughtsman Spix and a number of Indians, visited this forest near Topinambarana, south of the Amazon. He calculated that these trees dated from before Christ. Their girth is exceptional in all respects; their name is unknown. Published 1841

Marius Jacobs

The Tropical Rain Forest

A First Encounter

Edited by
Remke Kruk et al.

With a Chapter on Tropical America by
Roelof A. A. Oldeman

With a Foreword by Vernon H. Heywood

With 157 Figures and 8 Plates

Springer-Verlag
Berlin Heidelberg New York
London Paris Tokyo

Marius Jacobs†

Remke Kruk
Witte Singel 52, NL-2311 BL Leiden

Roelof A. A. Oldeman
Generaal Foulkesweg 76, NL 6703 BX Wageningen

Title of the Dutch edition:
M. Jacobs, Het Tropisch Regenwoud, een eerste kennismaking.

ISBN 3-540-17996-8 Springer-Verlag Berlin Heidelberg New York
ISBN 0-387-17996-8 Springer-Verlag New York Berlin Heidelberg

Library of Congress Cataloging-in-Publication Data. Jacobs, Marius, 1929–1983. The tropical rain forest. Translation of Het Tropisch Regenwoud, een eerste kennismaking. Bibliography: p. 273 Includes index. 1. Rain forest ecology. 2. Rain forests. I. Oldeman, Roelof A. A., 1937- . II. Kruk, Remke. III. Title. QH541.5.R27J3313 1987 574.5'2642 87-20653

Printed in Germany

Typesetting: Overseas Typographers, Inc., Makati, Philippines.
Printing: Druckhaus Beltz, Hemsbach/Bergstr.
Binding: J. Schäffer GmbH & Co. KG., Grünstadt
2131/3130-543210

Foreword

In recent years, tropical forests have received more attention and have been the subject of greater environmental concern than any other kind of vegetation. There is an increasing public awareness of the importance of these forests, not only as a diminishing source of countless products used by mankind, nor for their effects on soil stabilization and climate, but as unrivalled sources of what today we call biodiversity. Threats to the continued existence of the forests represent threats to tens of thousands of species of organisms, both plants and animals.

It is all the more surprising, therefore, that there have been no major scientific accounts published in recent years since the classic handbook by Paul W. Richards, *The Tropical Rain Forest* in 1952. Some excellent popular accounts of tropical rain forests have been published including Paul Richard's *The Life of the Jungle,* and Catherine Caulfield's *In the Rainforest* and *Jungles,* edited by Edward Ayensu. There have been numerous, often conflicting, assessments of the rate of conversion of tropical forests to other uses and explanations of the underlying causes, and in 1978 UNESCO/UNEP/FAO published a massive report, *The Tropical Rain Forest,* which, although full of useful information, is highly selective and does not fully survey the enormous diversity of the forests.

So there is still a need for a scientifically sound text which could be used by students, conservationists and other biologists who want to obtain a balanced, well-written assessment of the problems and issues involved in this unique ecosystem.

When Marius Jacobs wrote the Dutch version of this book, it was widely acclaimed. Recently, it has been given the Kluwer Award in the Netherlands, for a work which makes an area of science more accessible to a wider public. Not surprisingly, there have been demands for the translation of the book into other languages and I am delighted that this has now been arranged through the devotion of Dr. Remke Kruk and publication made possible by Springer-Verlag.

Marius Jacobs was a thoughtful, dedicated tropical botanist. He had a deep concern for the conservation of genetic resources and was a champion of the forest's right to survival. This book represents a personal statement of his views and beliefs about the significance of the tropical forest for science and mankind. Although several colleagues have generously contributed additional material for this English edition to make it more geographically comprehensive, this remains Marius Jacobs' book and is a fitting memorial to him. It will surely inspire students to follow his example and this would have given him great pleasure.

Vernon H. Heywood

IUCN Conservation
Monitoring Centre

Kew 1987

Note of the Editor

The history of this book is a complicated one. The original Dutch version appeared in 1981, and when Marius Jacobs – whose life I shared for many years – suddenly died in April 1983, he was about halfway with the preparation of an updated and slightly enlarged English version.

Many people expressed their regrets that the book would now remain unpublished, and several of Marius Jacobs' colleagues offered their help should I decide to try and get the book finished along the lines already set out by the author. Accordingly, the translation and – when necessary – revision of the remaining chapters were undertaken by such colleagues as were most expert in the subject concerned. Chapters 13 and 14 were translated unchanged. In one case (Tropical America), a new chapter was written in accordance with the author's original plan.

As a result, Chapter 3 of the present English version (Climate) was translated by Dr. L.R. Oldeman; Chapter 4 (Soils and Cycles) by Dr. P.M. Driessen; Chapter 5 (Trees) by Dr. E.F. de Vogel; Chapter 9 (Tropical America) was newly written by Prof. R.A.A. Oldeman; Chapter 10 (Malesia) was translated by Dr. M.J. van Balgooy; Chapter 11 (Tropical Africa) was rewritten, on the basis of the original chapter, by Prof. H.C.D. de Wit; Chapter 12 (Relationships of Plants and Animals) was translated by Dr. H.D. Rijksen; Chapters 13 (Evolution) and 14 (How Species are Formed) by Dr. Remke Kruk; Chapter 15 (At the Fringes of the Rain Forest) by Dr. Brigitta E.E. de Wilde-Duyfjes; and Chapter 18 (Protection) by Dr. H.P. Nooteboom. Advice on special points was given by the late Prof. C.G.G.J. van Steenis, Dr. A.J.M. Leeuwenberg and Dr. F.J. Breteler.

The English translation was corrected by Mrs. Wendy Blower (author's name Wendy Veevers-Carter), who had already worked on the book with Marius Jacobs, and continued to give her invaluable help after he passed away. Another driving force in the project was Mrs. Erna Prillwitz, Marius Jacobs' former secretary for conservation matters. Without her consistent help and encouragement, the project would never have been finished.

The final time-consuming job of preparing the illustrations, checking biological and bibliographical details, weeding out inconsistencies and preparing the – much enlarged – bibliography was undertaken by Dr. P.H. Hovenkamp. The Index was compiled by Dr. Annie Huisman-van Bergen.

The cooperative attitude of the Leiden Rijksherbarium, where Marius Jacobs was a staff member, has been an invaluable help throughout the project; moreover, its director, Prof. C. Kalkman, was never appealed to in vain for advice concerning the many obstacles which cropped up during the preparation of this book.

I want to thank all those people – any many others not mentioned by name – for their friendship, their help and their readiness to give their time to this book despite their own overburdening quantities of work.

Marius Jacobs dedicated the original version of this book to his future grandchildren. They may also be mine; and may the majesty of the tropical rain forest that is described in this book indeed be there for them to admire.

Leiden, December 1987 Remke Kruk

Preface

My first experience of a tropical forest was a sensation of mass, and of tranquil, timeless indifference. There I stood – in the Cibodas forest of West Java – amidst something at least as important, as evolved and in its way as powerful as a human being. Even now, I breathe more freely when, after a time in the forest, I come again into the open.

Speaking frankly, I don't believe that many people feel happy in a rain forest. It just stands there. There are strange noises. From the ground, monotony prevails. There are few visible events. A butterfly visits a flower, a leaf comes down, a big bird flies over. If such happenings are to have significance, it can only be over centuries, and thus gradually it becomes clear that the scope of time in a rain forest differs very much from our own. To appreciate this fully, long observation is required, and much thought. The forest then reveals considerable orderliness and systems. The more one looks, the more one sees. Trunks, branches, leaves, seedlings as well as animals are effectively distributed in the space available. Everything that happens can, after patient observation, be related to other events and assigned a significant role in the functioning of the forest as a whole, and thus in the continuing existence of each species.

But what meaning has the rain forest itself? One could as well ask about the meaning of mankind's existence on earth. There is, after all, some resemblance between the forest and mankind. Both are summits of creation. Fine comparisons can be made, for instance, between forests and human cities, the rain forests finding their best counterparts in old cities with rich cultural histories. Both are highly differentiated super-organisms full of singular creatures. Both are the result of long development, on which work is still going on, if at a different speed: for 1 year in a city's life must be the equivalent of some 10,000 years in the life of a rain forest.

Does this mean that beside Amsterdam, London, Paris and Rome we must place Andulau, Cibodas, Pasoh or El Verde? The idea is irritating. Few human beings like the notion that marvels exist which owe nothing whatever to man. Yet the rain forest is such a marvel, and everyone has heard something about it and felt something about it, however vague. Some possess certain knowledge of it, depending on interest, way of life and intellectual grasp. Let me give two quotations here as examples. First, one by Corner (1963, p. 1000) on Malaya: "... I measured my insignificance against the quiet majesty of the trees. All botanists should be humble. From trampling weeds and cutting lawns they should go where they are lost in the immense structure of the forest. It is built in surpassing beauty without any of the necessities of human endeavour; no muscle or machine, no sense-organ or instrument, no thought or blue-print has hoisted it up. It has grown by plant-nature to a stature and complexity exceeding any presentiment [sic] that can be gathered from books, and it is one of the most baffling problems of biology". And one by Lam (1927, 1945, p. 174) on New Guinea: "For some impressions are worthy of remembrance and, in spite of some disagreeable ones, make us always desirous of returning again to this immense and grand country. Magnificent with respect to its dimensions of forest and river, even when one looks upon them, grand in its virginity and its proud silence at the intrusion of people who will at some time conquer the country, it will persist until man also shall have vanished and his influence shall have been lost forever. Many memories bind us to each small place that we have visited: this is stronger still in the mountains, where so many more influences cooperate to that end. The Mamberamo navigators will ever see before them the broad muddy river with its numerous curves, the still, high forest walls, the translucent morning mists which hang without motion in the treetops until 8 or 9 a.m. They will feel the heat vibrate above the river banks at mid-day and again sniff the heavy damp odor of the forest, evidence of incessant progressive decay of organic material. Again the melancholy cooing of doves will be heard at regular intervals high up in the treetops, frequently also the noisy cry of the hornbills, which are hardly seen amongst the mass of leaves and branches until, with the harsh flapping of their wings, they fly away. Then we see again their dark silhouettes before us as they depart, frequently in pairs, over the river, and the screaming cockatoos with their blunt heads and swift wing-beat, which in the evening at the fall of twilight come in large flocks to sleep in the trees. Some will bear in memory the distant cry of the cassowary. Sometimes too it is the distant howl of a Papuan dog that comes hovering over the water by the river".

The second quotation uses the future tense. It conveys the belief, the hope that this source of experience will remain. Will that be possible? On this question, human beings have to make decisions, and stand by those decisions. To do so, they must have knowledge, and understanding, of exactly what is at stake. In this book, I have tried to explain, if one can, what a rain forest is, and to clarify the relationship between forests and humanity. My hope is that the result will be an increasing love and respect for the tropical rain forests of the world, for then there will be more of them left standing.

Leiden, 1982 Marius Jacobs

Contents

1 A Matter of Public Awareness

Tropical Rain Forest Characteristics

The tropical rain forest is such a diverse and complex system that a precise definition of it is hard to give. But its characteristics are clear. The tropical rain forest consists mainly of evergreen woody vegetation, forming a thick *canopy*, so dense that it is impossible to see the ground from above the trees. Consequently, little plant life grows at ground level. This clear area is several metres high, and is easy to walk through. No word for it exists; but it might be called the *hall of the forest*.

The canopy is usually 30 to 50 m from the ground; only the crowns of the very highest trees, the *emergents*, protrude partly or entirely above it. As they are 45 to 70 m high and stand rather isolated, they give the forest, as seen from a distance, a peculiar, uneven aspect (Fig. 1.1).

What also strikes the eye, in viewing the forest canopy from above, are the differences in crown shape, density and colour shades. Among the many tones of green, some crowns seem very pale, others, bronze; an occasional one is pinkish. It is obvious that the forest is thoroughly mixed in composition, even though the overall impression is a solemn dark green. Indeed, the forests of western Malesia are, in general, dominated by the Dipterocarpaceae, a family of many species, no one of which is regionally dominant, and nowhere does any one species account for more than 10–15% of the trees.

The trunks of the mature trees occur in any diameter up to 3 m, and occasionally more, although most trees are slender in proportion to their height, and bear comparatively small crowns. Only the emergent crowns can spread. Among them, a crown diameter of 22 m is not at all rare.

Many trees are buttressed at the base, the so-called buttresses being several inches thick and several square metres in area. In other trees, the trunk is divided into a collection of thin individual stems above ground level. Trees with prop roots are common, especially in swamp forests, and these roots take many bizarre forms. In addition, a number of thick-stemmed, poorly-branched, large-leaved plants like palms are always present, which often have spiny stems or leaves.

Lianas are also a conspicuous feature: these are woody climbers with flexible stems sometimes a foot or more in diameter, often forming big, disorderly curls.

Another characteristic is the phenomenon of *cauliflory*, in which both flowers and fruits sprout from the trunks or branches of trees and lianas. There is, however, no abundance of flowers or fruits except on single, isolated plants.

Trees and lianas – and their seedlings – dominate the scene: half or more of the plant species are woody. Where herbs do occur (orchids and ferns being conspicuous among them), they occur mainly high up in the trees, as *epiphytes* on the branches. Growth on the forest floor itself is sparse, as long as the canopy remains closed.

Altogether, characteristic features combine to create a structure of considerable complexity, yet all its elements are distributed within the available space in quite an orderly

manner, mostly vertically: it is often possible to discern several layers or storeys in the canopy (Fig. 1.2). A complex structure and a highly mixed composition characterize the tropical rain forests of all four continental regions, notwithstanding the fact that the species they contain are usually – and the genera and families often – completely different.

Fig. 1.1 Aerial photograph showing an area of ca. 1.5 × 1 km rain forest in North Sumatra: the study area Ketambe, where orang-utans occur. The river Alas runs from left to right, into which the Ketambe disembogues. On the far (eastern) bank a road is visible, and more to the *right*, a ford leading to the field station which is not visible (Photograph Rijksen)

Geographical Occurrence of Rain Forests

A map of the world's main vegetation types can be found in any good atlas. The tropical types shown will be: (1) rain forest: evergreen; (2) seasonal forest: deciduous; (3) savanna: trees among grass; (4) steppe: only grass; (5) (semi-)deserts: grass in patches on bare soil; (6) swamps; (7) coastal regions, under influence of the sea; (8) high mountains: near and above the timber line.

This does not mean that the tropical rain forest fills all the areas shown as type 1. It may have been destroyed or damaged; the colour merely indicates that conditions are suitable and that a tropical rain forest might have once grown there, and still may exist. The necessary conditions for its growth are:

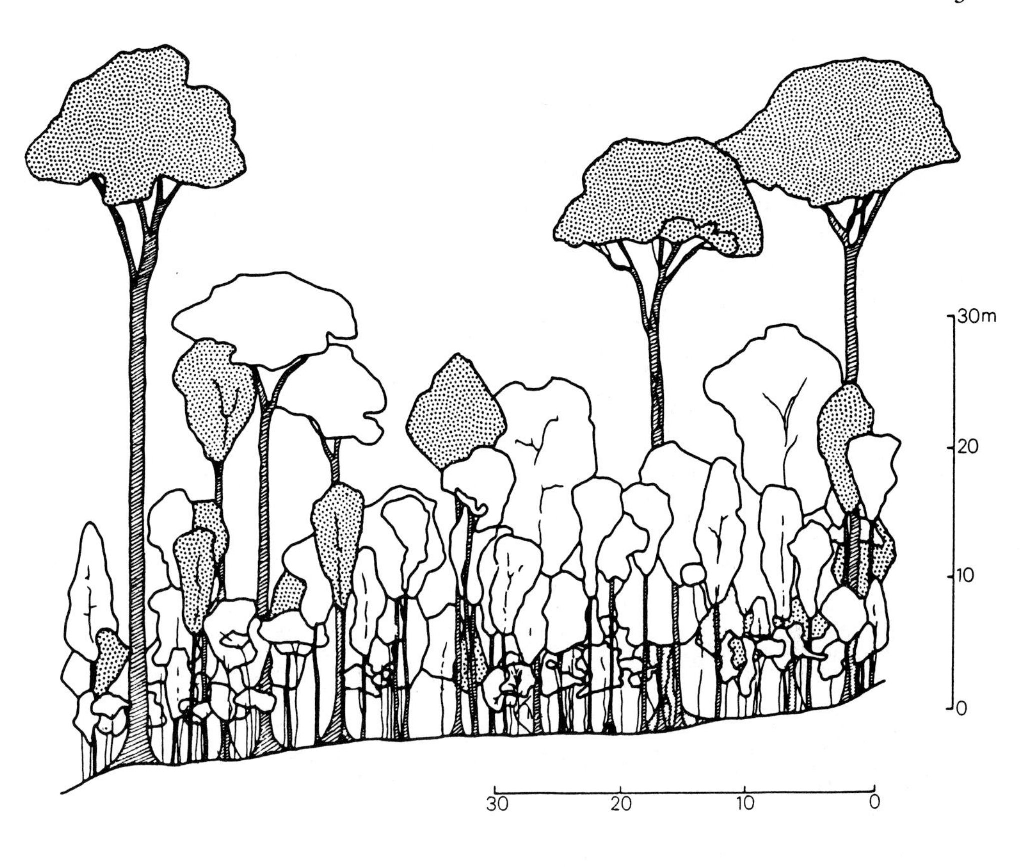

Fig. 1.2 Profile diagram of rich rain forest in Brunei, Borneo. Only trees of 5 m or more are shown, lianas are omitted. The canopy is closed, except for the emergents. Under the canopy is the 'hall of the forest', where walking is easy. Many trees have buttresses at their base. *Shaded*: Dipterocarpaceae (7 species). Total number of tree species: 45; list given in Ashton (1964) and Whitmore (1975)

1. A 'tropical' temperature, relatively even throughout the year, such as is found wherever between 23°30′ N and 23°30′ S latitudes the land is below 1000-m altitude.
2. Rainfall must be high: at least 1800 mm a year and evenly distributed throughout the year. Wherever such an 'ever-wet tropical' climate prevails, this indicates that the earth was once covered with this type of forest, with the exception of those lands into which rain forest plants were for some reason unable to spread. But it is not easy to outline on a map the precise extent of former rain forests; it is equally hard to delineate precise boundaries between rain forests and those which, under the influence of a drier season, shed part of their leaves periodically. It is also hard to show on any small scale map exactly how much of the world's rain forest has been destroyed.

The main range of the tropical rain forest is well known, however. Listed below are the three major rain forest regions in order of magnitude; each will be more extensively discussed later.

1. The *American* Rain Forest Region (Fig. 1.3). On the South American continent, there are three rain forest subregions. By far the biggest is Amazonia; because it is so vast, it must be considered as still largely intact. A smaller region lies west of the Andes and north of the equator, extending, intermittently, to Mexico. This region has already suffered badly. The smallest region is a narrow strip along the Atlantic coast of Brazil between 14° and 21° S; only a few remnants of it are still in existence. It is uncertain if true rain forests ever occurred on the larger Caribbean islands; those still present hardly deserve the name.

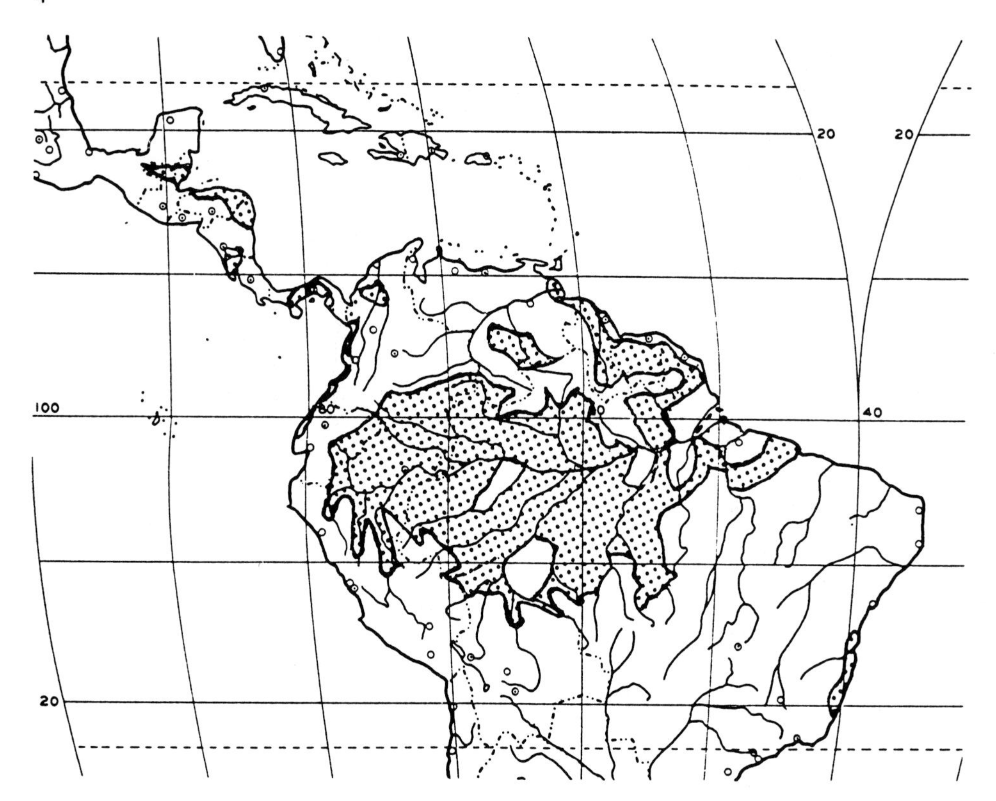

Fig. 1.3 The American rain forest area, below approx. 500 m altitude. Central America after Gentry (1977), see also Schmithüsen (1976). South America after Hueck and Seibert (1972), not including the 'development areas' in Amazonia as given in a map by Prance (1977a, p. 201). Near cities and roads the forest has disappeared or is doomed to disappear, on many other sites it is damaged. Along the rivers and on the borders with drier areas, other types of forest occur

2. The *Malesian* Rain Forest Region (Fig. 1.4), often called the Southeast Asian region, coincides almost entirely with the former Malaysian Archipelago (New Guinea included), to which the Malay Peninsula belongs botanically; the true rain forest is almost absent from continental Asia, except in Malaya, southernmost Thailand and southwest Cambodia.

In this region, two closely related nuclei can be distinguished. First, Borneo-Malaya-Sumatra and The Philippines, with outliers in the Andaman Islands, Ceylon and the poorer forests of southwest India (Fig. 1.5). Second, New Guinea, including the less species-rich forests on the islands to its west and east, with one outlier along the eastern coast of tropical Australia, where some pockets of rain forest survive. In the Malesian region generally, the larger the island, the richer the species count. But Malesia as a whole is mountainous, hence on most islands true rain forest can only occupy relatively narrow strips. Forests in Malaya, Sumatra, Borneo and The Philippines have been heavily depleted in recent times; New Guinea forests are still largely intact.

3. The *African* Rain Forest Region (Fig. 1.6) consists of four regions, all partly destroyed, along the Atlantic coast between ca. 10° N and 5° S, and in the Congo Basin stretching east to the mountains. Outlier pockets in East Africa may be regarded as more or less true rain forests, but not the forests of Madagascar (or what remains of them).

Peculiarities of the Rain Forest

In addition to their physical and geographical constraints, tropical rain forests are peculiar ecosystems, for the following reasons:

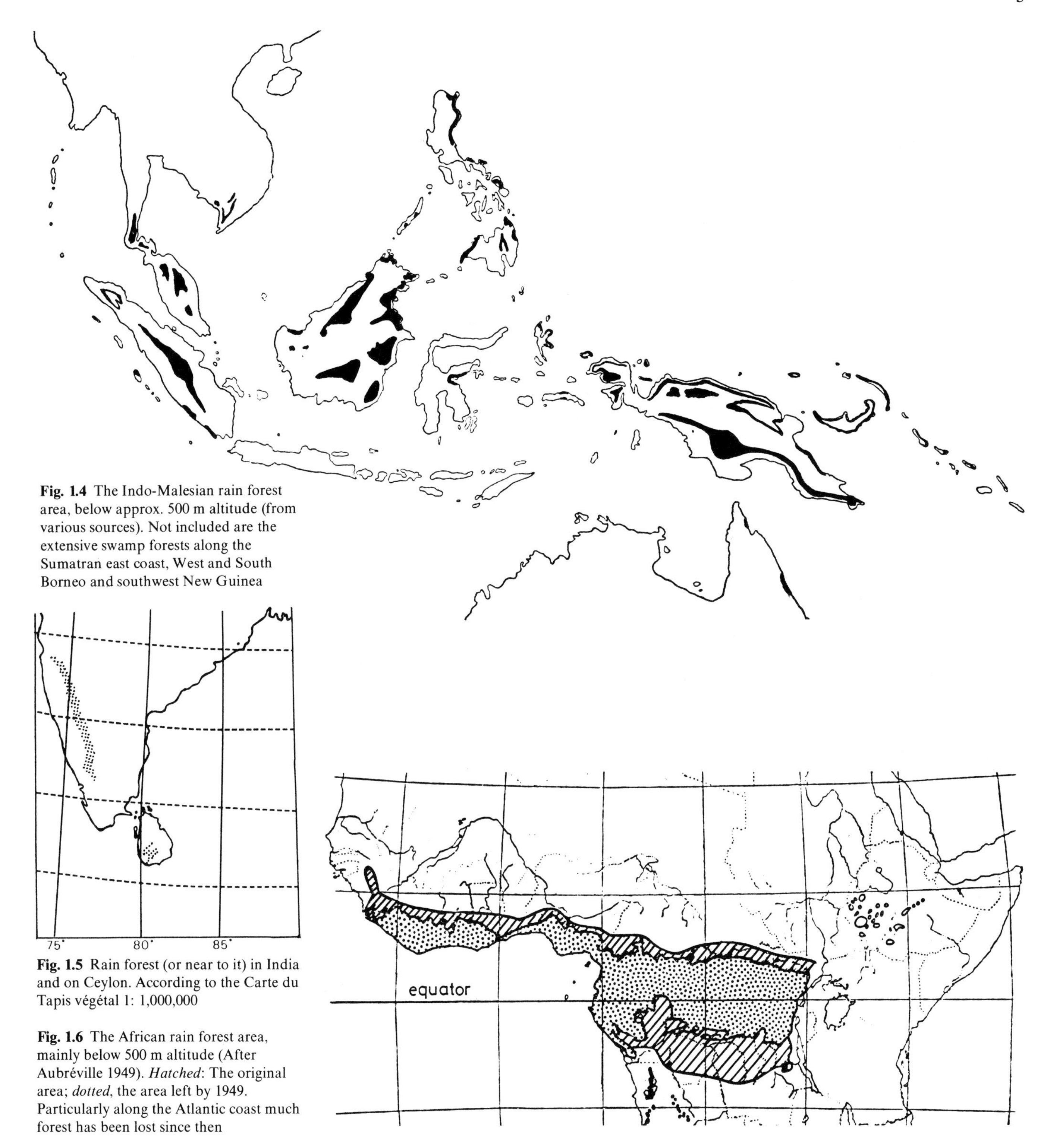

Fig. 1.4 The Indo-Malesian rain forest area, below approx. 500 m altitude (from various sources). Not included are the extensive swamp forests along the Sumatran east coast, West and South Borneo and southwest New Guinea

Fig. 1.5 Rain forest (or near to it) in India and on Ceylon. According to the Carte du Tapis végétal 1: 1,000,000

Fig. 1.6 The African rain forest area, mainly below 500 m altitude (After Aubréville 1949). *Hatched*: The original area; *dotted*, the area left by 1949. Particularly along the Atlantic coast much forest has been lost since then

1. They contain the largest numbers of species, of animals and plants, of any known ecosystem;
2. They occur generally on very poor soils;
3. Their potential for utilization is greatest in terms of quality, but smallest in terms of quantity;
4. They cannot easily be cropped in quotas, unlike some animal populations;
5. They contain huge capital assets in the form of timber which cannot be harvested without serious damage;
6. They cannot be 'managed' without the loss of large numbers of species;
7. They are unusually fragile, and, once damaged, do not recover, or recover too slowly for any human planning;
8. Nearly all of them are situated in 'developing' countries with shaky economies, and little or no power to implement laws on land use, if and where such laws exist;
9. They are now being logged rapidly by mechanical means since for a few decades machinery is available;
10. Yet they are so complex that even their main ecological features are poorly known even to professional decision-makers and misunderstandings about them abound.

Thus, the outlines have been sketched of the problem field: the tropical rain forest, which in this book we will explore.

Rain Forest Picture Books

Photographs of rain forests occur in many books and articles, but not many works succeed in conveying a fairly all-round impression of the subject. Three are mentioned here in order of their appearance.

Paul W. Richards, *The Life of the Jungle*, 232 p., illus. (1970, McGraw-Hill; New York), 26 cm.

After writing his classic 1952 handbook, *The Tropical Rain Forest*, Richards here describes in layman's language, but with remarkable profundity, the forests as "a world in harmony". All the essentials of structure, rhythms, plant-animal relations, circulation, succession and influence of man dealt with, while lightly touching on other relevant aspects. There are notes on reserves, endangered animals, exploration and insect collecting, as well as a glossary, a list of books and an index. The photographs are of good quality, most of them in colour; the diagrams are instructive. There is some emphasis on the New World. It is an expert introduction.

K.B. Sandved and M. Emsley, *Rain Forests and Cloud Forests*, 264 p., 229 illus. (1979, Abrams; New York), 36 cm.

The photographs (nearly all in colour) are the finest ever printed, showpieces in themselves. The text sets forth the wonders of the forests, often under romantic titles. Geological history, area, structure, energy and nutrients, epiphytes, animals of the forest floor and of the canopy, ants, flowering and pollination and nocturnal activities are discussed. Man's role is briefly explained. Some images from outside the forest have crept in, however, and not all features are well-connected or clarified. There is a bibliography and an index.

Edward S. Ayensu (ed.), *Jungles*, 200 p., many illus. (1980, Cape; London), 31 cm.

On each pair of pages, one aspect is discussed, 87 in all. There are a variety of figures, diagrams, colour photographs and some text, executed very tastefully. Each of the 18 chapters has good introductory material, and coverage is quite broad, although composition and evolution are under-emphasized, while too much emphasis is placed on animals. Man's influence is well explained; many forest products are described. There is an index, but recommended further reading is not given.

These books serve as popular, colourful introductions. In the next chapter, we will name some landmarks in the development of our scientific knowledge and views. Works of regional scope will be mentioned under Amazonia, Malesia/Australia and Africa.

Tropical Forest Inventories

As we remember from our glance at the atlas, there are various types of non-rain forest in the tropics. Since all forests are regarded as natural resources, and governments want to know their extent and potential, efforts have been made to explore, map and assess the various national forests. The resulting reports differ enormously in presentation, coverage, scale, thoroughness and vegetation classification, for example, botanists only began to distinguish clearly between peat swamp forest and freshwater swamp forest in the late 1950s.

Changes in vegetation cover will render data obsolete sooner or later. Yet, a good vegetation map is a priceless document, worth intensive study in combination with other maps, and vital for any form of field work or land-use planning. The general book by Küchler, *Vegetation Mapping* (1967) is a delight to read. And the same author's four-volume *World Bibliography of Vegetation Maps* (1965–1970) is a most useful tool in locating maps of rain forest regions as well. More recent information can probably be obtained from the *Institut de la Carte Internationale du Tapis Végétal*, Université Sabatier, Toulouse, France.

A forest inventory in the stricter sense often consists of descriptions and maps estimating amounts of harvestable timber. The Food and Agriculture Organization of the United Nations has compiled a draft list of the larger inventories made in the years 1955–1977 (FAO 1982).

FAO has also sponsored the compilation of inventories of tropical forests on a worldwide scale; one by Persson was published in 1974, another concise one by Sommer in 1976. Thereafter, FAO undertook a much more detailed study, *Tropical Forest Resources Assessment Project*, published in four massive volumes in 1981; two are on Africa, one is on tropical Asia, the other on Latin America. It was based on satellite images and local information, in the framework of the Global Environmental Monitoring System (GEMS). For each forest country there is a 'brief'. The main facts and conclusions are given in the 'regional synthesis' and the text is interspersed with clarifying remarks and documenting notes.

The sheer amount of factual information is stupefying. Forests are classified into Natural and Plantation, Hardwood and Softwood, Closed (crown cover more than 40%) and Open, consisting of trees or scrub, productive or not, intensively managed or not, in all sorts of combinations. Areas covered by each type are estimated in thousands of hectares (= tens of square kilometres) for 1980 and then for 1985. It has been hailed as 'the most authoritative estimate', and as 'the precursor of a whole generation of tropical forest surveys'. One might easily overlook one more fact: that this study, so complete in coverage and in detail the most advanced ever, is at the same time utterly obsolete.

These volumes remain a first-class feat in forestry fact compilation, however, and show a deep concern about worldwide deforestation, as well as excellent insight into its causes. On balance, the outlook for the future where hopeful is responsibly worded. Plantation forests make up an important portion of the study, with notes on their success, failure or planned extension, but nature reserves have not been forgotten. The long parade of figures, clarifications, trends and conclusions cannot conceal, however – nor does it want to – that this elaborate study is primarily concerned with a single goal: *the production of wood for industrial purposes.*

"The undisturbed forests predominate in Irian Jaya with a standing volume of 6335 million m^3" (p. 66 of the volume on tropical Asia) is one example. As a careful distinction has been made, throughout the study, between accessible and inaccessible forest, we may take it that this figure represents the exploitable stock of timber. The exploitable stock? Traditionally, the lands of New Guinea are owned by the native tribes. The terrain is steep in many parts: the erosion alone that would result from harvesting such an amount of timber is beyond imagination.

Nor does the study say much, if anything, about the value of the forest other than for timber: for instance, what about the forest products which support the indigenous way of life in that part of the world? Nor does the text examine the possible consequences of its numerical calculations. As Schumacher says in *Small is Beautiful*, the FAO approach, as it is expressed in this study, amounts to a view of nature as a mere quarry.

The real purpose of this monumental study is, therefore, ambiguous. While its declared long-term objective is "to assist the world community to formulate appropriate measures to avoid the potentially disastrous effects of the trends in the depletion and degradation of tropical forest cover" (p. 2 of the Asia volume), it estimates all growing stock in millions of cubic metres, the extraction of which is bound to bring about just such disastrous effects.

It is evident that the FAO estimates take no account of the true nature of the rain forest, but instead stem from the ancient mechanistic view of nature as a quantifiable collection of isolated things. Those who base exploitative programs on such data are bound to blunder immediately into environmental and social problems, and to add to these, problems which only tunnel vision could ignore. They were already well formulated when the FAO study was undertaken. We will enumerate them.

An Expression of Concern

In June 1978 a 'Strategy Conference' was held in Washington by the Department of State and the USAID, on Tropical Deforestation. In the Proceedings (*Tropical Deforestation 1978*) these were the principal conclusions:

1. Tropical forests are disappearing at a rapid and alarming rate. A lack of information exists about:

a) Deforestation rates and trends;
b) The nature and impact of forest removals at the local, regional and global levels;
c) The character and value of the species of trees, wildlife and other genetic materials being lost;
d) Whether and how tropical forests can be effectively and economically sustained.

Continuing devastation of forest resources in the face of such serious deficiencies in our knowledge of its long-term economic, ecological and environmental importance is both short-sighted and potentially tragic.

2. Although the cumulative national and worldwide economic, social and environmental *costs* of the loss of tropical forests are not well documented, they are already sizeable, and growing rapidly. Mankind's ability to meet basic human needs and quality-of-life objectives is critically dependent upon proper stewardship of tropical ecosystems. The consequences of forest mismanagement are:

a) The loss of sources for wood products and fuel wood;
b) The inability to maintain soil and water systems;
c) The disappearance of plant and animal genera and species;
d) The forfeiture of certain aesthetic and cultural values.

Global climatic changes associated with large-scale deforestation have often been postulated, but the exact role of forests in the carbon dioxide cycle is at present insufficiently known. Our inability to identify and quantify it and the preoccupation of many governments with more immediate matters, are limiting efforts to deal effectively with this aspect of deforestation.

3. Without greater recognition of the value of the forest resources and without improved stewardship, large areas of the world's tropical forests will be lost by the end of the century.

In the absence of better information about their nature and their value, and without significantly upgraded management performance as well as the intent to maintain and protect them, tropical forests must be regarded as *non-renewable* resources.

4. *Exponential population growth*, coupled with a lack of alternative economic development opportunities, is the basic cause of loss of tropical forest cover. Worldwide, this has meant the clearing of vast tracts of forest land for small-scale extensive agriculture, for large-scale intensive agriculture, for cattle grazing and ranching, to acquire fuel and building materials and to establish human settlements. The results of both shifting cultivation and commercial forestry are of lesser magnitude, even though the latter involves road building, which opens up forests to colonization by the land-hungry poor.

5. *Demand for fuel wood* by families in the semi-arid regions and for charcoal making has created critical economic and social problems for tens of millions of people, principally in Africa, Asia and the Caribbean region.

6. *Institutional ability* to address and control deforestation needs strengthening at all levels. In those countries most affected, forest management planners and practitioners need training and support, and international organizations and donor countries also need to expand their expertise in solving the often unfamiliar problems of tropical environments. Forestry in the tropics has, to date, been greatly neglected when compared to the collective level of effort that the international community has mobilized in other natural resource fields (water; minerals).

7. There is *little systematic monitoring* of tropical forests in progress. Consequently, the accuracy of current estimates of the rate of deforestation is extremely variable.

8. Because the deforestation situation requires urgent attention, *remedial action* will necessarily have to be taken on the basis of *incomplete information and data*. There is, however, enough knowledge of tropical forest management principles presently available to enable environmentally and politically committed countries to make significant inroads on their deforestation problems. It is, therefore, essential that all available information and guidance be presented to local decision-makers.

9. Major scientific and technological *challenges* remain. These include:

a) To establish and implement deforestation monitoring and assessment systems at the global, regional, national and local levels;
b) To identify and quantify the range of economic, social and environmental *costs* associated with deforestation, including possible climatic impacts; as well as the benefits to be expected from proper forest management. The latter would include economically important drugs, genetic materials and wildlife;
c) To improve the understanding of basic *physical, chemical, biological and ecological rain forest processes*, as a prerequisite to predicting the impact of large-scale clearing of vegetation and to making the best decisions on the best use of the land;
d) To develop *alternatives* to the present use of forests (e.g. fuel substitutes, agroforestry, firewood lots), as well as to improve techniques for reforestation and afforestation, including the use of faster-growing species.

10. *Responsibility for proper forest management*, and for the implementation of reforestation and other protective measures cannot be left solely in the hands of professional foresters, particularly since (as agents of an often distant central government) they are sometimes viewed as antagonists by rural people who feel unjustly prevented from utilizing an essential resource. *Local citizenry* must be intimately involved both in the planning and the implementation phases of forest and land management programs.

11. *International organizations and governments* have become increasingly aware of tropical forests since problems associated with forest disappearance more and more frequently have undercut or offset development gains made by programs in other sectors. Examples include:

a) Rapid siltation of reservoirs (as a result of erosion following deforestation upstream);
b) Reduction of the water supply for human and agricultural use;
c) Decrease of electrical power-generating capacity;
d) Intensified flooding;
e) Loss of badly needed wood products and firewood;
f) Loss of valuable plants and animals;
g) Intensification of basic social and economic problems in general (including the relocation or abandonment of new agricultural projects).

To date, however, the *collective effort* of these organizations and governments has been comparatively small and is clearly inadequate to meet the rapidly expanding need.

12. Countries and institutions which seek to provide assistance and guidance with respect to the improved management of tropical forests must be sensitive to the fact that almost all such forests exist in *developing countries* which legitimately exercise sovereignty over the resource.

Together, these points fairly well summarize the problems connected with deforestation. Most of them will be examined more closely in the course of this book. Discrepancies between problems and solutions in 17 countries as revealed in a report, *Forestry activities and deforestation problems in developing countries* (1980) will illustrate the above conclusions. First, however, we must learn more about the rate of deforestation.

The Speed of Forest Destruction

Destruction of natural forest in tropical countries was already a familiar phenomenon in the 1920s and 1930s – witness the forestry journals of those days. But many foresters have been saying for an equally long time that it was nothing to worry about. In the 1970s, however, (the Second Development Decade), even the non-professionals became alarmed about the dwindling of tropical forest, and about the unreliability of the data. A striking example concerns The Philippines. In 1974, Persson calculated an estimated forest cover of 51% of the land area. Shortly thereafter, satellite images revealed an *actual* cover of 38.2% at most.

To estimate the loss of tropical forest, differences in reliability notwithstanding, two studies were made. The first, by Lanly and Clement (1979), called *Present and future forest and plantation areas in the tropics*, was an offshoot of the FAO assessment study we just discussed. Although much more concise in scope, it gave estimates for the years 1980, 1985, 1990, 1995 and 2000. The figures for 1980 differ somewhat from those in the larger study for 1981 and are probably more accurate. More important is an understanding of the scope of Lanly and Clement's projections.

This scope was explained in their stencilled report, but not in its printed, widely circulated, but heavily abridged version. Lanly and Clement define a forest as a forest when the trees are 5 m high, and when the crown cover is 40%. This is a genuine catchall definition, and indeed: "forests temporarily unstocked are considered as 'forests' if they are under a process of natural or artificial regeneration and are not alienated for non-forestry purposes" (p. 3). According to the 40% criterion, the finest rain forests of western Malesia still just qualify as 'closed', when after logging they are virtually beyond repair. Species-poor secondary forest is pooled together with species-rich primary forest, and for the latter the important altitudinal limit of 300–500 m has been ignored. Forests on limestone and in swamps are, of course, included; mangroves which, like secondary forests, are renewable, are lumped together with dry land primary forests, which are not; all differences in composition and potential are statistically ignored.

Small wonder, then, that the report concludes: "The overall figure for the depleting of closed forests in the tropics (140 million ha in 25 years) appears less alarming than those generally quoted (like the one suggesting that 50,000 ha are cleared every day)" (p. 46). After the report outlines the concept which includes the most valuable rain forests in all (im)possible manners, it then suggests that there is plenty. What may be the result of such optimistic, yet ecologically untenable figures? Unrealistically high quota setting beyond false expectations, delaying of remedial measures and wrong planning will all conspire to increase pressure on the rain forests. Plants, animals and people in the tropics are bound to suffer if the figures of this report are uncritically accepted and used as the basis for 'development' policy.

The second report is by Norman Myers (1980) and was published by the National Academy of Sciences in Washington. In accordance with its title, *Conversion of Tropical Moist Forests*, it does not include the forests and savannas in regions with a strong dry season. Myers does include some rather seasonal forests, however, that should not be regarded as 'tropical rain forests' in the (fairly strict) sense adopted in the present book. This results in somewhat high estimates of forested land on his part, but this does not significantly extend the date the forests will be finished, given a certain rate of destruction. Myers has calculated this rate, worldwide, at 240,000 km^2 $year^{-1}$, or 65,750 ha day^{-1}, or 2740 ha h^{-1}, or 46 ha min^{-1}. This is in the same order of magnitude as an estimate given by the director of FAO at the 8th World Forestry Congress in Jakarta in 1978: 210,000 km^2 annually, or 40 ha every minute. In a personal letter dated 2 June 1982, Dr. Myers communicated his latest estimate: "that the rate of deforestation, in the sense of virtual elimination of forest cover, now amounts to 115,000 km^2 per year; and deforestation in the sense of gross disruption of forest ecosystems, with major loss of natural values, amount to another 85,000 km^2 per year". This represents a physical deforestation of 31,507 ha day^{-1}, and an ecological disruption of 23,300.

In his book Myers explains his method of calculation, then discusses all rain forest countries by region: 18 in tropical America, 15 in Asia-Malesia-Australia, 13 in Africa. He mentions areas with margins of accuracy, and discusses causes and rates by country. Four-fifths of the world's moist forests are concentrated in nine countries: Bolivia, Brazil (with one-third of the total), Colombia, Peru, Venezuela; Indonesia (with one-third) and Malaysia; Gabon and Zaïre.

In his cautious style, after weighing all the factors involved, he prophesies that at the current destruction rates, the exploitable moist forests will have disappeared in eastern Brazil, in Central America, in Colombia, in most of Ecuador on the Pacific side and altogether in Australia and Melanesia, Bangladesh, India, Sumatra, Sulawesi, Malaya, The Philippines, Ceylon, Thailand, Indo-China, Madagascar, East Africa and West Africa by 1990 or sooner.

By the year 2000 or earlier, according to Myers, another portion of Brazil will be deforested, particularly along the roads and rivers, and more of Ecuador and Peru on the Amazonian side. As for Indonesia, he expects Borneo to be depleted by 1995; but in Irian Jaya (and in Papua New Guinea) the forests may last until 2000; this also applies to the drier forests of Burma.

This leaves us, in or around 2000 AD, with forests in the western parts of Brazil, in the Guyanas, and, in Africa, in Gabon and a part of Zaïre.

Particularly endangered now are the remnants on the Atlantic coast of Brazil, the coastal forests of Ecuador and Colombia, Malaya, northwest Borneo and Ceylon; in New Caledonia (these are not rain forests, but very ancient and of exceptional value), in Madagascar (to which the same applies), and in the southwestern parts of the Ivory Coast.

These projections do not apply to legally protected areas of forest; Myers concerns himself with "forests . . . outside the reserves". But while the list of rain forest reserves is growing, a total of 5% of the land area of any country is exceptionally high. Many reserves are leftovers from logging schemes, or are in mountainous terrain, not in the species-rich lowlands. Many are comparatively small, or are already in danger of disappearing nevertheless. The establishment of more, larger and better protected rain forest reserves at low altitudes is a matter of top priority everywhere.

Some Misunderstandings Corrected

It seems strange that in the face of so much concern, a concern that has made ripples in so many disciplines, so much hardheaded effort is still made to cut down whatever can be cut down, under the most far-fetched pretexts. An exploitation dogma seems to prevail. Of course, we will investigate all the causes of forest destruction. But it seems appropriate, here, to correct a few common misunderstandings about rain forests and the effect of human impact.

1. It is a *mistake to regard a tropical rain forest as a source of timber*. There are too many other values in it, which are annihilated by logging.
2. Tropical rain forest is widely thought to be a *renewable resource*. This is only the case where minimal quantities of 'biomass' are withdrawn from a forest, and no plants are killed, i.e. *next to never*.
3. Damaged tropical rain forest *does not regenerate quickly*. Structure and composition are too complex. Recovery after substantial damage takes centuries.
4. *Reforestation cannot compensate* the loss of the majority of the values of natural forest.
5. Height and luxuriance of a rain forest are *no proof that the soil under it is fertile*.
6. Natural forest also *does not need to be exploited or managed* 'to keep it in good condition'. All age classes have their function in the forest ecosystem; all dead materials are recycled.
7. *Methods of exploiting* a tropical rain forest, in such a way that all species in it can be retained, *have not yet been discovered*.
8. So-called *selective cutting is not harmless*. On the contrary, it adversely affects the condition of the forest soil and the quality of the remaining tree population, since the best individuals are removed.
9. *Protection of single 'endangered species'* is *ineffective*.
10. Important species of the rain forests *cannot be saved in botanical gardens or seed banks*: there are too many of them, the seeds are short-lived, and the animals needed to maintain their life cycle, are not conserved simultaneously.
11. Reduction of tropical hardwood consumption *will not save any rain forest*. Only direct protection in nature reserves can be effective.
12. Other attempts to relieve the exploitation pressure, laudable as they are, will not save any rain forests either. *Destruction is proceeding too fast*.

A Difficult Message

In this first chapter, the focus has necessarily been on the disquieting relations between man and forest. The relationship is, at best, ambiguous. Even many a conservationist wavers in the face of 'the real world' of politicians and economists who claim forests for the benefit of the poor. We will revert to this later. But first we must look into the ambiguity. Of what does a tropical rain forest actually consist? And need it be subject to political conflicts of interest?

Our efforts to answer these questions will reveal another 'real world'. A complete answer is out of the question, though our knowledge of the rain forest is growing every day. But we can try to explain the main features, and the most important interconnections. Even so, the message of this book is difficult. This seems natural in view of the complexity of the subject. But most people have no idea of the diversity in nature, and yet such diversity is the key to understanding the functions and problems of all rain forests.

A word of apology may be in order to those many readers who have never had to plough through massive handbooks describing the plant and animal kingdoms, have never spent time in zoological museums or herbariums identifying long series of specimens, nor ever wanted to learn thousands of scientific names, and have no friends smitten by these fascinations. But one can imagine the long years of training, the mental stamina and exceptional memory required to be able to recognize, at a glance, the 380 species of Dipterocarpaceae in Malesia, or the 480 species of wild figs, and to identify them, sometimes by a single leaf.

But such is scientific life, confronted by life's diversity. In the Entomology Department of the British Museum of Natural History there are rooms full of cabinets, and cabinets full of drawers, each drawer filled with insects, neatly labelled and stuck on their pins, each an example of nature's diversity. The rain forest has perhaps shed its reputation of menace and impenetrability, but human dread of its overwhelming diversity has persisted. This is a very hard feeling to combat and great humility may be required to resist the impulse to change, simplify and, in the end, destroy this complex ancient creation of the natural world.

2 How Rain Forests Are Studied

Outline of the Chapter

The study of tropical rain forests has advanced brilliantly in recent years, but theories about the rain forest still differ markedly from expert to expert. Several traditional approaches can be discerned, however, and in any case we are interested here in the general picture.

I will first give an outline of the development of our knowledge of, and concepts about, the rain forest, and then briefly review some important books. Something must also be said about the collaboration between botany and forestry in tropical rain forest regions, and about those 'footholds' of ecology, the research plots, the foci of so many scientific pilgrimages.

The study of rain forests can be said to have four special aims or divisions: (1) the in-depth understanding of its biological diversity; (2) the analysis of the structure of its vegetation; (3) the discovery of what happens in the canopy at various heights; and (4) the study of the forest from above, from various heights. This chapter will discuss these four aims and finish with a brief enumeration of the methods and disciplines exemplified by the famous study of El Verde, though we must end with a warning against all oversimplifications.

A Handful of Terms

The key words to denote these forests are tropical and rain[1]. They derive from Schimper (1898), who introduced the German term *der tropische Regenwald.* In Latin America, derivations of pluvii sylva are often used.

The word tropical alludes to the constantly high temperatures usually prevailing between the tropic of Cancer and Capricorn below about 1000 m in altitude. Above 1000 m lies the montane zone. Forest growing between 1000 and 300–500 m is generally called hill forest; below 300–500 m is the lowland forest, the richest in terms of species and tallest in height.

The word rain suggests that these forests grow in ever-wet climates. A fairly common synonym, derived from the Greek word for rain, is ombrophilous forest. Since moisture in the soil and air determines plant life rather than rainfall, however, many authors prefer to use tropical moist forest as a term rather than tropical rain forest. The respective abbreviations are TMF and TRF.

In tropical forests not growing under ever-wet conditions there is, depending on the length and severity of the regular dry season, a gradual increase in the number of trees that seasonally shed (part of) their leaves. This telltale sign will also indicate many changes in

---

[1]Although Baur (1968, p. 1) makes a good case for the spelling of rainforest as a single word, "as indicating the community's status as a fully independent plant formation and to avoid undue emphasis on rain as the sole determining environment factor", we follow Richards (1952) in keeping the words separate, but spell them with small initial letters.

forest stature and structure, and in the composition of its species and life forms. Starting with the most moist, examples of which are found in Africa and tropical America, the evergreen forest is replaced by semi-evergreen, semi-evergreen by semi-deciduous, which, as seasonal dryness increases, becomes deciduous forest. In still drier climates, closed forest will give way entirely to open woodland, and this to still more open savanna. To understand precisely what is meant by these terms in a particular region, however, the local literature must be consulted.

Primary is the generally used term for virgin or untouched forest. Secondary is that for the fast-growing, species-poor, and at first impenetrable forest that grows on bared soil, especially in tropical ever-wet climates. This distinction, which is a vital one, will be dealt with in a special chapter. As secondary forests will, if not re-cut and if near or within a primary forest, be colonized by the nearby primary forest species and will eventually be replaced by them, such forests in transition are called successional forests. There are, however, other, very different kinds of secondary forests, depending on the climate in question.

In this connection, the term jungle is confusing. It means wilderness. But in a tropical moist climate, all 'wilderness' with its connotations of an impenetrable scrubby vegetation of thickets of low trees and bushes is secondary. But writers like Rudyard Kipling have given 'jungle' a romantic aura, and jungle has become the popular English word for any untouched forest, untamed and, therefore, potentially hostile to man. In German, the popular word for 'tall majestic forests' is *Urwald*, which is botanically correct; the English 'jungle' is not.

As for the French, they have one basic word for forest: *forêt*, whether used in a romantic or a scientific context, and add to it an array of qualifying adjectives as do the English. We must, therefore, remember that the words 'tropical forests' include all the forest types of the tropics, of which the lowland tropical rain forest is only a portion. It is, however, the richest of all in composition, the most ancient, the most interesting and the most fragile, and for these reasons it is with this type of forest we are concerned in this book.

The Growth of a Vision

Through the ages, tropical rain forests have been thought remote and inhospitable. If it was possible to work there at all for a prolonged period, it was a rare privilege. The pictures which Martius had drawn after his journey to Amazonia in 1817–1820, published later (1840–1869) in his monumental *Flora brasiliensis*, were the first really informative illustrations of the tropical rain forest that the botanists of those days could examine. Because they brought out important features, they were far superior to photographs. A few of them have been reproduced here.

Other scientists, too, began to explore and collect specimens in the rain forests, and to describe their travels in often, fascinating books. A selection of books on the Neotropics has been listed and annotated, in Verdoorn, *Plants and Plant Science in Latin America* (1945, p. xxiii–xxviii). The *Notes of a Botanist on the Amazon and Andes*, by Spruce in 1908, deserves special mention. Two fine travel books on the Malesian rain forests are Beccari's *Wanderings in the Great Forests of Borneo* (1902, Italian; 1904, English), and Lam's *Fragmenta Papuana* (1927–1929, Dutch; 1945, English). The African rain forest region was also well explored and described by, for instance, Mildbraed (1922).

It took a long time, however, before a reasonably comprehensive vision of all tropical rain forests could be arrived at since it had to be based on comparisons of the three main rain forest regions. Besides, the species richness being so great, sampling, comparison, identification and description all took a very long time. It was many years before even the merest outline of the incredible biological diversity of rain forests was established.

The first great synthesis of knowledge was given by Schimper (Strassbourg 1856 – Basel 1901); his biography is in Ber.deut.Bot.Ges. 19, Gen.Vers.Heft, pp. 54–70, 1901. Schimper travelled in the West Indies, Brazil, Ceylon and Java. His book *Pflanzengeographie auf physiologischer Grundlage* (1898) is still a classic, and one of the pillars of modern ecology; the English translation *Plant-Geography on a Physiological Basis* was published in 1903 and reprinted in 1964. 'Physiological basis' means the influence of climate and soil, though plant-animal relations, too, received due attention. Although worldwide in coverage, this important book describes tropical conditions and vegetations extensively, particularly those of the rain forest.

The second great synthesis was that of Richards. Richards was born in 1908; his biographical details can be found in *Flora Malesiana Bulletin* 33: 3365–3373, 1980. His book, *The Tropical Rain Forest*, appeared in 1952. There have since been many reprints with small corrections, but it has not been materially revised. His findings were based primarily on field work in Guyana, Borneo and Nigeria, supplemented by a wide and scholarly knowledge of the literature available at the time. Richards was the first to work with profile diagrams, and his book is still a mine of information. Many quantitative data, especially those relating to (micro)climate, soil, composition and ecological niches are systematically examined. A rain forest is presented as a self-renewing climax vegetation, from which, under various limiting conditions, or after various kinds of damage, different forest types can be derived. Cycles and plant-animal relationships, though hardly present as separate subjects here, are subsequently well described in the same author's picture book, *The Life of the Jungle* (1970).

The concepts of evolution were introduced into rain forest thinking by Corner (born 1906; biographical notes in *Fl. Males. Bull.* 29: 2536–2538, 1976, and *Gard. Bull.* 29, 1977). Corner was resident botanist in Singapore from 1929 till 1945, but also studied in Amazonia, and became one of the first botanists to develop a really 'tropical-centered' vision of the rain forest. He is well known as a taxonomist of fungi and of the Moraceae family; in the chapter on Evolution we will discuss his famous 'durian theory' of 1949. In his book *The Life of Plants* (1964), he discusses the evolution of the entire plant kingdom from marine algae, and views the tropical rain forest as the matrix of terrestrial vegetation types. The relation between plants and animals is given a temporal perspective, the modern rain forest being a product of co-evolution. Disdainful of plodding laboratory research, Corner describes the forest in action, in magnificent prose, and from simple, obvious facts he draws many daring and far-reaching conclusions. A difficult man, but an inspiring botanist and teacher.

Peter S. Ashton (born 1934) was one of his pupils. For biographical notes on Ashton, see *Flora Malesiana* (i 5: cclv–cclvi. 1958, i 8: xi, 1974). After a brief period in Brazil, Ashton was a forest botanist in Borneo for 9 years. Besides specializing in the taxonomy of Dipterocarpaceae, he was the first to do prolonged ecological work in tall, species-rich lowland rain forests. The very detailed study which resulted (1964), contained greatly improved profile diagrams and revealed the great influence of soil properties on forest composition. In a remarkable paper on *Speciation Among Tropical Forest Trees* (1969), Ashton used modern genetic concepts to speculate on the transport of chromosomes (in pollen and seeds) in a rain forest, and on its influences on evolution below the species level. He has continued to work along these lines towards a better understanding of the factors which influence or control rain forest composition, concentrating on western Malesia.

Daniel H. Janzen, to some the *'enfant terrible'* of modern tropical botany, has worked predominantly in Central America, concentrating at first on the biological interactions between ants and plants and on insects which consume seeds. In the 1970s, he developed into a dashing, brilliant theorizer on the ecological effects of various plant-animal relationships, interpreting many phenomena in terms of biological defence and survival value. His booklet, *Ecology of Plants in the Tropics* (1975) brims over with ideas about plant behaviour and suggestions for further research. We shall be discussing several of his challenging ideas elsewhere.

Sedate by comparison, the book by Hallé, Oldeman and Tomlinson, *Tropical Trees and Forests* (1978) is a potential classic in the field. Hallé (born 1938) is a French botanist whose main experience is in Africa, although he has recently spent 2 years in Indonesia; Oldeman (born 1937) is a Dutch silviculturist, who worked in the Ivory Coast and then in Guyana and Ecuador. Tomlinson (born 1932) is a British-American tropical plant morphologist with experience in Florida, Malaya and Ghana.

The three authors view a forest as a collection of individual trees, arranged in fascinating structural and dynamic patterns. Trees are classified according to their architectural models, 23 in all; these models remain constant throughout the life of a tree, renewing themselves by repetition during each of the three main growth phases. Viewed as a whole, the canopy, too, is subject to dynamisms, and there are different distributions of meristem concentrations (Tomlinson's main contribution), characteristic of the mature forest as opposed to various succession stages. In this book, the art of making profile diagrams has been elevated to unprecedented heights to the point which makes refined quantification possible. The concept of the existence of natural units in the forest at different levels of organization reveals much about changes in forest structure; consequently, better understanding of its natural processes may create the possibility of manipulating some of them, if this is desirable.

A precursory book by Hallé and Oldeman in French (1970) was translated into English (1975).

Many others have, of course, contributed to the very considerable body of rain forest knowledge; the choice of the six landmarks above is only my personal one.

Rain forest expeditions from temporarily established headquarters within the forest continue to be an obviously practical idea, even if over the years the reports have become less romantic and more technical. One delightful example: the book by Hanbury-Tenison, *Mulu/The Rain Forest* (1980), an account of a British expedition to a conservation area in Northwest Borneo. During the 15 months of its duration, 115 scientists participated, each pursuing his or her specialty. And in a highly instructive (and almost lighthearted) manner, the essential methods by which rain forests are studied are revealed. That, at the same time, much consideration was given to the tribal peoples who live in and off the forest there, and to broad conservation aspects makes it one of the finest works of its kind. Here is a sample passage: "Rafting is the best way to travel a river, even though the Medalam had a few daunting rapids where we were nearly swamped; silently we drifted for two hours beyond the last boundary of the park at Long Mentawai seeing things a motor would have disturbed: a five-foot monitor lizard draped asleep in the sun over a submerged log, a green heron standing motionless gazing into the water, striped squirrels playing on a branch" (p. 57).

Regional Differences

The three main rain forest regions of America, Malesia and Africa differ markedly in extent, coherence, composition and problems. These differences will be discussed in later chapters. But, in addition, separate traditional methods of study have developed in each region. The following influences have caused this:

1. Influence from northern countries. Latin America owes its epithet to the Latin-derived languages of Portugal and Spain, countries which conquered and christianized the Neotropics as early as the 16th century.

During the 19th century, when the Latin American nations became autonomous, the scientific establishments set up in each country operated in relative mutual isolation, while scientific ties with Spain and Portugal did not persist. Contacts with scientists from other western European countries and the United States have existed for long periods, but have never led to long-lasting, structural relationships. Verdoorn's *Plants and plant sciences in Latin America* (1945) contains much interesting information on this subject.

In the Guyanas and the West Indies, however, the situation is more comparable to that in Malesia and Africa.

In Malesia, The Philippines were colonized by Spain (until 1898, when Americans took over), Indonesia by The Netherlands, Malaya by Britain and New Guinea by Germany and Australia. Organized scientific work was established and expanded in the course of the 19th century, until 1940. After World War II, scientific ties were vigorously resumed in new forms, essentially between the former colonizers and the newly independent nations.

In Africa, the same thing happened, the European countries most involved here being Belgium, Britain, France, and, as a newcomer, The Netherlands.

2. The location of the main institutes follows the local historical pattern. Rain forest study in the Neotropics is concentrated in Central and South America; the United States has only become involved in recent times. In Malesia and Africa there are regional institutes as well as tropic-oriented institutes in the former home countries in Europe. All these institutes and the relations between them have their own character, formed according to the languages and personalities involved.

3. State of exploration. Although spectacular early work was done in Amazonia in the tradition of Humboldt and Martius, the exploration of this vast rain forest continuum with its many species lags far behind that in the other regions.

4. Organization of the scientific effort. Scientific strengths and weaknesses also stem from traditional historical associations. In Malesia, regardless of national boundaries, all data regarding exploratory and taxonomic work have been carefully scanned, digested and published as parts of the *Flora Malesiana*, while an annual information issue of the *Flora Malesiana Bulletin* keeps everyone involved posted on the scientific state of affairs.

In Africa after independence, the former colonizers continued scientific work on a nation to nation basis, though many scientists soon realized the limitations of such an approach and joined forces in the AETFAT (*Association pour les Etudes Taxonomiques de la Flore d'Afrique Tropicale*). At its regular meetings valuable information continues to be exchanged and published.

In Latin America, a single coordinating centre does not exist. The *Flora Neotropica*, an important series of monographs, was not preceded by a strong bibliographic effort, and its board cannot be compared with a vigorous organization like AETFAT.

5. State of the taxonomic knowledge. This is more or less in inverse proportion to the number of species involved, and is also closely related to the state of organization outlined above. African and Central American rain forests are relatively poor in species, but these are well known because of the many monographs written on them. The Amazonian rain forest is the richest in species, but the least known. Malesian taxonomy is in an intermediate position.

It makes a great difference whether the taxonomic knowledge of an area has been built up from species descriptions eventually forged into a Flora, or if it has been synthesized by critically comparing all the material from the entire region. Time and again such syntheses have resulted in considerable reductions in the number of species thought to be confined to the area. In Malesia, locally endemic species do occur, to be sure, but only work of the monographic sort can bring out this kind of 'hard' data. In Latin America, such work has only begun to explore the plant world, though a start has been made with the *Flora Neotropica*.

6. State of ecological knowledge. When in the mosaic of all these differences a few favourable conditions coincide, it then becomes possible to build up a scientific infrastructure, a prerequisite to the prolonged, high-quality ecological work necessary for a real understanding of an area of rain forest. Research plots like Pasoh have thus become

established through the fulfillment of a whole series of conditions in the past. Scientific conditions are now also excellent for intensive ecological work in Surinam, while several parts of Africa are now well enough known to spearhead ecological research in the area. But the pattern is a chequered one, and unfortunately the feasibility of planning further research is entirely dependent on it.

Collaboration Between Botanists and Foresters

Tropical forest biology owes much to the collaboration between foresters and botanists. Tropical foresters occupied with natural forests have to find their way among vast numbers of species, many times more than those which constitute temperate forests. To distinguish these species, they depend heavily on taxonomic knowledge, and for this reason every respectable forestry service in a tropical country maintains a Herbarium staffed with forest botanists.

Forest Herbaria organize exploratory expeditions which result in the collection of specimens, tens of thousands over the decades per institution, and in the description of rain forest tracts. Such descriptions often surpass in value the more prosaic stock inventories in cubic metres, though this depends on the interest and ability of the forest botanist involved, and of his superiors. In many cases, forestry services have acquired broad spectra of data, and from them have compiled highly informative maps and handbooks, or chapters in handbooks on the country concerned and on its forests in particular. Such works are worth searching for as they are not always widely distributed.

Forestry services, often continuously maintained and having access, over the years, to considerable sums of money, will usually have facilities which far surpass those available to 'pure' scientists. One advantage is that a good forestry service always maintains a network of regional, district and local officers, labourers and transportation which are usually generously made available to the visiting scientist, insofar as they can be spared.

Also, since foresters remain in their stations much longer than a 'pure' biologist can stay, some of them acquire great personal knowledge of the local terrain, wildlife and biological phenomena, and quite a number of them have done much to establish rain forest nature reserves, and for conservation in general.

Forest research institutes are also extremely important. They usually maintain plots where routine observations are made, on the weather, soil movements, tree growth, flowering, fruiting and germination, often over many years.

The role of a tropical forestry service is, of course, forest management, a word that stands for a blend of exploitation and protection, with great differences in emphasis from one forest service to another. In the tropics, biological diversity is the forester's real challenge, and any good forester will want to understand it in order to reduce it advantageously.

The 'scientific' management, thus understood, of mixed species-rich tropical forests is now about a century old. Its founder was Sir Dietrich Brandis, who worked mainly in India (see Hesmer's 1975 biography). Since Brandis' time, the state of the art has been written up with great knowledge and wisdom by George N. Baur, *The Ecological Basis of Rain Forest Management* (1964).

How to Recognize Plants in the Forest

.The painstaking and skilled science of identifying a plant by a detailed examination of its flowers, in which herbarium taxonomy specializes, is of little use in the forest. The trees are tall and flower infrequently. The knowledge of families, genera and species obtainable in a herbarium is only applicable in the forest after it has been adapted for use with living trees.

This is the job of a forest botanist. He commutes, so to speak, between forest and Herbarium, to which, if he is lucky, a good 'arboretum' may be attached. He carries a pair of field glasses, a hand lens, a bush knife and a notebook. Every plant in flower or fruit which he does not know he collects for identification, even though years may elapse before a name has been attached to a specimen beyond any doubt (see p. 26: The Work in a Herbarium). He regularly visits marked trees to secure fruits as well as flowers and flush. If a beginner can work with a more experienced colleague, after a year he may know precisely what to look at, and will be able to name a fair proportion of the plants at sight. Several forest techniques will help him to do this.

For instance, a whole array of characteristics can be discovered in a 'slash', that is, by making a cut through the bark to the wood, and, a hand's breadth above it, another cut, so that a chip of bark can be removed and examined. In the Guttiferae, the first thing one notices are the drops of yellow sap which ooze from the slash; the Connaraceae, Myristicaceae and some Leguminosae, exude a red resin; Apocynaceae, Moraceae, Sapotaceae and some others, a white sap. The Thymelaeaceae, on the other hand, can be recognized by the long silky fibres in the bark; the Fagaceae slashes reveal vertical bluish lines in the wood; and when you cut the bark of a *Dillenia* (Dilleniaceae) and listen closely, you will hear an effervescent sound as in a glass of soda water. The 'slash' of one liana (Vitaceae) in New Guinea has the typical smell of corned beef. Less spectacular, but very revealing are the gross anatomical characters of the inner bark, and of the sapwood, which can easily be seen with a hand lens.

The outer bark may also have many distinctive features: for a collection of photographs and descriptions, see Rollet (1980). In the genus of ebony, *Diospyros* (Ebenaceae), the bark is often blackish. *Melaleuca* (Myrtaceae) owes its vernacular name kayu putih (Malay for white wood) to its very light-coloured bark. *Tristania* (Myrtaceae) sheds broad scrolls of thin outer bark, which pile up below the tree, similar to the bark shedding of *Eucalyptus* in the same family. Corner (1940, pp. 11–12), distinguishes seven main types of bark; more data are given by Whitmore (1975, pp. 256–257).

Fallen or collected branches can also display, even when sterile, many features which alone or in combination are characteristic of a family or genus:

1. The growth of twigs may be continuous or in flushes;
2. The distance between leaves may be equal, or, if very unequal, will result in the arrangement of the leaves in tufts;
3. The twigs may be 'hairy' or 'smooth' and, if hairy, the habit and the structure of the hairs will vary;
4. The twigs may be round or angular in section;
5. The surface of the twigs may be smooth or ribbed;
6. The buds may or may not be covered with scales;
7. The leaves may be simple (i.e. with a bud in the axil of each blade) or compound (i.e. many blades on one 'rachis');
8. The leaves may be opposite, scattered or in two rows along the stem;
9. Stipules may be present or absent near the leaf base, look first at the young leaves in case they are shed with maturity;
10. The leaf margin may be entire or incised (look near the apex);
11. There may be glands in certain places, on the underside of the leaf, for instance;
12. There may be other differences in the upper and lower surfaces of a leaf;
13. Vein patterns differ widely; and
14. If leaf stalks are present, they may have swollen 'joints' at the base and/or apex, or the leaf may be 'sessile'.

In several countries, forest botanists have compiled useful pocket guides to identify the common trees by such vegetative characteristics.

Fallen flowers and/or fruits, and sometimes the leaves, can often with care be matched with the proper tree or liana. If possible, one should always try to ascertain the position of the ovary with respect to the sepals and petals: is it inferior or superior? This is very often a family characteristic. Some fruits (in genera of the Apocynaceae, Elaeocarpaceae, Lecythidaceae) have hard shells or cores, which remain exposed on the forest floor for a long time after falling, and these may provide useful clues.

Establishing a plant's architectural model (this will be discussed in Chap. 5) will often reveal telling structural differences between similar-looking trees, especially between saplings.

The base of the trunk, too, may have interesting features such as buttresses or prop roots, or may be spectacular in size or shape; such characteristics, however, depend greatly on the age of the tree.

Seedlings of trees may also differ so strikingly from the twigs of adults, that special publications have been prepared to recognize them. The classic books on seedlings are those of Troup (1921) and De Vogel (1980); see the latter for a bibliography on the subject. Accurate seedling identification is essential when making estimates of natural regeneration.

Faced with the many problems of plant identification in the forest, workers with insufficient knowledge are prone to place too must trust in locals who pretend to know the plants by their vernacular names. By this means, a list of local names is compiled and then, back in camp, the botanical equivalents are looked up. Horrendous mistakes creep in here! Many foresters in Indonesia rely on this practice when making inventories of timber stands, usually by the process known as strip surveying. Using a compass, a trail several kilometres long is blazed in a straight line. All trees above a certain diameter at breast height (dbh) or a certain girth, and within 5 m to the right and left of the trail, are listed according to size and local name. According to Dr. A.J. Kostermans, so many errors are made in these identifications, that the results are completely unreliable.

A good 'arboretum', where trees are labelled with their proper names, is very useful for getting to know them. Such an arboretum should ideally be a plot of 'civilized' forest, with trails. Retaining the natural plant densities of a natural forest is essential: trees planted in the open as in a park or garden will not develop a long slender bole and globose crown, but will usually develop atypically a conical crown on a short trunk. Many forest research institutes maintain good forest arboreta; the Semengoh Arboretum near Kuching in Sarawak is a fine example.

Research Plots

Though there are many tales of lost or destroyed research plots, there are still a few demarcated bits of tropical rain forest which have been staked out and studied by various research workers over a span of years in an effort to build up scientific data.

One example is the Pasoh Forest Reserve in Malaya, 140 km southeast of Kuala Lumpur. Although some timber cutting was done in the 1920s, and although as a result the canopy is still somewhat open and uneven, it is still a very nice 650-ha tract of lowland dipterocarp forest (LDF) between 75 and 150 m in altitude. It is surrounded by a similar area of depleted forest, as a buffer zone. Under UNESCO's International Biological Programme, 70 persons did research there between 1970 and 1974 (see Pasoh 1978). It has been calculated that the money spent on this work – US$ 460,000 – equalled the value of the timber the reserve contains, which could lead us to speculate on how expensive timber is, or how cheap the work of scientists! Thanks to this work, the area has now become priceless and the natural focus of further study.

The staff and students of the University of Malaysia at Kuala Lumpur analyzed the flora in five plots of 2 ha each, and in five plots of 0.2 ha; they measured, mapped, identified and labelled a total of 5907 trees 10 cm in diameter or thicker, belonging to 460 species. From a three-storey wooden tower, 33 m high, 170 trees were marked for regular observations within a radius of 50 m, and twice a month, observations on flush, flowering and fruiting were recorded. Instruments measuring light quantity, wind velocity, temperature, rainfall and humidity were also placed on the tower, at different heights. Soil profiles were made; and the distribution of dipterocarp trees turned out to be closely correlated to the eleven different soil types identified (Ashton 1976a).

A newcomer might walk the trails of this highly varied forest, where Dipterocarpaceae represent 7–14% of the trees, Euphorbiaceae, 5–16%, Burseraceae, 5–9%, Myrtaceae (nearly all *Eugenia*), 4–7%, Leguminosae, 4–7%, Annonaceae, 4–7%, Sapindaceae, 2–5%, Myristicaceae, 1–5%, and Guttiferae, 2–4%, and be struck only by the uniformity of the undifferentiated mass of green, for only knowledge and practice will reveal the rain forest's diversity. It is only very gradually that a student comes to notice diagnostic features, to grasp relationships and to form an understanding of the vegetation, and even so can still be daily

Fig. 2.1 A biologist's camp in the rain forest. Shelters have been made of poles cut nearby; tarpaulins give protection from rain. Note the hammocks: in the lowlands, like here, it is much too warm to sleep in a closed space. Some plant presses lie in the foreground. North of Manaus, Brazil. (Photograph MJ 1982)

more puzzled and intrigued. The richer one's knowledge of plant and animal species, the fuller one's theoretical background, the sharper one's eyes and the more open one's ears, the more such a forest will reveal. The rain forest makes more demands on a student's personal qualities and training than any other ecosystem of this planet.

Plant Collecting for Taxonomic Purposes

Botanists in Herbaria, like zoologists in museums of natural history, produce the taxonomic knowledge that is to be applied in the forest, but the material for their study must first be collected in the forest. A cycle thus exists: forest–specimen–herbarium–taxonomic literature–name–forest. In the previous section, we discussed one part of the cycle: name–forest. Now it is the turn of forest–specimen–herbarium. In the next section, we will complete the cycle: herbarium–taxonomic literature.

Specimen collecting is essentially a kind of random sampling, good collections are, therefore, few and mediocre collections abundant. Only fertile material is desirable, i.e. flowers and/or fruits, and every specimen should be in duplicate. All the material from one species in one location and taken, if possible, from one plant, is then registered by the Herbarium under one collector's number, mounted on paper if possible and incorporated as 'specimens'; if duplicates have been taken, for instance, branches of the same tree, these can later be sent to different Herbaria.

Because the 'harvest' of specimens is bulky, a collecting expedition is always expensive in both assistance and transport. Ideally, the trip should be planned by or with a local Herbarium, whose staff know good areas and how to get there. Arrangements for transportation must be made, personnel recruited, foodstuffs purchased, on the basis of x days times y men, everything packed and then transported by boat, car or aircraft as far as one can obtain these various means. To reach undisturbed forests these days, zones of various degrees of destruction will have to be traversed, and the last (tens of) kilometres generally negotiated on foot, and the supplies carried by porters at the rate of 20 kg or more per man, depending on the terrain.

Once in the forest a camp or a series of camps (Fig. 2.1) are made from which the botanist can make daily excursions, accompanied by half a dozen native workers to climb or cut trees as necessary. He carries the botanist's indispensable field equipment: binoculars, pruning shears, a linoleum knife, a hand lens and a notebook. With the binoculars he will try, from far below, to locate flowers and fruits in the canopy. Some which have fallen to the forest floor can help in this search, but fresh material must be obtained, and in any case 'detached' parts picked up from the ground can be a source of confusion. From below, it is often impossible, of course, to see whether a tree carries young foliage, buds or open flowers. But if a botanist glimpses anything moving just above a crown, he will climb himself or send a man up to collect specimens, for movement indicates that insects or birds are visiting the tree.

If the tree cannot be climbed, it may be possible to catapult a line over one of its branches, having previously tied to it a cable which can be used to help a climber, or which can be pulled to break off a branch. If this fails, the tree must be cut down (Fig. 2.2). If buttresses make this difficult, the helpers can quickly make a scaffold from poles chopped in the neighbourhood, tying them with liana stems. The tree is then felled above buttress height and once down searched for its fruit or flowers as well as for any climbers and epiphytes in the fertile state which it may have been supporting. In this search it is particularly important to be aware of variability: a leader shoot may differ in structure from the lateral shoots, and sterile branches bear much larger leaves than fertile ones. And, of course, duplicates must be exact duplicates, including those involving large fleshy fruits. (In the field, it is wise to keep such fruits in a 'potato net' in plastic vessels containing alcohol.) The job of selecting and then reducing the size of specimens depends on what parts or sections a taxonomist will later need for his work.

Fig. 2.2 Chopping down a tree for botanical collecting. First, a scaffolding made of small trees has been built around the trunk; this particular trunk is 'fluted', it has a number of rounded ridges with narrow grooves between them

Another complication, frequent in the rain forest, is that the plant bears only female or only male flowers, so that another plant with flowers of the opposite sex will have to be found; and for plant groups of deviating structure or delicacy, many precise collecting instructions will exist. Wood samples, often extremely difficult to dry, should also be taken if their transport is possible. Ideally, every part essential for study must be taken, but redundancies avoided. The collector must also write down in his notebook all observations that cannot afterwards be made from the dry specimen in the herbarium. Most of these comprise the data later to be inscribed on the label: place, date, altitude, habitat, habit and size of the adult plant, colour of the flower parts, shape of the fruit, state of maturity and colours generally. Vernacular name(s) and local use(s) are added if known. And, of course, each separate item must be tagged with the collector's number.

Once back in camp, the work of preserving the selected material begins (Fig. 2.3). To dry specimens, they must be carefully arranged between sheets of newspaper about 45 × 30 cm, interspersed with ribbed cardboard, and pressed between boards or a lattice frame tightly lashed together. To dry them quickly, exposing the press or presses to hot dry air is the ideal

Fig. 2.3 Collecting equipment for taxonomic purposes. In the *back*: a packet of plants in alcohol, sealed in plastic, plastic containers with alcohol of various sizes. *On the blanket, back row*: binoculars in carrying case, axe, chopping knife, adhesive tape, tin containing camera; *front row*: fruits, branches (all are duplicates with the same number), bow saw, secateurs in case, gardening gloves, notebook with pencil attached and labels, hand lens on a piece of string; *behind* that: wood sample with number written on a piece of adhesive tape; in *front* of it: long scissors for the cutting of plastic and tape, altitude meter, field compass. New Guinea. (MJ 1973)

way, usually achieved by placing the press(es) high above a fire so that the heated air can pass through the ribbed cardboard. After a couple of days the plants are dry; they no longer feel cold to the touch, and they are brittle. But spreading and drying takes expedition time, and it is very difficult to maintain dried specimens in an ever-wet climate.

Rain forest botanists therefore generally prefer what they call the wet method. All duplicates of one 'number' are just folded together in a newspaper; a bundle of these 1 ft high is placed into a sleeve of firm polyethylene. Half a litre of 70% alcohol is poured into this and the vapour will keep the material preserved, if the sleeve is then sealed airtight with tape. Such packages can be kept for months, and the material can be dried afterwards at a convenient time and place.

Once dried and back at the Herbarium, the material is tentatively named as to family and genus by those few enlightened botanists who have a working knowledge of the flora of the region. Only then can be labels be made, and the duplicates distributed to other Herbaria, as part of an important exchange relationship by which a collection can be safeguarded, for if material is lost in one place, it can then be consulted in another. The specimens are now mounted on sheets of thin cardboard to prevent breakage, and bulkier items such as the wood samples, dried fruits and alcohol-preserved fruits or fleshy parts, are properly stored and identified, so that although the specimens are incorporated into the general collection, they can always be retrieved. A Herbarium is thus an archive of preserved plant specimens, and it is in these archives that plant taxonomists work.

As one can imagine, the problems of collection, preservation and storage are much increased in the case of large plants like palms, or of plants with very spiny parts, hand-sized flowers or large fleshy fruits. Consequently, such plants tend to be poorly represented in collections and specialized collecting expeditions must be made to obtain adequate material for the taxonomists. This is the more necessary since these difficult plant groups often include many plants important economically or nutritionally. Fortunately, more finance for such work is becoming available.

As for forest animals, they are collected, preserved and 'filed' in comparable ways, mutatis mutandis, in zoological museums, but to describe the very different techniques used to preserve animal specimens is far beyond the scope of this book.

Botanical Gardens

To make possible the study of individual living plants is the main function of a botanical garden. The oldest tropical garden, established in 1795, is in Calcutta. The one in Bogor, in western Java, dates from 1817. Particularly in Bogor, there has been outstanding research work done on rain forest plants by a veritable galaxy of biologists. An extensive summary of this work was published by Dammerman in 1935. In recent times seedlings of rain forest trees have been planted and observed in this garden by De Vogel (1980).

Modern air transport has made it possible to send plants for further cultivation in the greenhouses of institutes where specialists are working; aroids, ferns and orchids are the groups favoured for this sort of treatment. When the plants become fertile, they are described and illustrated; their chromosomes are counted, they are identified and parts of them are, belatedly, preserved. Such back-up work by botanical gardens greatly increases the effectiveness of any expedition.

The Work in a Herbarium

A herbarium is a collection of dried plants. A Herbarium (with a capital letter) is a botanical institute devoted to the study of such plants, and which to that end also maintains a library of the relevant literature. The collections are kept in condition by qualified technical personnel, and studied by taxonomists or systematists. Taxonomy or systematics is the classification and scientific naming of plants and animals.

A Herbarium can therefore be characterized as a device to ingest specimens and produce Floras. A Flora with a capital F is a book or books in which all plant species of a certain region are described, i.e. it is the end product of the taxonomists' efforts. All the available knowledge about plant families, genera, species and varieties is, in condensed form, eventually channelled into regional Floras. The main corpus of botanical knowledge, however, is to be found in the many specialized series and books written by scientists for other scientists. Besides full treatments of families or genera that cover the whole of their chosen specialty, called monographs, and similarly detailed studies reworking some section thereof, called revisions, botanical literature contains descriptions of newly discovered species and genera, lists of species collected by an expedition, changes in the definition of certain genera which may necessitate re-naming species, records of taxa in regions in which they hitherto were unknown, treatises about the correct name for a taxon (i.e. any taxonomic unit, like genus or species) and data on wood and pollen, for which a Herbarium is also a source.

Fig. 2.4 Label with data. See main text

M. JACOBS 9225 . EASTERN NEW GUINEA, 14.X.1973

VITACEAE det. C.E.Ridsdale 1974

Leea coryphantha LAUT.

Papua, near Waro airstrip 20 km SSW of Kutubu, 6°31'S 143°10'E alt. 500-600 m.

Habitat: limestone country, primary forest on level land, sometimes partly flooded.

Notes: treelet 4-5 m. Juvenile shoots brown--flushed. Young leaves beneath with reddish nerves. Stipules green. Leaf rhachis cavernous, inhabited by ants. Buds dark yellow. Corolla inside pale pink-yellow, disk reddish, anthers pale yellow. No fragrance. Fruits sordidly green, unripe.

Duplicates 2.

Botanical literature now contains data on every taxon in the world, all of which must be retrieved whenever scientific work is to be done. All such data are derived from Herbaria specimens, which must be cited with the collector's name and number, and of special importance in this connection are the so-called type specimens, to which a scientific name has been officially coupled (Fig. 2.4). Wrongly interpreting the identity of such a type specimen will lead to the wrong use of its name and then all sorts of confusion are possible. For instance, a species may have been described from different locations under different names. Or the same name may have subsequently come into use for a quite different species. Or a name that was never published in accordance with the accepted international rules of nomenclature may have none the less come into common botanical usage.

The only way to disentangle such confusion is by a personal critical scrutiny of all the Herbaria specimens of that genus, and of all the relevant literature published after 1753, the year of Linnaeus' work. To collect enough specimens, a taxonomist must borrow material from the 10, 20 or 30 best-stocked Herbaria, and request certain specimens on loan from others.

An important stage of any taxonomic examination is the dissection of the flowers. To do this, a dried flower is boiled in water for perhaps half a minute until all parts have softened and air bubbles have been driven out. In a petridish, in water, under a 10–30 magnification, all parts of the flower are studied: sepals, petals, stamens and pistil(s), then each is drawn to scale and identified on the drawing. When a flower has thus been depicted two-dimensionally, it can be compared more easily with the drawn flowers of other specimens.

Such comparisons between specimens will reveal clearly both the resemblances and the differences, which will be found to be either essential or accidental. They are essential if they constantly recur, and are correlated to others in fixed combinations. It is the existence of these combinations which enables a taxonomist to sort out his specimens, at first by putting them in different piles. Each pile will differ from all other piles in a number of constantly recurring features, which we call characters or characteristics. Each pile thus represents a taxon. Each taxon must be assigned a rank by the taxonomist: order; family; tribe; genus; section; species; variety; forma; with subranks in between.

He tests his distinctions by compiling a key for identification, which he applies to the specimens, to see if the characters are indeed as constant as he presumed. This done, he looks up which type specimens are contained within the various taxa, as these furnish the name. If there are more than one, a choice must be made, in accordance with the rules of nomenclature. The correct name is accepted, the others are 'reduced' to synonyms.

The final results of all these investigations are written up in a condensed, technical language, rigidly prescribed in form, which has been developed over centuries of taxonomic work. All important taxonomic publications dealing with the taxon must be cited as references. The taxon is described botanically, its range of distribution outlined, its 'ecology' or habitat described and notes are added on its uses and vernacular names (Fig. 2.5). Errors made by previous authors are corrected, and all important collections must be mentioned together with the collector's name and number. The resulting monograph or revision should also include identification keys, illustrations, and indeed, all the important facts about the family or genus. The species must be sorted out, correctly named and all doubts or difficulties cleared up. All the examined specimens are then labelled with the correct name and the borrowed ones returned to their respective Herbaria. In all such revised taxa it should be possible to identify all new incoming material without a problem.

This summary may suffice to demonstrate the fundamental nature and value of monographs and revisions. Only a skilled taxonomist can discover and describe the reliable, constant differences between taxa and assign to them the proper scientific names. Once the entire flora of a country or region has thus been written up, the knowledge is used in innumerable ways by all who have directly to do with the plant world, in science, exploration, cultivation or trade. The scientific name of a plant is thus the key to all the available knowledge about the species to which it belongs. Those who still doubt 'the value of Floras to underdeveloped countries' should refer to Brenan's 1963 paper so titled.

A reasonably experienced full-time taxonomist is able to 'work up' an average of 15–20 species a year. In the case of large, well-marked genera with not too much material and literature, he will score higher, but a single, widespread, polymorphic species may engage him for a whole year. Administrative or teaching duties, of course, slow down the process. The same happens, in terms of revised species, if he goes out on field work: a 5-month expedition may easily take up 1 year of his professional time.

The State of Inventory

From the foregoing, it is evident that even the simple-looking species list of a research plot in a forest is the result of three laborious tasks: (1) specimen collecting, over decades or even centuries in the region, (2) taxonomic revision, also often over long periods, and (3) locating the (initially) unknown species in the plot in both herbaria and literature in order to identify them beyond doubt.

As an example of just how time-consuming and laborious these tasks can be, the three-volume definitive and reliable *Flora of Java*, by Backer and Bakhuizen van den Brink, was not published until 1963–1968, that is, after a century and a half of organized botanical work on this island. And Java, with 4598 wild species, is botanically the poorest part of western Malesia! It is understandable then, that most epecies lists of rain forests are only partially complete and partially reliable, even when perfectly up-to-date scientifically.

Just how 'partially' is the question we will now briefly examine. Owing to historical circumstances, considerable differences exist in our knowledge of the three main rain forest regions. As was mentioned in the section on regional differences, Africa is the best known, but

Fig. 2.5 Page from *Flora Malesiana*. Revision of the genus *Euthemis* (Ochnaceae) with name, references to literature, description, distribution and ecology, then key to the species, treatment of one species with name and synonyms together with references to the literature, then description, distribution, ecology, uses and vernacular names

108 FLORA MALESIANA [ser. I, vol. 7[1]

limestone in the Langkawi Islands a race '*microphylla*' is found with small leaves and very reduced inflorescences. Similar, but less reduced forms are found elsewhere along the coasts of the Malay Peninsula.

Excluded

Gomphia magnoliaefolia ZIPP. *ex* SPAN. Linnaea 15 (1841) 186, *nom. nud.* = *Pycnarrhena longifolia* (DECNE) BECC. (*Menispermaceae*), *fide* DIELS, Pflanzenreich, Heft 46 (1910) 51.

4. EUTHEMIS

JACK, Mal. Misc. 1, 5 (1821) 15; KANIS, Blumea 16 (1968) 62. — **Fig. 4.**

Shrubs or shrublets, sparsely branched. Stipules free, caducous. *Leaves* coriaceous, glabrous, denticulate, nerves numerous, parallel, from the midrib curving sidewards, straightly ascending to the marginal veins at an angle of *c.* 80°; petiole ± winged. *Inflorescences* terminal, many-flowered, compound racemes; bracts small, caducous. *Flowers* ☿ or polygamous. *Sepals* 5, turning purplish red in fruit. *Petals* 5, white or pinkish. *Staminodes* 0(–5), filamentous. *Stamens* 5, free; anthers subsessile, rostrate. *Ovary* 5-celled; ovules 2 per cell, pendulous, axile; stigma minute. *Fruit* a berry with 5 pyrenes. *Seeds* 1 (2) per cell.

Distr. SW. Cambodia, in *Malesia*: Sumatra, Malay Peninsula, Borneo.

Ecol. Everwet tropical forest below 1250 m, in kerangas forests, on low ridges in peat-swamp forests, and in open ridge forests, on poor, mostly sandy soils. Dispersal probably by birds because of conspicuous, white, rose-pink or red berries (RIDLEY, Disp. 1930, 410).

KEY TO THE SPECIES

1. Inflorescence a panicle, branches well developed with scattered flowers. Leaves 8–40 cm long, margin distinctly denticulate. Mature fruit white. . . . . **1. E. leucocarpa**
1. Inflorescence a very slender, often cernuous raceme, nearly all branches reduced with conferted flowers. Leaves 4–15 cm long, margin faintly denticulate. Mature fruit red. . . . . **2. E. minor**

1. Euthemis leucocarpa JACK, Mal. Misc. 1, 5 (1821) 16; WALL. in Roxb. Fl. Ind. 2 (1824) 303; JACK in Hook. Bot. Misc. 2 (1830) 69; PLANCH. in Hook. Ic. Pl. II, 4 (1845) t. 711; MIQ. Fl. Ind. Bat. 1, 2 (1859) 675; Sumatra (1860) 208, 533, *incl. var. latifolia*; SCHEFF. Nat. Tijd. N. I. 32 (1873) 411; BENN. in Hook. *f.* Fl. Br. Ind. 1 (1875) 526; KING, J. As. Soc. Beng. 62, ii (1893) 234; RIDL. Trans. Linn. Soc. II, Bot. 3 (1893) 285; BARTELL. Malpighia 15 (1901) 167; HALL. *f.* Beih. Bot. Centralbl. 34, 2 (1917) 30; MERR. J. Str. Br. R. As. Soc. *n.* 86 (1921) 388; RIDL. Fl. Mal. Pen. 1 (1922) 368; DIELS, Bot. Jahrb. 60 (1926) 311; BURK. Dict. 1 (1935) 987; AIRY SHAW, Kew Bull. (1940) 249; MERR. J. Arn. Arb. 33 (1952) 224; VIDAL, Adansonia 1 (1961) 60; KANIS, Blumea 16 (1968) 62; Fl. Thail. 2 (1970) 29. — *E. robusta* HOOK. *f.* Trans. Linn. Soc. 23 (1862) 163; BARTELL. Malpighia 15 (1901) 168; HALL. *f.* Beih. Bot. Centralbl. 34, 2 (1917) 32; RIDL. Fl. Mal. Pen. 1 (1922) 368.

Shrub up to 6 m. Branchlets stout, green. Stipules ovate, 4–6 by *c.* 2 mm, acute to acuminate, ciliate. *Leaves* oblong to linear oblong, 8–40 by 2–10 cm, acute at apex, tapering at base, margin distinctly and irregularly denticulate, nerves 1–2 mm apart; petiole 2–5 cm. *Panicles* erect, 8–20 cm; pedicels 4–10 mm, articulate at base; bracts 8–10 by 2–4 mm, lanceolate, acute. *Flowers* ☿, erect, often in pairs. *Sepals* obliquely ovate to elliptic, unequal, 4–7 by 2–3½ mm, ciliate. *Petals* obliquely obovate to spatulate, 4–10 by 2½–5 mm. *Anthers* 3–5 by *c.* 1 mm, yellow. *Ovary* ovoid to bottle-shaped, 2–4 by *c.* 1 mm, style 1½–3 mm. *Fruit* globular, up to 1 cm ø, fleshy, via red turning white. *Seeds* like sectors of a sphere, *c.* 4 by 2 mm.

Distr. SW. Cambodia; in *Malesia*: Sumatra, Riouw & Lingga Is., Banka, Billiton, Malay Peninsula, Anambas Is., Borneo.

Ecol. From sea-level up to 1000 m, on poor soils, preferably in moist, shady places (see also under the genus).

Uses. Medical applications of the roots is reported from Malaya. In Brunei the fruits are used against eye-diseases.

Vern. Sumatra: *bĕlusung putih*, *kaju padang*, *mata pĕlanduk*, Banka; *balong*, Billiton; Malaya: *pĕlawan bĕrok*; Borneo: *tambu*, Sarawak, *ranggas hutan*, Sabah, *iur iur*, W. Borneo.

2. Euthemis minor JACK, Mal. Misc. 1, 5 (1821) 18; WALL. in Roxb. Fl. Ind. 2 (1824) 304; JACK in Hook. Bot. Misc. 2 (1830) 69; MIQ. Fl. Ind. Bat. 1, 2 (1859) 675; Sumatra (1860) 209, 534; SCHEFF. Nat. Tijd. N. I. 32 (1873) 412; BENN. in Hook. *f.* Fl. Br. Ind. 1 (1875) 526; KING, J. As. Soc. Beng. 62, ii (1893) 235; BARTELL. Malpighia 15

has the poorest forests, while tropical America is the least known, but probably has the richest forests, and Malesia's position is intermediary. Prance's paper *Floristic Inventory of the Tropics: Where do we Stand?* (1977b) made a first attempt at an overview.

Collecting is, as we said, a kind of random sampling. When through historical research all collecting efforts made in a region have been unearthed and recorded, as has been done for Malesia by Van Steenis-Kruseman (1950, 1958, 1974), a picture emerges of the collecting density. The 'density index' is expressed in the total of numbers gathered in an average 100 km^2 in each area, or, in the case of Indonesia, on each island. For the larger islands, this density index is quite a reliable guide to the possibility of completing a Flora. The surprising fact is that a seemingly low index of about 50 collections per 100 km^2 of original vegetation is already a good basis for a complete Flora, and even half that number carries us quite far. From Borneo, 739,175 km^2 in area, a total of about 194,200 numbers had been collected by 1972, a density index of 26, an average of 20 numbers of each of the estimated 10,000 species of the island. New Guinea collecting is about on the same order.

Of most species, 97–98% in Malesia, we have at least some material in our Herbaria. But what low densities do not reveal is the variability and geographic distribution of species. To obtain such information, highly necessary for any further work, there must be an even coverage of collecting. Many blank spots on the rain forest map are left to be filled, as in Indonesian Borneo (about 73% of the whole island), where the density index in 1980 stood at 10.5; in the adjoining Malaysian parts it was 87. To add one point to the density index of Indonesian Borneo requires 5364 numbers; the latest large expedition brought 4000. But it is only in the Malesian area that we are so well informed. In Africa and especially the Neotropics, the picture is hazier.

The state of taxonomic knowledge as a whole is chequered. The plant families are well known. There are (very roughly) five to ten times as many genera; we have a fairly clear idea of them, with several unsatisfactory omissions. There are again (very roughly) five to ten times as many species; in Malesia, only about one-third of these rain forest groups have ever been taxonomically revised, as they are always the largest and the least well collected.

The literature provides, as we have seen, a rather uneven mixture of published knowledge, in which only professionals can find their way. In the *Flora Malesiana*, one-third of a century after its beginning, only some 19% of the 23,000 species of spermatophytes or seed plants have yet been written up. Monographs, revisions and manuscripts have clarified matters in at least another 30–40% of the species, but this is as far as we have gotten to date. The cause? At a rough estimate, 60–70% of all Malesian species are tropical rain forest species.

Knowledge is inevitably weakest in most of the larger families, though there are also small groups which have never been touched. Publication of a major monograph or revision dealing with, let us say, 300 or more species, is indeed a triumph. Progress in taxonomy has been better during the last 40 years than ever before, yet is far too slow if measured against the demand for applicable knowledge, and against forest degradation.

The identification of rain forest plants at species level is for all these reasons an arduous task, and here I am only speaking of the vascular plants, i.e. the flowering plants and ferns! In some families, it is not even possible to determine the genus of a plant with certainty, though in a fair number of genera the species can be identified with reasonable certainty (more in Africa, less in tropical America). But it all takes time.

Any new collection is first sorted out as to family. If there is a specialist working on this family at the time, specimens are sent to him, and in a couple of weeks (or months) he answers. The other families are tackled one by one, with varying success, depending on the quality of the material and of the literature available. The more time spent on a collection, the higher the percentage of reliable species names. One may reckon that an all-out effort to identify 100 rain forest specimens consumes one taxonomist-month, which will result in species names for about two-thirds of them, and in generic names for most others. Yet the accuracy of the inventory of a tract of rich forest will depend on such all-out efforts. Unfortunately, there is no way to compile reliable lists of typical rain forest species from taxonomic literature or Floras. The information they contain on occurrence is too scanty and too incidental.

It will be evident that the factors limiting rain forest inventory are fewer than in taxonomy, since field work can be organized and carried out on relatively short notice in the form of projects. But the taxonomic backlog in every Herbarium can only be reduced by employing more taxonomists. It takes 1 year to train a plant collector, but 5 years to make a taxonomist. Many more of both are needed to understand the rain forests as they deserve.

Books Describing Rain Forest Trees

Of course, ways have been found to divulge much wanted information on important rain forest plant groups in a useful form. There are, for instance, several books of practical scope confined to identifying trees, of 20 cm dbh or more, or a selection of the most common species. The fine introduction in Brandis' *Indian Trees* (1906) explains the purpose of such books, and they make an excellent introduction to understanding the diversity of the forests, since they often have illustrations and usually a glossary. Here is a tentative list of titles.

Africa:

Aubréville A (1950) *Flore forestière Soudano-Guinéenne*
idem (1959) *La flore forestière de la Côte d'Ivoire,* 2me éd., 3 vol
Dale IR and Greenway PJ (1961) *Kenya trees and shrubs*
Eggeling WJ (1952) *The indigenous trees of the Uganda Protectorate,* ed. Dale
Irvine FR (1961) *Woody plants of Ghana*
White F (1962) *Forest Flora of Northern Rhodesia*

America:

Lindeman JC and Mennega AMW (1963) *Bomenboek voor Suriname*
Little EL et al. (1964, 1974) *Common trees of Puerto Rico and the Virgin Islands,* 2 vols
Pennington TD and Sarukhan J (1968) *Arboles tropicales de México*

Australia:

Francis WD (1951) *Australian rain forest trees,* 2nd ed

India:

Brandis D (1906) *Indian trees*
Gamble JS (1902) *A manual of Indian timbers,* 2nd ed

Malesia:

Corner EJH (1952) *Wayside trees of Malaya*
Browne FG (1955) *Forest trees of Sarawak and Brunei*
Meijer W (1974) *Field guide to trees of West Malesia*
Whitmore TC (ed.) (1972) *Tree Flora of Malaya*

How to Name Plants

No biological work done in a rain forest has any value unless the proper identity of the species involved is known or can be determined, and the specimens preserved for documentation. Without these precautions, the work is impossible either to verify or to repeat. Many poorly prepared scientists come up with pretexts to circumvent this responsibility, and try to conceal their fault when it is too late.

It is doubly sad, therefore, when leaders of rain forest research projects neglect their duty to teach all the participants how the plants and animals of the groups singled out for study are to be collected, preserved and identified. Like species knowledge and taxonomy, this discipline must be acquired early in any botanical career and scrupulously maintained, especially in the face of the enormous biological diversity of the rain forests.

If an amateur botanist contemplates collecting in the forest, he or she should first consult a knowledgeable botanist, listen carefully to the advice given and then, with the aid of published guidelines for collecting and preservation, painstakingly follow them. If the procedures are correctly followed, the consulted specialist will then give advice on whom to approach to identify the collection. Some help can usually be expected from a local Herbarium (there is an international, regularly updated directory, the Index Herbariorum). If the collected material is fertile, and of good quality, with twigs and leaves, well-labelled and in duplicate, a specialist will always welcome it. Dealing with poor specimens consumes both time and goodwill.

The Profile Diagram

One way to grasp the structure of a plot of forest vegetation is by making a diagrammatic section of it, called a profile diagram. The customary size of a plot involved is 7.5×60 m.

The making of a good profile diagram will keep a field crew busy for a week or longer. First, a reasonably level strip of forest has to be selected, which has not been obviously damaged by storms or other causes. This may already be hard to find, and if longer strips are wanted, it may be practical to stake them out in adjoining, but more or less zigzagging portions. The forest in front of the plot is generally cleared, very carefully so that no tree falls into the plot, and also the undergrowth in the plot is removed after collecting data on it and saving small representative sections. On grid paper, the plot is then meticulously mapped to scale, including the horizontal projections of crowns, for which instruments exist, and fallen trunks. Stems and crowns of all trees and lianas within the limit of the plot are measured with a 'dendrometer' and their vertical projections drawn. Corrections are made for distances and slopes. Photographs help, of course, but are not essential. Then herbarium material is taken from every tree, even if it is sterile, for identification purposes. Sketches made in the field in colour are worked out later, just as a surveyor's job is finished by a cartographer.

A modern, detailed diagram (Fig. 2.6) reveals:

1. The height of the trees;
2. The shape of the crowns;
3. The height and shape of the trunks, and the direction of the main branches;
4. The larger buttresses;
5. Distances between the trees, in projection;
6. The area of each crown, in projection;
7. The trees belonging to one species, and their relative size;
8. (Sometimes) the architectonic models of the trees;
9. The larger lianas and larger plants of the undergrowth;
10. The 'life phases' (of the past, present or future) of the larger trees;
11. The lower limit of the crowns of the emergents;
12. The position of each tree in relation to light and shade;
13. The effects of previous damage.

It is also possible to derive from a fine diagram (in combination with the field notes):

1. The diameter of each tree;
2. The 'basal area' of all the trees in the plot;
3. The relation between tree diameter and height;
4. The volume of the stems;
5. The volume of the crowns;

Fig. 2.6 Profile diagram of a piece of lowland rain forest, 60 × 7.5 m, in Andulau, Brunei, Borneo. Only trees of over 4.5 m are shown, all in all 53 species in 42 genera in 21 families; 9 species in 5 genera are Dipterocarpaceae (*dotted*) (Ashton 1964). This diagram was reproduced in several publications, probably due to the height of the trees, yet it is not typical. Oldeman (1974, p. 174) indicated spots where these large trees probably had been damaged (*stars*). He also drew the *three lines* that are reproduced here. The *uppermost* one connects the lowest branchings, the *middle one* indicates half the height of the trees, the *lowermost* one connects the tops of some small 'trees of the future'. It seems that a number of trees have grown up at the same time, after a gap in the canopy was made. The small 'trees of the future' of later date are apparently stunted in their growth

6. The stage of development of certain tree species;
7. The density of the canopy, at various heights;
8. The proportion between parts of the crown in the light and in the shade;
9. The proportion between the mature and developing parts of the canopy;
10. Important changes to be expected in the canopy during the coming decades, when decaying trees will give way to others, and emergents will dominate and cast more shade.

The earliest profile diagrams on a quantitative basis were published by Davis and Richards in 1933; a few have been reproduced in *The Tropical Rain Forest* (Richards 1952). Those by Hallé et al. (1978) show an enormous increase in sophistication. It is even possible, using their methods, to draw new conclusions from the earlier diagrams.

Up to 1975, no more than 30 ha forest in all had been drawn in diagrammatic form, according to Rollet (1978, p. 114), and these samples were still far from representative. Continued diagram construction will provide very welcome data for detailed comparison and better understanding of forest dynamics. See also Oldeman (1979), Richards (1952, p. 24) and Rollet (1978, p. 113).

Fig. 2.7 Several constructions used for the study of animal life in the upper levels of the rain forest (After Mitchell 1981)
(a) Hut for generators
(b) Light traps on pulleys for catching insects
(c) Sleeping platform
(d) Bat trap
(e) Observation platform
(f) Scientist being raised into the canopy, collecting samples on the way
(g) Camp hut

How to Look into the Canopy

The canopy of a tropical rain forest is a world of its own: climb the stairs of a 12-storey building to imagine its height. Most animals who live there never descend to the ground; and each species prefers to inhabit a different level. One way to observe them is to climb up a large tree oneself, as I often did (for botanical reasons) in eastern Java and in Sumatra, after exploring the way up from the ground with field glasses. Or better still, to attach a ladder to a big tree and to construct a platform a few square metres in size in its crown. In the eastern part of the Kutai Reserve in Borneo, Leighton and Thomas (1980) constructed one 39 m above the ground, the access being by a ladder made of nylon rope with wooden rungs. The whole thing cost less than US$ 250. Dr Leighton told me of his surprise at being heavily attacked by mosquitoes when he had installed himself to spend a night up there, while in the camp on the ground there were very few. Apparently the mosquitoes know well enough that most of the warm-blooded animals remain high in the canopy at night.

To get an idea of all that can be observed from such a platform, take the study by H.E. McClure (1966), an ornithologist. In 1960 in Malaya, he built a platform 43 m above the ground in a huge tree. An aluminium 'cover ladder' was fixed with chains to the trunk. He found it a fascinating place to be, since he had a clear view over more than 300 ha of magnificent forest. In the vicinity of his tree, about 1.6 ha, 60 large well-recognizable trees had

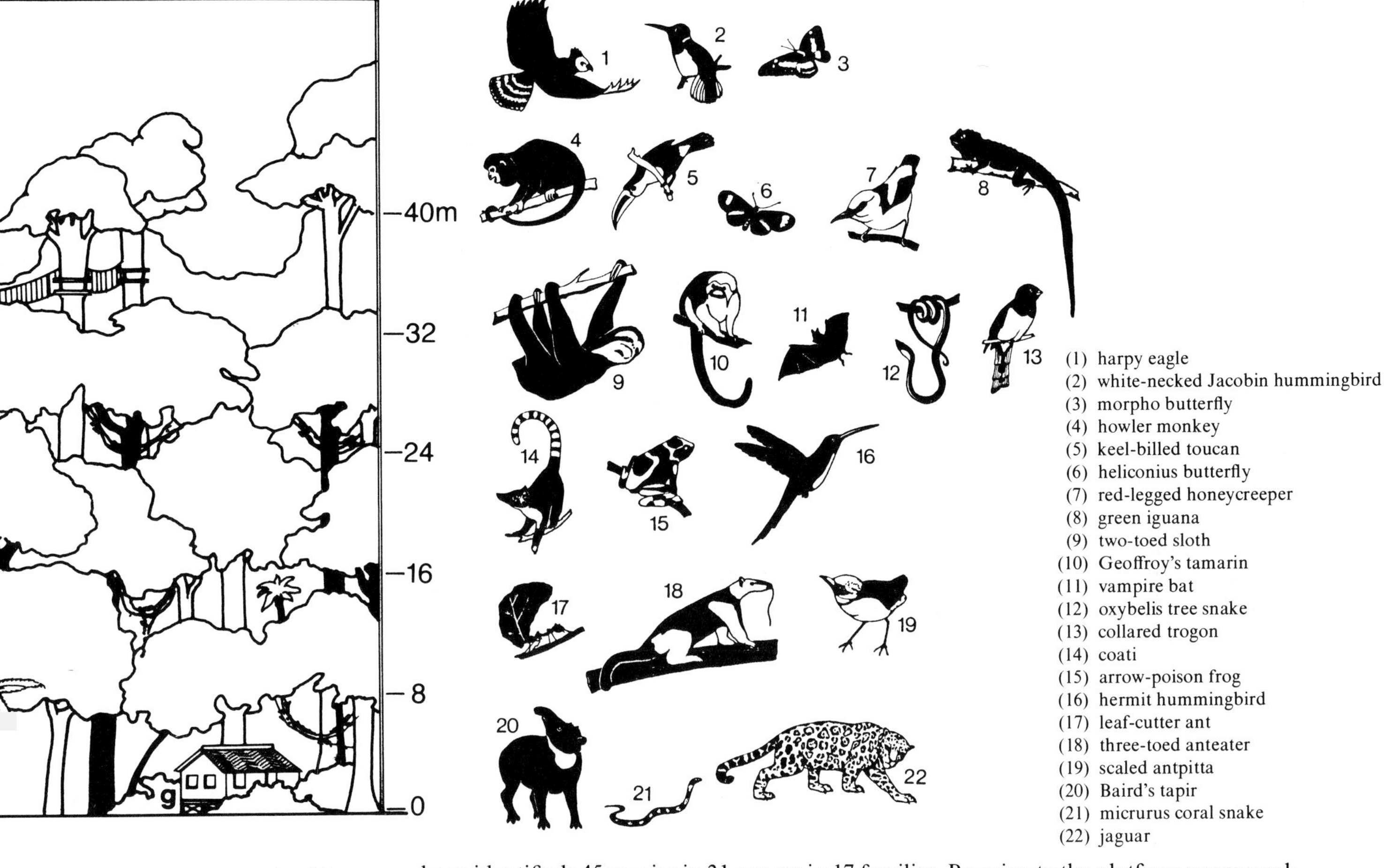

Different animals are found at different levels in the forest. Using the aerial walkway, Operation Drake studied the life in the upper canopy of tropical rainforests in Panama, Papua New Guinea and Sulawesi. This diagram shows the position of the walkway and some of the traps that were used. Larger animals such as tapirs and leopards obviously live at ground level, but even iguanas will ascend into the treetops where the climbers such as primates and arboreal anteaters live. Birds and many insects were also found to have their own favoured position in the forest.

been identified: 45 species in 31 genera in 17 families. By going to the platform every week, one could follow their phenology of flush, flowering and fruiting, correlating it with meteorological data, and could make observations on the birds, monkeys, squirrels and bats which were moving about and feeding. From 1960 to 1969 these phenological observations continued to be made with great regularity, and are summarized in a fine paper (Medway 1972). But since then, this precious tree has been sacrificed to road construction.

Towers or imitation trees have also been constructed in some research plots, often with several platforms to accomodate instruments registering light quantity, wind, rainfall; and in order to measure the temperature properly at different levels. Ideally, the tower should be made of wood, not metal.

Further improvements on the idea culminated in the invention of an aerial 'walkway', a kind of suspension bridge in sections, by which one can move horizontally through the canopy. An extensive installation of this kind was constructed as part of the scientific investigations conducted by 'Operation Drake' (Mitchell 1981, see Fig. 2.7). An informative booklet on this whole subject is A. Mitchell's *Reaching the rain forest roof* (1982).

Radio-Tracking Animals

It has become established practice to provide large mammals with a radio-transmitter on a collar. One of the more esoteric of these experiments was that of Montgomery and Sunquist (1978, p. 331) who applied the technique to sloths, the slow-moving edentates of the Neotropics.

On spotting a sloth, workers would climb up to the branch from which it was hanging by its long two-toed or three-toed claws, and cut the branch with a saw after catching the animal with a noose on an extendable aluminium pole. The gentle animal was then carried to the nearby field station, inspected, measured, weighed and fitted with a radio-collar and released. Its movements through the canopy could now be followed, and data collected about their selection of trees and on the leaves on which they feed. These sloths are interesting subjects with regard to social organization and competition: Montgomery and Sunquist speculated that sloths have been successful at keeping monkeys out of their range to a considerable extent. Several primates occupy similar niches in the Old World tropical forests where sloths are absent.

The scientists were also able to measure deep body temperatures by inserting a telemetry transmitter into a sloth's rectum. Since a sloth defecates only every 8 days, observations spanned a week, with luck. It is then seen that their body temperature goes up in the daytime and down at night, and that the animal moves to sunny spots when it is cool, and to the shade when it is warm (p. 343).

This does not exhaust the potential of transmitters, in the view of M.G.M. van Roosmalen. He told me that such devices of 1 cm^3 could be inserted in a seed, to trace its whereabouts.

By Canoe

Possibilities of studying the upper canopy by canoe often occur when tracts of forest have become submerged as a consequence of building a hydroelectric dam. D.C. Geijskes availed himself of such an opportunity in the 1960s when Lake Brokopondo in Surinam was filling up. When the leaves had fallen from the drowning trees, the nests of bees and other animals were fully exposed. Epiphytes, too, can then be inventoried, as they remain alive longer than the trees which have supported them.

This is Corner's lively description of an inundated swamp forest in Malaya (1978, p. 10): "During the three days at Danau we paddled in a canoe through the flooded forest. The force of the flood was lost among the trees, though impossible to stem on the main river. I could stand up at a height of twenty feet above the floor of the forest and collect from the tops of the undergrowth trees. Wherever we touched leaf, twig, trunk or floating log, showers of insects tumbled into the canoe. Everything that could had climbed above the water. Ants ran over everything. I bailed insects and spiders instead of water, even scorpions, centipedes, and frogs. All around there was the incessant swishing of the half-submerged leaves, the oblong-lanceolate form of which was eminently successful for the occasion, and the incessant honking of frongs and toads. Lizards clung to the trunks; earthworms wriggled in the water, with snapping fish.

I realized the importance of the hillocks in and around the swamp forest to animal life, for anything that could escape the flood must have fled there. We met no corpses. Pig, deer, tapir, rat, porcupine, leopard, tiger, monitor lizard, and snakes must have congregated on those hillocks in disquieting proximity".

Of course, on the Amazon river system where the annually recurring differences in water level may be 8 m and more, boats are indispensable. In the Anavilhanas Archipelago on the Rio Negro above Manaus, there is even a floating research station.

El Verde

Most of the study methods described so far consist primarily of observation, with the object of understanding the main features of the tropical rain forest. The only sophisticated items of equipment we have mentioned are the radio-transmitters used to investigate sloth ecology. There is little point here in describing all such apparatus, but the book on El Verde, *A Tropical Rain Forest*, well edited by Odum (1970), is a mine of information and ideas.

El Verde is a research plot on the island of Puerto Rico in the Caribbean. It is a rather high-altitude plot and supports a relatively poor, somewhat depleted, but quite interesting mixed forest. It is probably the best-analyzed forest in the world, and furthermore it was exhaustively examined both before and after an experiment with atomic radiation. The three-volume report gives excellent, detailed, well-illustrated accounts of the methods and techniques, down to the computer programs, of all the disciplines involved. Here is a synopsis in outline form:

1. A. The rain forest project (2 papers)
   B. The rain forest at El Verde (22)
      Aerial sensing and photographic study
      Relation of crown diameter to stem diameter
      Upper canopy crown closure
      Pollen analysis of Sphagnum bogs
      A system for representing forest structure
      How to describe the geometry of plants
      Keys for identification of seedlings
      Illustrated leaf key to the woody plants
      Effects of herbicides
      Climate at El Verde
   C. The radiation experiment (5)
2. D. Plants and the effects of radiation (20)
      Tree growth
      Seedling diversity
      Lichens
      Phenology
   E. Animals and the effects of radiation (14)
      Lizards and frogs
      Niche separation of tree frogs
      Population seasonal changes of birds
      Aquatic communities in bromeliads
      Termites
      Insects
      Microarthropods
   F. Microorganisms and the effects of radiation (10)
      Fleshy fungi
      Mycorrhiza
      Root, soil and litter microfungi
3. G. Cytological studies within the irradiated forest (6)
   H. Mineral cycling and soils (22)
      Rain forest structure and mineral cycling homeostasis
      Biomass and chemical content
      Hydrogen budget and flow
      Electrical conductivity and flow rate of water

Leaching of metabolites from foliage
The phosphorus cycle
Mineral retention by epiphylls

I. Forest metabolism and energy flows (10)
Metabolism and evapotranspiration of some plants and soil
Overview

Looking down from the Air

High tree-based or free-standing platforms, as we have seen, make detailed observations and data collection possible in a small area, and of a limited number of trees. But, and this is especially true in an even climate, the outbreak of leaves, the flowering and the fruiting is so irregular that a much larger area must be placed under observation if patterns in time and space are to be detected. Such patterns largely determine the movements of animals, and thus propagation, selection and, ultimately, composition.

In 1978, the Hladiks, a French couple (see Hladik and Hladik 1980), purchased a 12 m^3 balloon which they took to the forest, filled with gas and allowed to rise to 800 m. Fixed to the balloon was a camera, operated by radio signals, which produced 'instant' colour pictures 20×25 cm of about 4 ha forest from a height of 250 m above the canopy. At this height, individual trees are perfectly recognizable, and a series of valuable observations can thus be made on them, and the photographs used to guide a scientist to the places where changes have been spotted.

Gliders and mini-helicopters are variations on the balloon idea. Photographs at scales between 1: 20,000 and 1: 100,000 can be made from all airborne craft. Mounted and carefully juxtaposed, they can be viewed in stereo.

'Ordinary' air photos are made by keeping the camera angle perpendicular to the earth. Their success depends on clear weather, a rare condition in an ever-wet climate. Clouds can, however, be penetrated by radar waves, and considerable rain forest photography is done with SLAR (Side-Looking Airborne Radar) which records strips to the left and right of the aircraft. This method was used to map the vegetation of Colombia and for the great RADAM project of Brazil.

Monitoring from Satellites

But air photography is costly and is, therefore, hardly suitable for the continuous monitoring of forest cover.

Monitoring over long periods is a particularly important device to detect those changes which are only with difficulty detected by a field observer. To monitor the changes in soil cover, a grid is placed over the subsequent images to enable comparisons to be made in one square after another.

While the word remote-sensing technically means all surveying from a great distance, including that made from aircraft, it is most commonly associated with satellite imagery (Fig. 2.8). A satellite encircles the earth in an orbit well above the atmosphere usually at a perpendicular distance of about 900 km. Over a path 185 km wide, it picks up light reflections in four different wavelengths, known as bands: band 4 for green, band 5 for lower red, band 6 for upper red, band 7 for near-infrared. The reflections are transmitted to the earth, in units. The number of units per whole image determines the satellite's power of resolution. In the present satellite imagery (known as LANDSAT), these units or picture elements ('pixels') cover 0.44 ha or even 0.325 ha. Under magnification, objects on the photographs more than 79 m apart can normally be distinguished; roads and rivers can also be detected, the roads

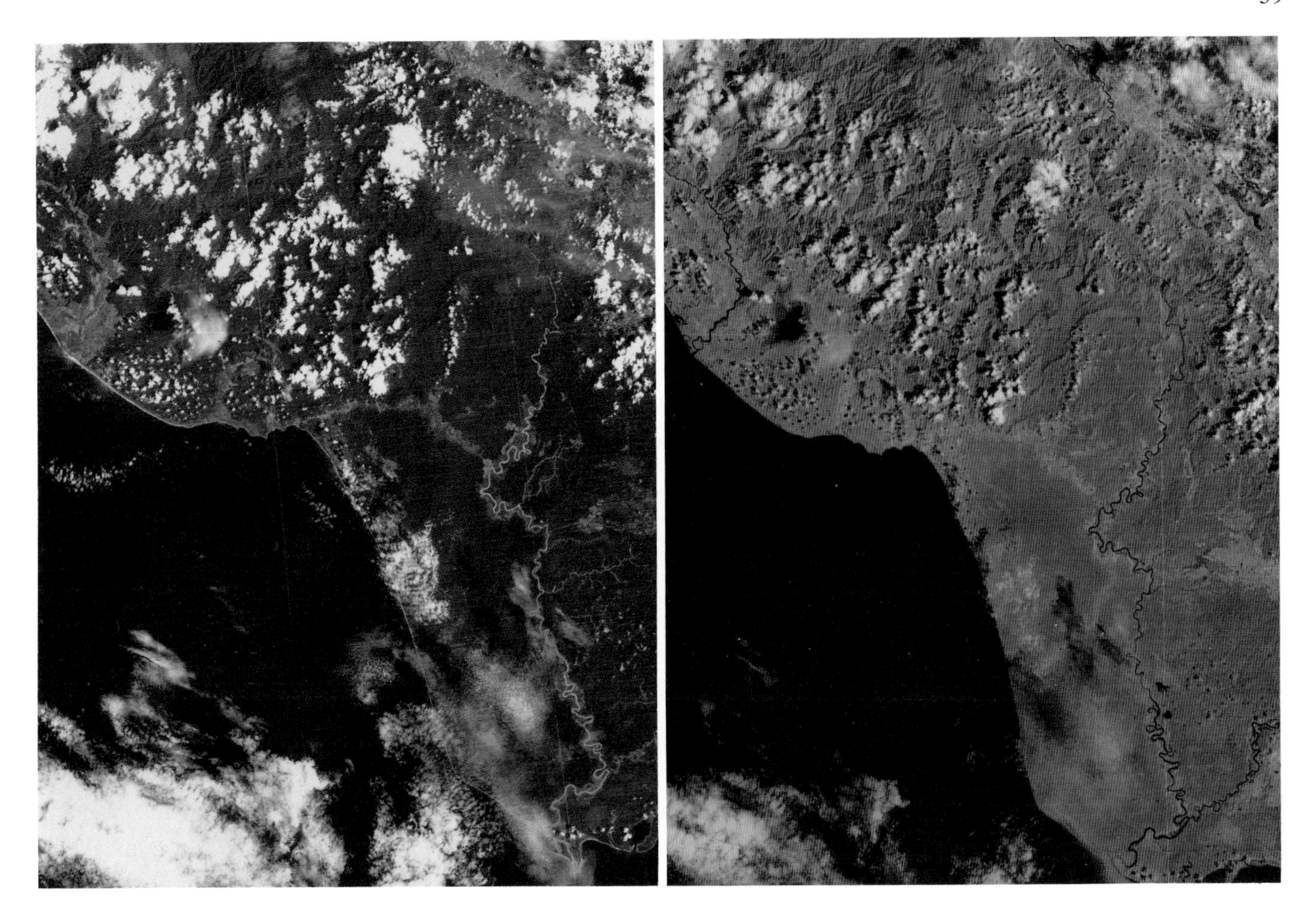

Fig. 2.8 Two satellite photographs showing the southern part of Aceh, North Sumatra, from ca. 97° to 98° E and ca. 2°30′ to 3°30′N. *Left* band 5; *right* band 7. The Alas river flows from the north, meandering under the name of Simpang Kiri to the Indian Ocean on the west side. *Light grey areas* (clearly visible along the river) are deforestated, *darker areas* are probably intact. Over the Gunung Leuser Reserve clouds are obstructing the view, particularly on band 5

made for logging operations usually showing up as a cluster of pale worms. Recent clearings are easily distinguished from intact forest by their lighter tone since they reflect more light. As a secondary forest matures, however, this difference vanishes, and an image with more than 10–20% cloud cover becomes almost worthless in every respect. To overcome this obstacle, the satellite is placed in an orbit at a speed synchronous with the earth's rotation, so that it passes over the same spot every mid-morning when the cloud cover is minimum. Scanning by computer permits the quick detection of (illegal) impacts on the forest cover. GEMS, the Global Environment Monitoring System, developed by UNEP and FAO, and set up in the late 1970s is an excellent idea and will no doubt be continued and perfected. Now it is at least possible to determine, worldwide, where the world's forests are being attacked and destroyed, and sometimes how.

Simplifications

So integrated and complex is the rain forest system that those dealing with it often ignore one or another of its separate components. The FAO inventories (FAO, 1981; p. 7) are an example of this. The only proper way to study, to collect in or to inventory our rain forests is to ask incessantly: what has been left out?

The rewards of such a questioning attitude are many. Sometimes the animals are ignored as a component, and thus their vital services as pollinators and seed dispersers are forgotten. Others ignore insofar as possible the factor of time, with the result that both evolution and future possibilities are discounted. Species diversity is seldom adequately treated; the forest's potential in terms of non-timber products is then naturally left in limbo. Traditional mechanistic thought encourages such tunnel vision; some aspects of physics and chemistry actually depend on it.

Yet, at the same time, there is a need to handle big chunks of 'total reality' in a scientifically responsible manner; modern science meets this need by using a variety of mathematical models. Holdridge's et al. impressive work along these lines is revealed in his book *Forest Environments in Tropical Life Zones* (1971). This is a remarkable work on three accounts: (1) he has amassed an enormous number of measurements, 110,000 from nearly 15,000 trees in 46 locations; (2) these measurements were all, seemingly, contemporaneous: time as a factor is absent from the calculations; (3) they represent a high level of abstraction. He converts a real profile diagram of a strip 60 m long into an 'idealized profile' projected over a length of 100 m, which differs more from the original than it resembles it.

The Holdridge approach is a typically deductive approach. A program is first written in detail and then carried out punctiliously; the sequence is material—method—results. The elements of observation, surprise, and, perhaps, of discovery, are pushed aside. It would be inconvenient if reality were found to be different from the model.

Oldeman's (1983) approach to forest complexity seems more appropriate. He distinguishes various sorts of units. An eco-unit is a piece of vegetation which has started to grow in one homogeneous area simultaneously, for instance, where a tree has crashed. A chrono-unit is the minimal area where all stages of one eco-unit occur (growing up, mature state, decline). A silvatic unit contains chrono-units of all eco-units at one site. We will come back to this system in Chapter 8 (Primary and Secondary Forest).

Simplifications according to area are, of course, common; so are aut-ecological studies of single species.

It also seems logical to confine scientific work to one life form, or one ecological niche, or one small group of biological relations.

Oldeman differs from Richards in his admirable emphasis on the dynamism of the forest. While Richards gives great attention to the forest as a whole, he portrays this whole as something essentially static and immutable, an outlook prevailing in all earlier publications as well. Oldeman emphasizes the changes in the canopy: growth, decay and replacement. Perhaps this explains why Oldeman fails to distinguish the three layers or strata in the canopy: in a structure which incessantly renews itself, however slowly, the significance of layers remains obscure. Yet they are apparent, and it is the existence of such strata which enabled Richards to discover some order in the green chaotic mass.

Another polarity is that between aut-ecology and syn-ecology. In the former, the organisms function essentially in isolation. It characterizes the views of Ashton and Oldeman. Both have a background in tropical forestry, and are deeply involved in problems concerning forest management for which an aut-ecological view is indispensable. Pollination and dispersal, in such a view, are mechanisms for the transport and exchange of genetic material. On the other hand, it is syn-ecology, in which plant-animal relationships come first, which occupies the conservationist. Plant-animal relations constitute the forest web of life: the durian and the bat, is one example we will discuss later.

My chosen manner of simplification, for of course one must simplify, is to try to make this immense subject comprehensible through selected examples. Each of them has its story to tell, often in connection with the others. If I succeed in my aim, a fairly coherent picture will

Flora Brasiliensis, tabula XXVIII. 'Forest, intractable because of roots and lianas.' Coastal area of Brazil, drawn by B. Mary, 1836, published 1847. The tree is a *Ficus* (Moraceaea), the development of the roots resembles the freshwater swamp forest. Aerial roots (branching below) and stems of lianas well visible. On the ground, as well as epiphytic on the tree trunks, are several orchids (Martius 1840–1869)

gradually emerge, consisting of a body of knowledge with which to answer most questions that may come up.

This method of illustrating the subject leans closest to that of Corner's. His personal knowledge is, of course, much wider, his strokes are firmer, his palette richer in colour; and his evocations are always magnificent. As he is the best writer on the tropical plant world ever, perhaps I should have kept silent! But there are new things to relate.

3 Climate

The Decisive Factor: Heat

The very phrase, 'tropical rain forest', indicates that the existence of this type of vegetation is primarily conditioned by climate. The tropics are generally considered to be located between latitudes 23° N and 23° S, the 'tropic of Cancer' and the 'tropic of Capricorn' respectively. At the equator, the mean temperature at sea level is around 26°C, and the prevailing 12-h day length shows little seasonal variation. Though hot, the tropical region therefore never becomes excessively hot compared to desert regions at higher latitudes, where days become longer in summer time, and these high, but not too high, temperatures permit the existence of a true tropical rain forest. In other permanently wet areas with cooler temperatures, for example in New Zealand, a kind of rain forest exists, but not with all the characteristics typical of a tropical rain forest growing between the tropics. There is a close relationship between heat (plus moisture) and the typical structure of a tropical rain forest with its giant trees and buttresses. But what is it? This is an intriguing question. In the humid tropics the sky is usually overcast: another factor in tempering the heat. The diurnal

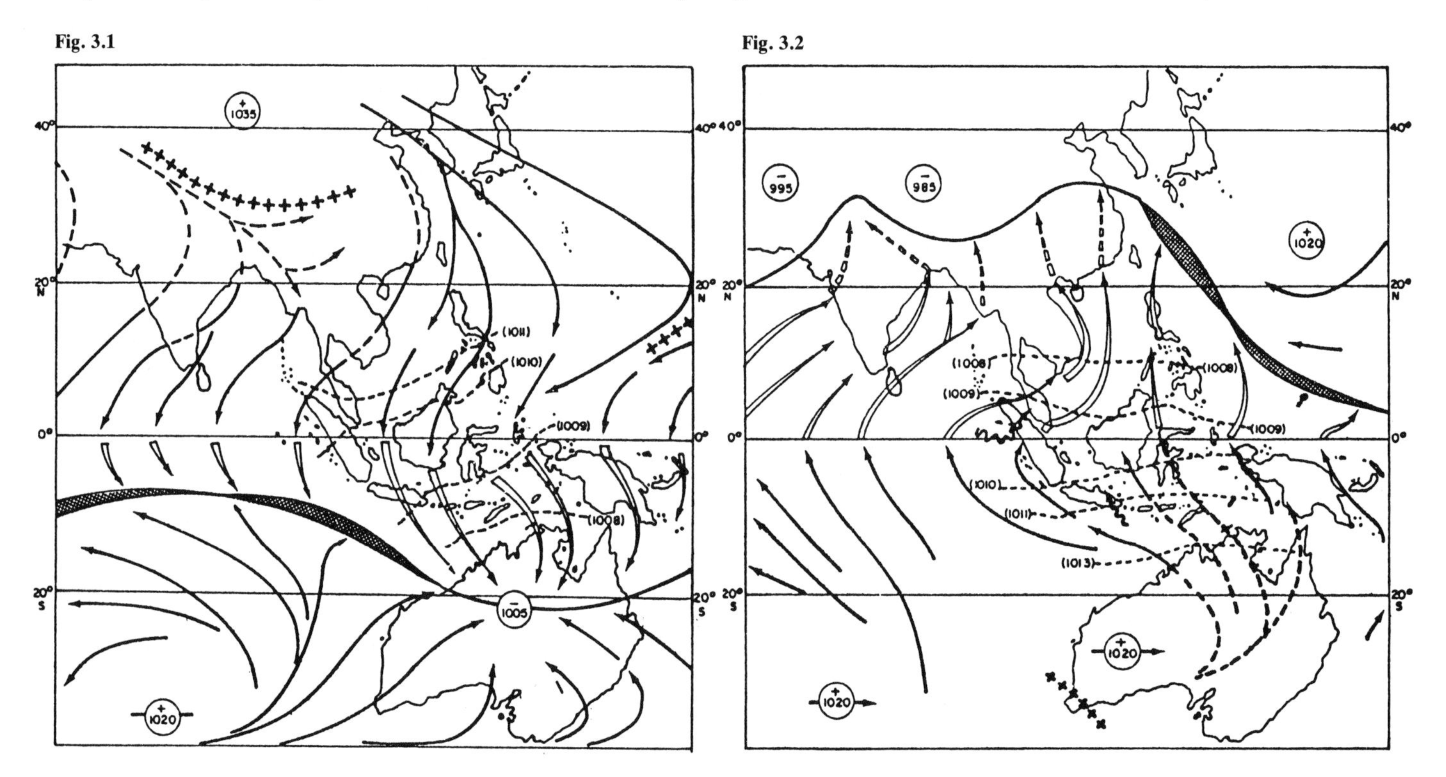

Fig. 3.1 Atmospheric circulation over Southeast Asia in January (After Fontanel and Chantefort 1978)

Fig. 3.2 Atmospheric circulation over Southeast Asia in July (After Fontanel and Chantefort 1978)

temperature fluctuation is only 3° to 6°C in equatorial regions, and around 7–12°C near the tropics of Cancer and Capricorn. The maximum temperature seldom exceeds 34°C. A mean temperature of 20°C in the coolest months of the year marks the boundary of the tropical rain forest. This boundary is indeed near the tropics of Cancer and Capricorn. Temperature, however, is not only related to latitude, but also to altitude. In the equatorial region the 20°C boundary is usually reached at an elevation of 1000 m, since temperature decreases by 0.61°C per 100 m increase in elevation. At higher latitudes, the altitudinal boundary is correspondingly lower.

Precipitation: The Second Limiting Factor

While a temperature boundary of 20°C or above forms the first requirement of the tropical rain forest, it can only grow, as the word 'rain' indicates, if there is sufficient moisture. But this is not always the case. In the tropics the extremes of wet and dry seasons are much more pronounced than temperature variations between seasons. In the equatorial zone, where air pressure is low, the water evaporating from the oceans, rivers and forests ensures that the ascending air has a high moisture content and is unstable. As it cools, rain clouds are formed and when the saturation point is reached heavy precipitation and thunderstorms occur. Energy is released as the air ascends to altitudes over 10 km, then spreads out in northerly and southerly directions meeting at 30° N and 30° S respectively, the so-called westerly jet stream, which forces the air to descend, picking up moisture like a sponge. At an altitude of 1500 m the air streams back to the equator. These trade winds have a northeasterly direction north of the equator and a southeasterly south of the equator due to the earth's rotation. These air streams meet in the zone of low pressure near the equator, called the Intertropical Convergence Zone. Characterized by low wind speeds, the 'doldrums', widespread cloudiness and precipitation, this zone is not stationary, but from January, when it is located 5° to 10° S, moves northwards to its most northern position (20° to 25° N) around August. It then moves back towards its most southerly position. Figures 3.1 and 3.2 illustrate the atmospheric circulation over Southeast Asia in January and in July. Figure 3.3 shows some climatic diagrams in Thailand, Malaysia and Indonesia.

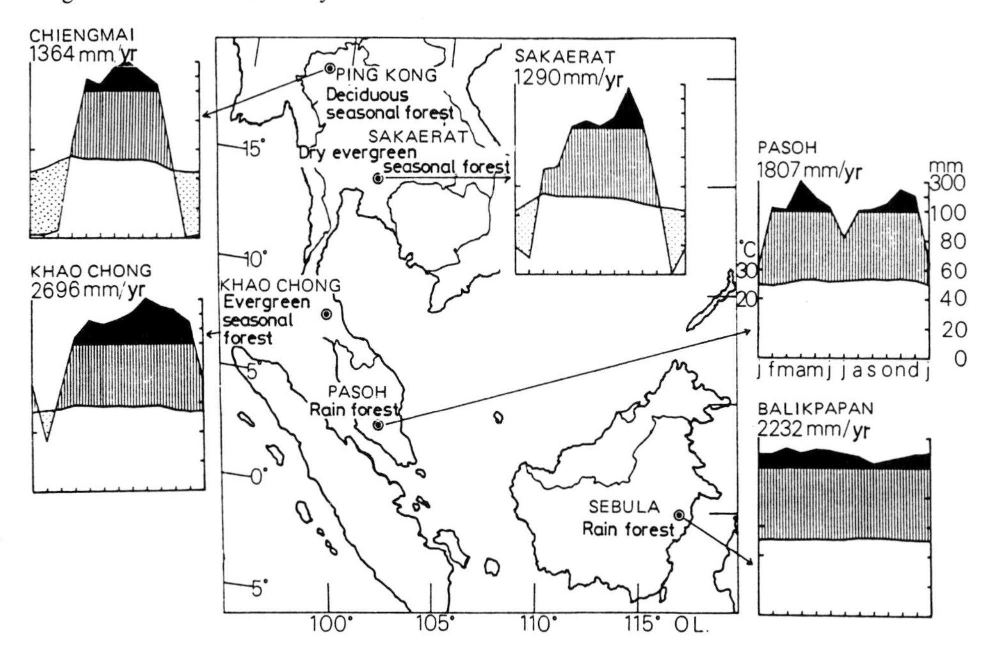

Fig. 3.3 Climate diagrams from Thailand, Malaysia and Indonesia. The *lowest*, almost horizontal, *line* indicates the mean monthly temperature. The other *curve* indicates the mean monthly precipitation. Surplus rainfall is *blackened*, deficit in water is *dotted*. Compare Pasoh in the ever-wet region where 1807 mm sustains a rain forest, with Khao Chong, where with 2696 mm only monsoon forest can grow (Kira 1978)

During the northern hemisphere winter, an area of high pressure develops over Siberia and over the North Pacific, while in the southern hemisphere we find a high pressure cell between Africa and Australia, but a trough of low pressure over New Guinea and Australia. As a consequence, northerly winds bring relatively cool and dry air over India, Burma, Bangladesh, Thailand, Malaysia, the western regions of The Philippines and as far as the northeastern coast of Sumatra. The northeasterly winds, however, pick up moisture passing over the South China Sea and cause extremely heavy rainfall along the east coast of peninsular Thailand and Malaysia and also along the east coast of The Philippine islands. These airstreams eventually cross the equator and become northwesterly due to the earth's rotation, bringing the humid air towards Indonesia.

During the northern hemisphere summer, the conditions are reversed. Low pressure cells develop over India and Pakistan, due to the intense heating of the continent, while a cell of high pressure develops over Australia. Southeasterly winds, cool and very dry, originating from Australia, sweep over the southern part of West Irian, the Lesser Sunda Islands and East Java. As they blow over water they gradually become more humid, and their effect is therefore less pronounced in West Java than in East Java. Cold air flowing north from the southern Indian Ocean warms up, becomes more humid and after crossing the equator and reaching the Indian continent from the southwest, marks the onset of the rainy season there, and later in Bangladesh, Thailand and The Philippines. Long, dry seasons therefore occur at higher latitudes, while near the equator the dry seasons are less pronounced. In areas subject to long, dry seasons, tropical rain forests will not grow, since it is not the total amount of annual rainfall that counts, but its annual distribution. A guaranteed monthly minimum of 60–65 mm is more important for a perennial vegetation than a peak of rainfall in the wet season followed by a 3-month rainless period. In his book about the rain forest of Surinam, Schulz (1960) works with 'moving totals' of rainfall over the preceding 30 days to incorporate the aftereffects of drought and wet spells. He defines as a wet month any month with more than 100 mm rainfall, as a dry month one with less than 60 mm and as a "not-wet" month one having 60–100 mm rainfall.

Averages and Exceptions of the Rain Forest Climate

For more than one reason climate is a difficult subject. Despite the fact that a humid tropical climate is generally considered very uniform, there are many factors, physical and biotic, that interact every day to create changes. In the rain forest the strongly buffering influence of the forest canopy is particularly interesting. One must also always bear in mind that although average conditions determine the usual course of events in an ecosystem, and it is therefore useful to measure and correlate long-term climatic averages, it is the exceptions – the longest dry spell, the strongest wind gusts, the heaviest volcanic ash rains of the century, the largest earthquake of the millennium – that will limit what is possible in an ecosystem. Schulz considers that very dry years have large and lasting effects on a rain forest, and that the trail of a typhoon can be traced decades later.

Wind, Thunderstorm, Downpours

In forest areas on hilly terrains, the presence of wind is difficult to ascertain. Inside the forest and especially at or near ground level, it is usually calm. Sometimes one notices a single large leaf, perhaps of *Curculigo* (Hypoxidaceae), slowly moving on its vertical stem, blown by some very local airstream, while the rest of the vegetation remains deadly still. "A ghost", whisper the helpers. Then, on emerging from a forest near the open sea, one can be surprised by the

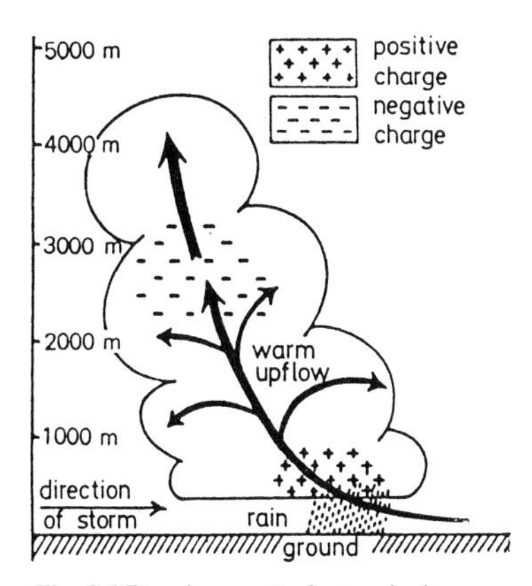

Fig. 3.4 Development of a tropical thunderstorm. The *arrow* indicates the upward movement (up to 160 km^{-1}) of hot air (After Dobby 1960)

wind's sudden force. (Yet, it does not appear that the proximity of the ocean influences either the habitat of the forest or its composition.)

During thunderstorms, strong wind gusts can occur that can break branches or uproot trees, particularly after heavy rainfall when the soil is soft. In this way, gaps in the forest roof top can become wider. And although the long-lasting storms common at higher latitudes are rare in the tropics, tropical typhoons can leave a trail of destruction, broken branches and uprooted trees, destroying the regular development of the roof of the forest canopy (see Fig. 3.4). Even ordinary rainfall can be a destructive force.

After each shower, only about 3 mm water is retained in the forest canopy (Mohr 1946), while about 80% of the precipitation penetrates through the canopy and reaches the ground. Although the forest canopy breaks the force of the rain, big drops of water form again on the leaves, and after falling 7–9 m regain 95% of the kinetic energy, thus causing erosion wherever the ground is bare.

Vertical movements of air masses high in tropic skies can be terrifying. Cumulus clouds develop in no time, thousands of metres high. The sky becomes cloudy and rainstorms of unbelievable intensity (up to 30 mm p h^{-1}), but relatively short duration pour down. Lightning often accompanies these sudden rainstorms (see Fig. 3.5), and one encounters occasionally badly damaged trees; Whitmore (1975) tells of a 36-m tree in the Botanical Garden in Singapore that was completely decapitated. Sometimes a forest fire will occur, even in a rain forest, when excessive logging and unusually dry climatic conditions have combined to make it vulnerable.[2]

The Treacherous Climate

Therefore, the climate can be rough, even in a tropical rain forest. In the cities of the tropics this rough climate is generally not encountered, and until recently, meteorological stations have often been located near such population centres. The recording of weather data outside these centres in the tropics has often started only recently and a decade of observation cannot provide sufficient data to arrive at significant climatic averages. Futhermore, most remote weather stations have been installed primarily for aviation or water management purposes and only secondarily for vegetation studies.

Some climatologists are also tempted to quantify climates and to extrapolate climatic data for large areas from the comfortable conditions of their offices. The resulting schematic image often masks or omits small zones with strong bioclimatic differences. If one accepts that a rain forest climate is automatically optimal for all plant growth, one can make enormous mistakes, for example when planning regional economic development. Tosi (1975) has most strongly emphasized this point. It must be remembered that plant growth and agriculture are not identical. A continuous rainy climate leaches the soil of all its nutrients, and the optimal climate for a rain forest vegetation exists only under forest cover.

Vapour Pressure Deficit Inside and Outside the Forest

The total quantity of precipitation and its annual distribution does not tell us much about its effect on the vegetation. Plants maintain a sap flow from the roots to the branches, and water transpires through stomates in the leaves. A plant can press water from the ends of the veins

[2] There is evidence indicating that the disastrous fires in Borneo in 1983 were set deliberately (as a cover-up for illegal logging activities) and got out of hand because of the peculiar drought conditions.

Fig. 3.5 A 'storm forest' in Malaysia after 88 years. The two trees in the *centre* are *Shorea parvifolia* (Dipterocarpaceae). The *kink* in the trunk halfway is probably caused by the growth of lianas. Photograph Forest Research Centre, Kepong, in Whitmore (1975)

□ : at air temperature of 31°C
• : at air temperature of 22°C

Fig. 3.6 Relationship between relative humidity and vapour pressure deficit at two different air temperatures; one typical for the soil surface (□), the other for the canopy (*dots*)

Fig. 3.7 Mean daily fluctuations of the vapour pressure deficit at the canopy of the forest (at 30 m), at 5, 1.5 and 0.1 m above the forest floor. The deficit is zero at all heights at sunrise, whereas a maximum deficit occurs at 14.00 h. Surinam, dry season of 1957 (After Schulz 1960)

into its young parts; in this way a continuous flow is maintained even in humid air. This is important because the amount of water that can transpire depends on the vapour pressure deficit of the air, which is the difference between saturated and actual vapour pressure at a given temperature. Relative humidity, a better-known term, is the percentage of saturation at a given temperature. The saturated vapour pressure at 26°C is 33.61 mbar. At a relative humidity of 80%, the actual vapour pressure is 26.89 mbar; the vapour deficit is therefore 6.72 mbar. This relationship is illustrated in Fig. 3.6.

Water vapour in the air has two sources. First, the common evaporation of water from wet surfaces (e.g. on leaves). Secondly, the active evaporation from leaves through their stomates (transpiration). The quantity of vapour derived from both sources also depends, however, on the saturation deficit, which decreases as the amount of vapour in the air increases, but which increases at higher temperatures or with higher wind speeds. The wind speed and the temperature of the air also vary with the height above the ground, the forest canopy's tempering effect notwithstanding. The result is a gradient in the diurnal fluctuation

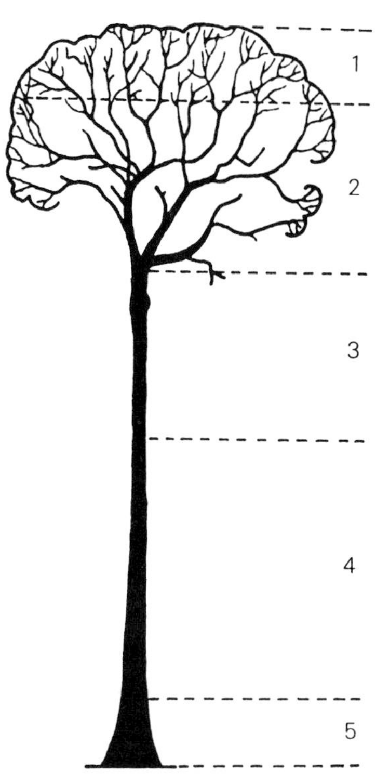

Fig. 3.8 A forest tree with five microclimatic zones (Longman and Jenik 1974)

of the deficit within certain limits that some plant species can tolerate better than others. Both high forest trees and pioneer vegetation in open terrain tolerate these deficit extremes relatively well, while other rain forest plants develop better where these fluctuations are small. Each species has its own optimum for transpiration, and is characterized by the saturation deficit limits it can tolerate.

The graphs of Fig. 3.7 show clearly the buffering effect of the forest canopy. At 1.5 m above the ground, the diurnal fluctuation of the saturation deficit, which indicates the regularity of the sap flow in the plants, is about 50% of the diurnal fluctuation at the upper surface of the canopy. At 10 cm above the ground, the diurnal fluctuation is only 25% of the fluctuation in the canopy.

Longman and Jeník's Tree

Each vegetation creates its own microclimate. The taller and more permanently closed the canopy, the greater the protection it provides; this can be expressed as a reduction in diurnal fluctuations. But since these fluctuations are externally caused, and since the protection given by the forest roof is only relative, the outside climate does eventually determine the climate inside. Because the forest canopy is high, at least 25 m, and generally fairly dense, the protection inside a tropical rain forest is large, yet between the outside climate and the microclimate inside one can identify a series of microclimates by examining the epiphytic growth on the trees.

Longman and Jeník (1974) distinguished five such levels on a forest giant (see Fig. 3.8):

1. Top of the crown, exposed to weather, with micro-epiphytes.
2. The protected part of the crown, dominant zone of epiphytes.
3. Driest upper part of the trunk, with flat, crust-forming lichens.
4. Moist lower part of the trunk, with lush growth of lichens and mosses.
5. Trunkbase with buttresses, with moist shadowy holes with abundant moss growth.

This classification is not related to the well-known stratification of the forest as defended by Richards (1952), and doubted by Hallé et al. (1978). We will discuss this later (p. 98).

Temperature Inside and Outside the Forest

With regard to the diurnal fluctuations of temperature inside the tropical rain forest, Fig. 3.9 illustrates a pattern similar to that of the saturation deficit. The temperature in the forest is usually 7–10°C less than it is outside. Since it is a fact that with an increase of 10°C all chemical reactions occur two to three times as quickly, the significance of this difference is enormous. Not only do biochemical reactions vary in plants at various heights above the ground, but also the oxidation of humus, and thus the nitrate concentration, essential for protein building in plants, is affected. As oxidation of humus increases at higher temperatures, many of the nitrates released will be washed away by rain. Inside the forest, the temperature of the so at depths between 5 cm and 40 cm is rarely lower than 23°C (Schulz 1960). Therefore, all reactions in the soil, such as the root functions, decomposition of organic matter or activities of the soil fauna, which proceed continuously, will be disturbed if normal conditions are altered, for instance when an opening is made in the forest cover. The greater the opening, the stronger the diurnal fluctuations, as illustrated in Fig. 3.10. At a depth of 75 cm the eoil temperature remains constant at 31°C, which is equal to the average air temperature inside the forest hall. Just above the soil surface inside the forest, the air temperature fluctuations are small (22–27°C). A maximum of 34°C may be attained high up in the canopy and there is apparently a gradient of increasing fluctuations.

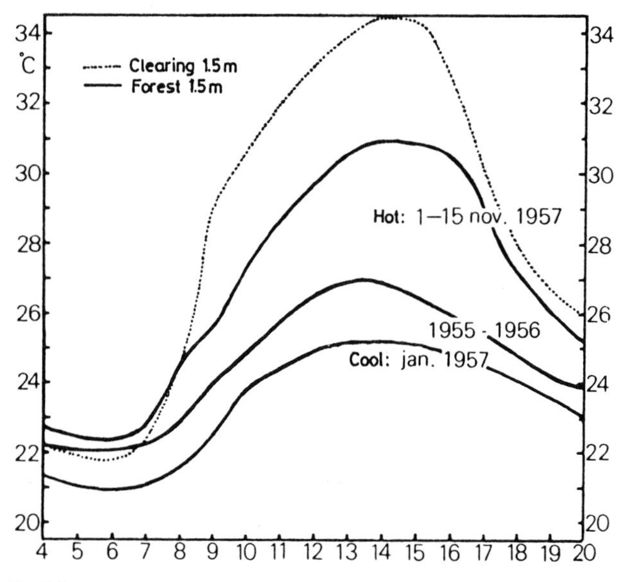

Fig. 3.9

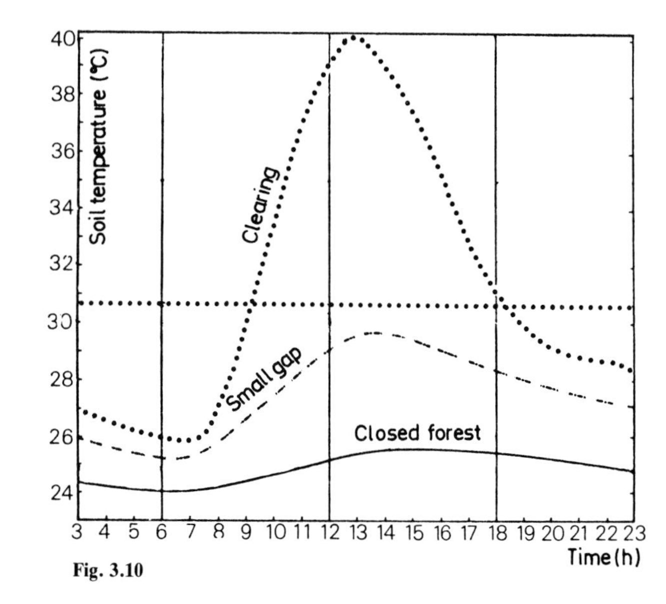

Fig. 3.10

Fig. 3.9 Mean daily fluctuations of the air temperature during the hottest and the coolest season in Surinam at 1.5 m above the forest floor; also the mean daily fluctuation of air temperature during the hottest season at 1.5 m in an open spot in the forest in Surinam (After Schulz 1960)

Fig. 3.10 Mean daily fluctuationof the soil temperature in Surinam, at 2 cm depth during the dry season inside the forest (*solid line*) at a small open spot (*broken line*), at a large open spot (*dotted line*) and at 75 cm depth at a large open spot (*horizontal dotted line*) (After Schulz 1960)

Light Intensity Inside and Outside the Forest

The effect of the tropical forest on the light intensity under the canopy is illustrated in Fig. 3.11, from Kira (1978). The reduction of light is enormous, although other sources indicate that at least 1% reaches the surface. There is always sufficient light to work in the forest, however, unless it rains late in the afternoon; and somewhere there is always an opening in the canopy through which a ray of sunlight will penetrate, illuminating a small area for a short while. Even illumination of only 40 min day^{-1} is of great importance for plant development, as Schulz found out. Light is the source of energy for growth, and while most forest herbs start as, and remain, shade-lovers, the typical forest tree will start as a shade-lover, but change as it grows into a sun-lover. The essential presence of light must be combined with a regular sap flow, determined by a constant soil temperature, a certain small vapour pressure deficit and an ample supply of water. If one of these variables changes more than the other, this variable becomes a limiting factor. The rates of the two biochemical processes in the plant, photosynthesis and respiration, have a certain optimum, conditioned by the leaf temperature. Both rates are graphically illustrated in Fig. 3.12. The point where the two curves meet is called the point of inversion. At higher temperatures, the photosynthetic rate becomes the limiting factor. It appears that the shape and size of a tree is strongly related to these inversion points. A giant tree from the forest, if planted in a park, will receive too much light, and the vapour pressure deficit will be somewhat too high. As a consequence, park trees develop with short stems and more spherical or conical crowns. In the tropical forest we see the reverse: a tall stem and a more elliptic to cauliflower-shaped crown.

From ground level to the upper canopy of the forest, there is an increase in light intensity, and an increase in vapour pressure deficit (or decrease in relative humidity). The shape of these two curves depends on the density of the forest canopy at different heights. The pictures in Fig. 3.13 illustrate these effects. The density is determined by the growth activity at a certain height.

The process of growth depends on the prevailing quantity of light, CO_2 and water. The tree manufactures bioproducts as assimilates. Part of these assimilates are used to fuel

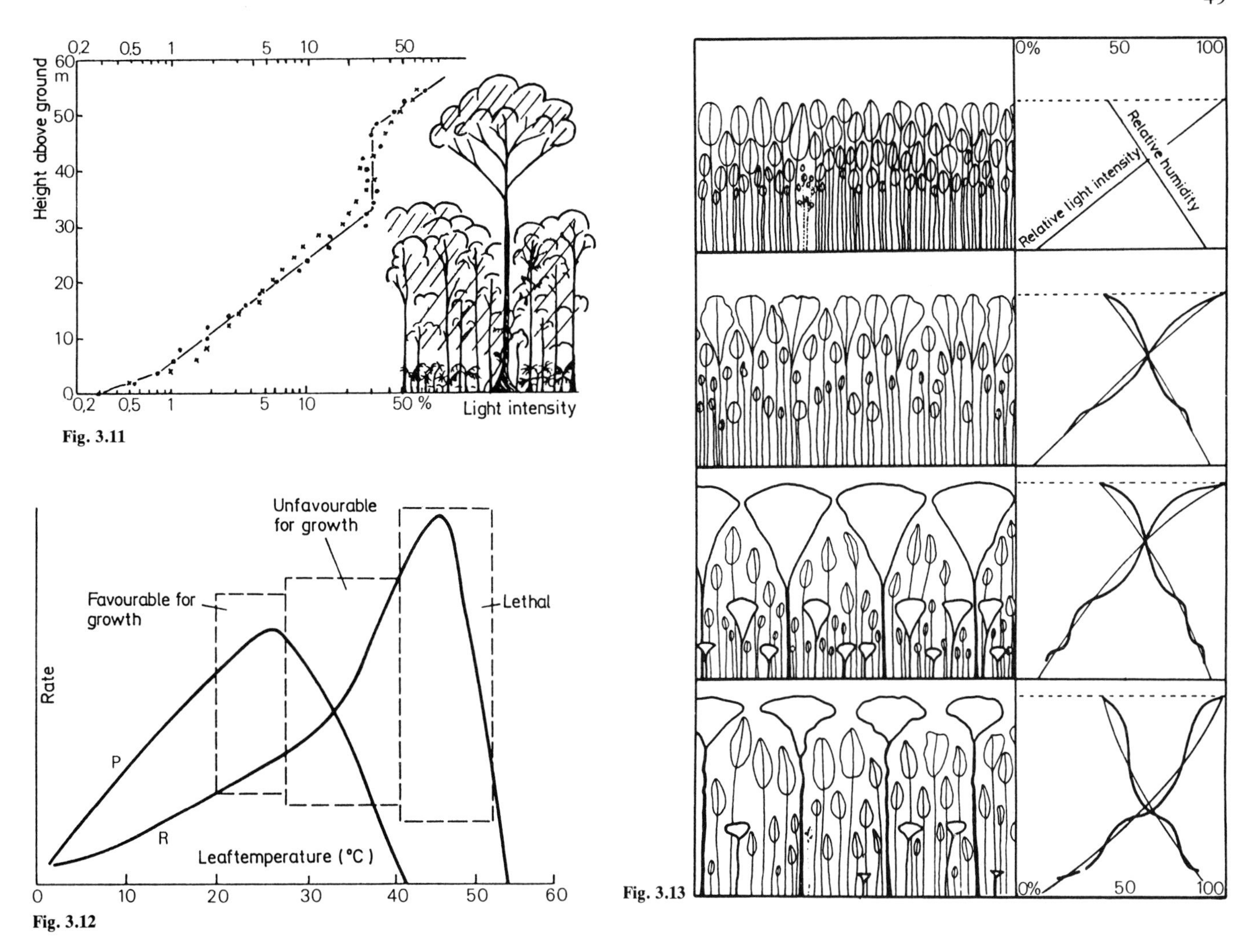

Fig. 3.11 Intensity of light penetrating through the leaves of a tropical forest at various heights above the forest floor (Kira 1978)

Fig. 3.12 Relationship between leaf temperature, rate of photosynthesis (*P*) and the rate of transpiration (*R*). The point where the two curves meet is called the inversion point (After Grubb 1974)

Fig. 3.13 Four phases in a forest, developing on an originally open terrain. The graphs on the *right* represent curves of the relative humidity and relative light intensity in relation to height. In reality, as indicated, the lines are not straight. The desintegrating crowns of the old forest allow more light to be transmitted, which is reflected in the curves (After Oldeman 1978)

respiration and growth processes, and a part will be transformed into biomass. Under a closed forest canopy, the water supply and relative humidity are ample, but light will be the limiting factor and in such cases the highest tree will have the best chance. As it grows, the distance that the sap flow must travel (from roots to leaves) becomes longer. The balance between the growth factors becomes more delicate, and any disturbance of this balance will be more noticeable. A giant tree will also hamper the development of other nearby trees because it will greatly reduce the available intensity of light below its crown. Once a forest giant has established itself, there will be little chance for a competitor to grow in its shade.

The Structure of the Tropical Rain Forest

This chapter started with the question of the relationship between the tropical rain forest climate and the typical structure of a tropical rain forest in the lowlands, characterized by the presence of giant trees, often with buttresses, and tree stranglers, which are to be described in more detail in Chapter 6.

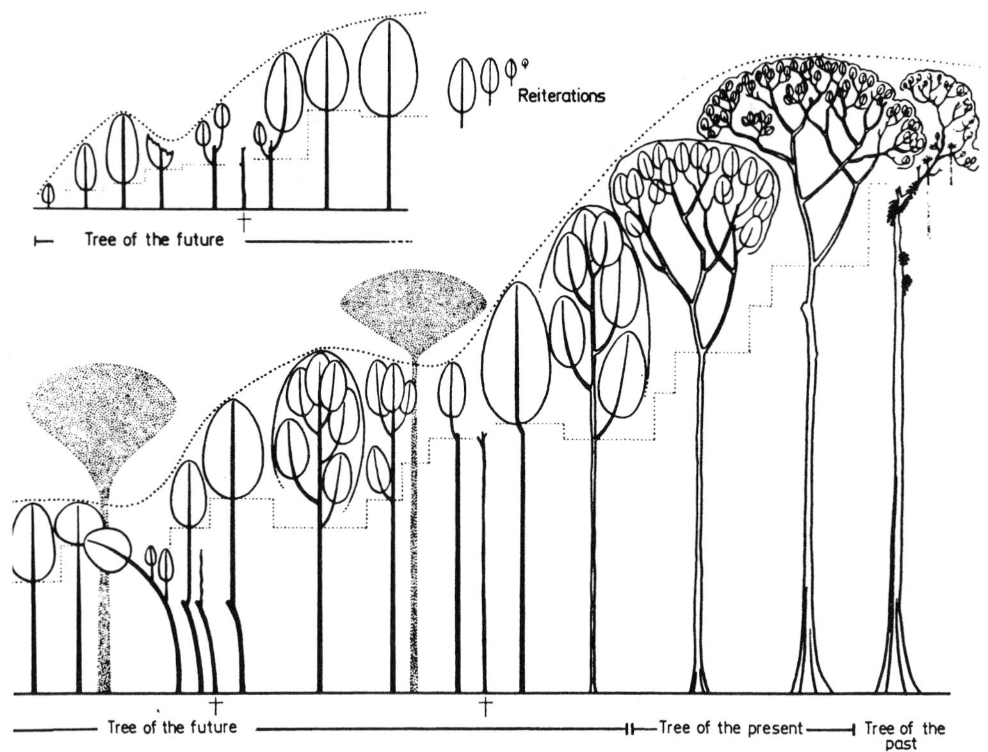

Fig. 3.14 The life cycle of a forest giant. Growth is initially hampered by shade from middle-height 'tree of the present' (*greyish*), occasionally death of small trees occurs (†). The *highest dotted line* shows the stagnation in their upward development. The other trees reiterate repeatedly, while their lower branches die off because of the shading effect (*lowest dotted line*). Buttresses develop only on tree of the present (the three trees on the *right*) (After Hallé et al. 1978)

The life cycle of a forest giant is illustrated in Fig. 3.14. Initially the crown has an elliptic shape; only when it has reached its final stage does the shape resemble a cauliflower. The crown is divided into smaller units by the process called reiteration. The smaller units arrange themselves on the inversion surfaces where a balance exists between the effects of light, photosynthesis and the sap flow rate, caused by vapour deficit. Where light is present in excess, but the supply of sap is insufficient, the trees will form widespread flat crowns, as they do in savanna climates.

In the rain forest, the incentive to grow in the upper canopy is apparently so high that such forest giants will occur in all forested areas, wherever there is ample moisture and a high and constant soil temperature that makes this particular forest configuration possible.

The majority of trees remain, by nature, relatively short. Hallé et al. call them "structural ensemble I" (see Fig. 3.15). Their crown development also follows the reiteration process, but does not require that special balance between light and sap flow rate that the giant trees need, since they grow under conditions of lower light intensities and higher humidity. Amongst these smaller trees we find the future giants. As long as they are small, they are shadowed by the mature, middle-high trees (shaded in Fig. 3.14). Some of them will die. Those that receive enough light will grow to the top of the canopy as indicated in Fig. 3.15, and will take the lead over other trees in the same generation. That there will be great differences between trees of the same age group is assumed by analogy with trees in the temperate zone (Hallé et al. 1978): tropical trees do not show characteristic rings indicating their age. Trees with typical buttresses or with almost partitioned stems, a characteristic known as 'fluted', and the tree stranglers have in common an enlargement of the surface area relative to volume, when we compare them with the single cylindrical shape of normal stems. Hallé et al. (1978) believe

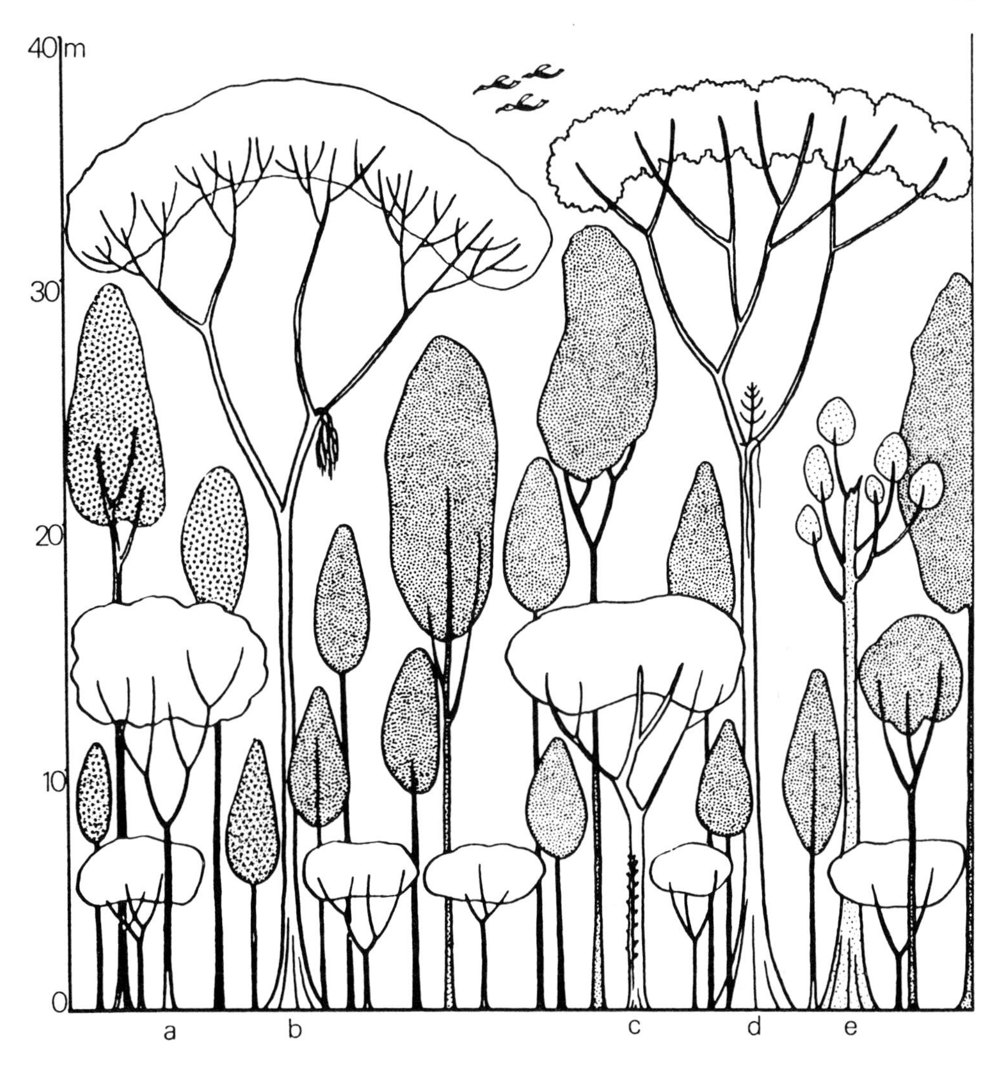

Fig. 3.15 Schematic diagram of a tropical rain forest with mature forest giants (ensemble II: *b* and *d* of the present, *e* of the past); mature trees of middle height (ensemble I: *a* and *c* of the present); below that a still smaller ensemble of trees remaining small when mature. In between (*shaded*) the trees of the future from ensemble I and II (Hallé et al. 1978)

that the ratio between outside surface and volume, made possible by special climatic conditions prevailing under the rain forest canopy, is an advantage for the energy management of these trees, particularly if they are big and tall. But so far we do not know the precise reasons for these typical trunk shapes.

4 Soils and Cycles

Geomorphology

Soils are an integral part of the landscape; regional variations cannot be understood unless the soils are seen in their physiographic setting. Landscapes form as a result of complex and interacting geological forces that, on the one hand, create different 'landforms' through volcanism, tectonics or sedimentation, and, on the other, destroy them through weathering and erosion. These latter processes are greatly influenced by climate, vegetation cover and by the actions of man. The science of landforms is called geomorphology, as defined in a delightful textbook by Sparks (1972). Geomorphology helps us to correctly interpret maps, air-photos, satellite imagery and field observations; it gives insight into the formation, age and stability of landscapes and to some extent also provides clues to their future. Leafing through Löffler's book, *Geomorphology of Papua New Guinea* (1977), one quickly becomes convinced of the importance of geomorphology in the study of rain forests. The book contains fine pictures of the Fly River and its surrounding crackle-like area of 140 × 125 km, made up of a disarray of ridges under dense primary forest. It also shows rain forests from which gruesome limestone peaks protrude ('broken bottle country' is the local name), the wide and tranquil lower reaches of the Fly River with levees under forest and wide swampy basins, forest on limestone reefs that emerge from the sea and wide barren strips that curve downhill; the landslides. There are close-ups of tree roots washed free of soil by erosion, where in some instances the eroded soil has formed natural steps, and pictures of earth structures, built by worms, which extend above the dead leaves on the forest floor and themselves are exposed to the forces of erosion.

Landforms determine to a large extent the type of vegetation that occurs. The nature of deeper substrata, the slope of the land and, of course, the water regime are all involved. The importance of water is obvious in the case of marsh or swamp vegetations, but in 'dry-land forests', the presence of water, and its dynamics, influence the floral composition just as much. For instance, the fact that certain tree species are found only on slopes might point to their preference for moving groundwater, as is known to be true of the giant horsetail, *Equisetum telmateia* (Equisetaceae) that occurs in The Netherlands. A species that grows in valleys but not on ridges may be sensitive to dry soil or require more nutrients than the soils of the ridges can supply. We shall pay some attention to this matter later. First, let us look briefly into the most important processes that shape a landscape: weathering, erosion, sediment transport and deposition.

Weathering

Any rock exposed to the air will weather sooner or later, and in the humid tropics, water is an especially effective solvent. Its relatively high temperature promotes the dissolution of solids because chemical reactions proceed faster. That is one important reason why the weathered surface layer is often deep: 20–30 m are not exceptional (Fig. 4.1). The climate and the parent rock determine to a large extent the nature of the weathering products. In the case

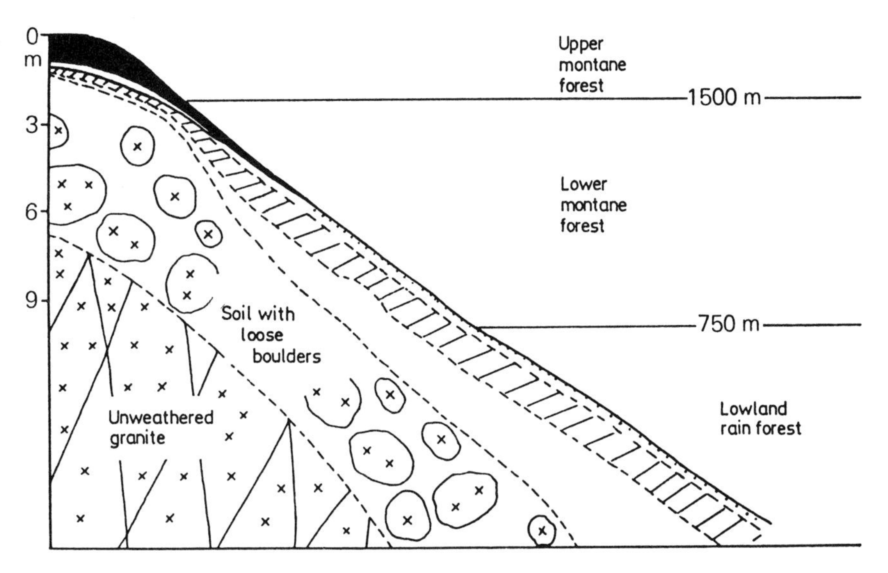

Fig. 4.1 The process of weathering on a slope with granitic bedrock, in Malaya. Thickness of the layers indicated at the *right*. The weathering is very deep. The upper end of the slope, where the soil is strongly leached, is covered with peat. On the slope itself, the soils are lateritic (Simplified, after Whitmore and Burnham 1969)

of feldspar-rich parent rock, percolating rainwater will remove silicates and aluminium hydroxides first and thus cause the relative accumulation of aluminium and iron oxides ('sesqui-oxides') and quartz in the remaining soil material. The ratio of silicates and oxides shifts gradually from 3:1 to 1:1 and can serve as an indicator of pedogenetic development over a period of, e.g., 5000 to 20,000 years. The final product is often a sticky, reddish mass that becomes as hard as rock upon drying and is known as laterite. Where the parent rock is acid and rich in quartz, weathering will produce a soil body of sand-sized quartz grains. Through this, oxides and any organic matter are leached downwards by percolating rainwater and accumulate at considerable depth where an indurated and virtually impermeable hardpan often develops. This podzolization process yields poor soils: the leached surface soil, in particular, has a low content of plant nutrients and poor physical properties.

The concepts of 'laterization' and 'podzolization' stem from the early days of soil science. In recent decades, new comprehensive soil classification systems have been developed with, at first sight, awesome concepts and terminology. The *Soil Map of the World 1 : 5,000,000* (FAO-UNESCO) 1979) has no less than 60 different colours on the Southeast Asia sheet alone, each representing a map unit identified on the strength of, among others, 22 'diagnostic soil criteria'. But this book is not the place to go into such matters in any depth; those who are interested should refer to the elementary – and understandable – introduction to modern soil science written by Buringh (1970).

Erosion

When rain drops fall on bare soil, their impact dislodges tiny soil particles: erosion begins on a very small scale. Rainwater which cannot instantly infiltrate into the soil runs off over the soil surface and carries some of the detached soil material with it. On level land, this 'sheet wash' proceeds at a relatively slow rate, but where slopes are steep, the water collects in rills and gullies and removes much more soil as it speeds downwards. Such erosion can be considerable (Fig. 4.2), even under a rain forest vegetation: Löffler (1977, p. 145) mentions annual soil loss rates of 0.5 to 3.5 mm. Many podzolized soils in wet climates show interesting erosion phenomena when exposed to direct rainfall, for instance in road cuts. After only a few showers, pebbles and leaves that were on the surface lie on top of tiny soil columns between

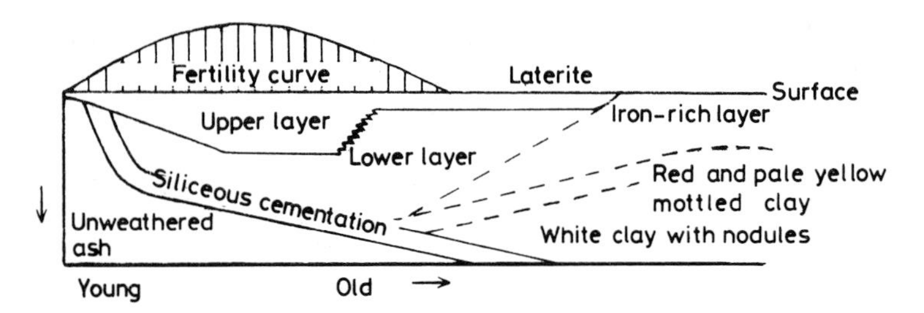

Fig. 4.2 The gradual weathering of volcanic soil, from *left to right*. Finally, laterite remains. A *curve* showing the corresponding fertility is drawn above the surface. (After Mohr and Van Baren 1954; Whitmore 1975)

which the unprotected soil has been washed away. This suggests incidentally one way to fight erosion: to maintain some sort of soil cover that absorbs the impact of falling rain drops. A cover of grass or gravel can provide adequate protection, but requires, maintenance. Forest cover intercepts rainfall twice, once in its canopy and subsequently in the layer of dead debris on the forest floor. That explains why many countries have passed legislation that a 'protection forest' must be preserved on slopes steeper than a certain angle. It also explains the importance attached to the use of cover crops in tropical agriculture. Many farming systems make use of low, dense and broad-leaved crops to minimize soil exposure and therefore loss.

In climates with a distinct dry season, where the vegetation withers or sheds its leaves and is often prey to fires, the danger of erosion is particularly grave and gullies of staggering depth and width form in only a short time. Rivers transport the eroded soil material to the sea where coastal waters are muddy and brown; that same soil material could have been a land resource for future generations.

Transport, Sedimentation

Transport of soil material can be large or small; it is sometimes spasmodic, but more commonly gradual (Fig. 4.3). Landslides are a particularly dramatic kind of transport. Anyone who travels in broken jungle land will occasionally notice a narrow or wide, short or long, vertical strip of land where the vegetation is clearly different from the surrounding forest. On some of these, the dominant colour is still that of the naked soil, others have the greenish tinge of an establishing pioneer vegetation. It is evident what has happened: the surface soil has slid downwards, vegetation and all, only coming to a halt at the foot of the slope. In some cases, the dislodged soil mass ends up in a stream or river which is temporarily blocked until the water removes the obstacle again and carries trees and sediment towards the sea.

In volcanic areas, gigantic mud flows, 'lahar' in Malay, are associated with volcanic eruptions. A blown-out crater lake, for instance, can instantly cover and destroy large tracts of land. Transport can also proceed at a very slow pace, hardly notices able while it happens, for example during torrential showers when water acts as a lubricant. The consequences of

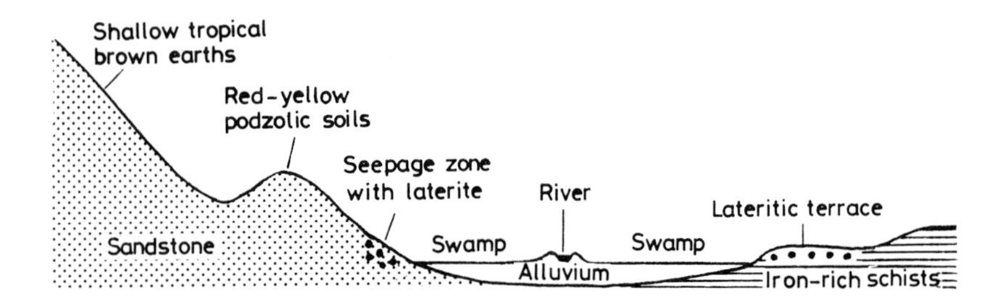

Fig. 4.3 The results of weathering, transport and sedimentation, on the *left* of sandstone, on the *right* of schists. The trough in the middle is filled with fine deposits, which carry a river with levees (After Whitmore 1975)

such 'soil creep' are clearly visible: trees lean over in every direction and one's feet sink easily into the soil. Transport of soil for which gravity is the major force results in heterogeneous strata of colluvium. There is also the transport mainly associated with flowing water. Such transport is often more constant and results in more homogeneous strata of well-sorted alluvial deposits, predominantly found in former and present fluvial and marine landscapes. Löffler (1977, p. 134) discusses a river catchment of 78 km^2 in New Guinea and estimates sediment transport there at 2.5 million m^3 over a period of 30 years. This corresponds with an average loss of land surface elevation of 1.1 mm each year! In the course of the process the riverbed widens and sediments are deposited on the valley floor. The average grain size of the deposited material becomes smaller as the flow velocity of the water decreases. Coarse material (gravel, sand and silt) is deposited close to the riverbed, particularly in the upper stretches of the river course; finely textured (clayey) deposits accumulate in the basin areas at some distance from the riverbed and in the downstream parts of the flood plain. Naturally, the newly deposited material is subject to renewed weathering. If weatherable minerals are still intact, they release nutrients from which a plant can benefit. That is one reason why coarse-grained levee formations often support a comparatively luxuriant vegetation: the varzéas along the Amazon River are a good example.

Soil Poverty

After this brief introduction to certain important geomorphological processes, some attention is due to the generally low natural fertility of soils in the perhumid tropics. These regions have an appreciable rainfall surplus, evenly distributed over the year, which results in continual downward percolation of rainwater through the soil. Plant nutrients in the soil go into solution and are removed from the root zone in what is virtually one-way traffic. This leaching of the soil is particularly prominent in the elevated parts of hilly land: a steady downward stream of nutrients is lost from there but can, to some extent, replenish nutrients lost by hill slopes and valley bottom soils. Therefore, the vegetation is often poor on the topmost parts of hills and ridges: clearly an edaphic, or soil-determined, phenomenon (Fig. 4.4). Even where trees stand tall, the soil material may be structurally deficient in nutrients. Early ecologists and land-use planners used to think that a luxuriant rain forest would indicate fertile soils. This notion is almost never correct. It applies only where soils are young, i.e. on some of the recent sediments mentioned in the previous paragraph, or on young volcanic deposits. When volcanoes erupt, they produce lava and/or ash which settles on top of the surrounding land. The molten lava solidifies to rock and the weathering products of this rock, and of the ash and other effluents, are carried off by water that seeps through them to enrich irrigation water. Wetland rice cultivation on the volcanic slopes of Java and in parts of Sumatra relied traditionally on this source of nutrients. Volcanoes become extinct eventually, however, and, then, natural fertility is no longer replenished, while leaching, laterization and so on continue (see Mohr and Van Baren 1954, Pl. A). Ultimately, volcanic soils can become just as poor as most other soils in the humid tropics.

The situation is completely different where wet and dry seasons alternate. The evaporation surplus during dry spells induces an upward flow of water, i.e. from the groundwater towards the surface. This flow brings minerals from deeper strata to the rooted surface soil where they accumulate. In extreme cases, salt crusts form on top of the soil; the same mechanism accounts for the white efflorescences on the soil surface in flower pots.

The chemical poverty of most soils under tropical rain forest has four important consequences:

1. It affects the nutrient regime of the system. The forest depends for its nutrient requirement almost entirely on a limited amount of recycled nutrients. At any one moment the majority of these are stored in the biomass, i.e. in roots, stems, branches, leaves, flowers

Fig. 4.4 Difference between height of vegetation on a ridge and in a gully. For the scale, note the man standing to the right of the large bare tree at the *right* of the photograph. A sharp boundary between forest and open land as shown here always indicates damage. In this case the forest that occupied the foreground was cut for short-lived agriculture. After depletion of the soil the grass alang-alang *Imperata cylindrica* (Gramineae) established itself. Regular burning has removed all traces of forest and inhibits the establishment of seedlings, but cannot penetrate far into the forest. Southeast Celebes (Photograph MJ)

and fruits, but also in the fauna that is indigenous to the forest. All these components of the ecosystem will eventually die, decay and be reabsorbed, a process that will be discussed later. Decomposing organic debris is the major source of plant nutrients. The importance of nutrient cycling is witnessed by the fact that the forest vegetation has shallow root systems only, irrespective of the depth of the weathered surface layer. If one examines a fallen tree or the decaying leaf mass on top of the mineral soil, one often finds a dense network of fine roots. Any nutrient that is set free is quickly taken up again in the finely tuned process of cycling that developed during the long process of evolution.

2. It follows that the forest, as an ecological community, is to some extent independent of the inherent fertility of the soil material. Majestic trees with heavy trunks thrive on land too poor to be fit for any agricultural use. If such forest is cleared and the land taken into cultivation, the soils rapidly degrade to such a degree that fertility is permanently lost. It is astonishing that forest plants have been able to evolve this intensive process of nutrient cycling, continually recovering minute quantities of inorganic compounds in defiance of structural soil poverty and ongoing leaching.

All this does not mean that soil conditions in general are of no importance to rain forest development. On the contrary: the floristic composition of the forest depends strongly on soil conditions, as we shall see later. And where the nutrient poverty of the soil is extreme, even rain forest cannot survive. On such land, only certain types of light or stunted forest (kerangas) can occur; these types will be discussed in Chapter 15.

3. There is a direct relationship between the quantity of nutrients available to a vegetation and the possibilities for forest exploitation. Any biomass component contains mineral elements. If biomass is removed from the forest, the nutrients contained in it are lost from the system. Though normally this loss will be replenished in due course, foresters know that forests on rich land recover more quickly than they do where soils are poor or chemically

exhausted. It should be possible to describe all relevant nutrient fluxes in quantitative terms, so that it can be calculated how long it takes before the nutrient regime of a certain tract of forest land has fully recovered from disturbances incurred in logging operations. Estimates available now suggest that this is a matter of centuries. The selective removal of small quantities of particularly valuable forest products such as rattan, resin, gums, honey, waxes, medicines and stimulants is much less harmful.

4. The possibilities for other forms of land use after clearance of the forest depend to a large extent on the inherent chemical fertility of the soils. Fertile soils can produce several crops of maize, rice, sugarcane, tobacco, etc., even though much biomass has been removed. Poor soils cannot stand such treatment. They only allow certain forms of permanent cropping where small quantities of high-grade products are won from slow growing species (commonly rain forest in origin); examples are coffee (the seeds), oil palm (the fruits) and rubber (the latex).

The Shade Community

The foregoing has demonstrated how all-important nutrient cycling is to a closed, self-reliant system such as a rain forest. Dead leaves, withered flowers and fruits fall on the ground throughout the year and, occasionally, pieces of bark, whole branches (with epiphytes growing on them) and trunks. The collective term for all this debris is litter. Its accumulated dry weight amounts to 5–13 t ha^{-1} $year^{-1}$ (Hladik & Hladik 1972, p. 167): a small percent of the standing biomass. All this litter is processed by an inobtrusive 'shade community' that breaks the litter down as soon as it is deposited on the forest floor. A botanist working in the forest may suddenly find himself without a crew if their attention is temporarily monopolized by a putrefying tree. They will rush to open its trunk with their bush knives in search of beetle larvae; these larvae are as thick as a finger, whitish and soft, the people eat them on the spot and quarrel over the fattest. Such larvae tunnel through decaying trunks and provide access roads to fungi which are responsible for further breakdown of the wood.

Termites do the same in a different manner. They are seldom bigger than a few millimetres and shy away from light and dryness, covering their passages across trunks and logs with earth roofs and building earth mounds to protect their nests. Most of the approximately 2000 species known today feed on dead wood, including unfortunately home timber and wooden furniture. Some also devour rubber and even plastics, but cellulose is their staple, digested with the aid of a fungus that is cultivated by the termites inside their nests (see Becker 1976).

Ants are everywhere in the rain forest. There are thousands of different species, in sizes between 3 and 30 mm. They handle also animal matter which they bite to manageable pieces before hauling it to their nests. Of course, other carrion eaters help them with the really big jobs; worms, millipedes and all kinds of microorganisms bring the material further in circulation. Ultimately, it is fully mineralized, i.e. broken down to ions that can again be taken up by plant roots. It is sometimes assumed that the soil microflora and fauna is essentially the same the world over. We should realize, however, that we know still very little about this silent community whose diligence keeps nutrient cycling going, inobtrusively but very effectively.

It is perhaps good to mention here once more how important the climate is to the functioning of a forest. In ever-wet forests where rainwater seeps continually into the soil, the shade community lives under a closed canopy in a permanently moist and relatively cool environment. In climates with a distinct dry period (where water and nutrient fluxes are alternately upward and downward), the foliage is periodically thinned because of drought stress. As this happens in the most critical period when temperatures are high and drought makes the fallen leaves crackle under one's feet, it is conceivable that decay may follow different pathways in ever-wet and periodically dry forests, and a comparative study would certainly further our insight into the dynamics of nutrient cycling and the possibilities of manipulating this process for optimum forest management.

Humus

Litter decays into humus. The difference is that litter can be picked up from the soil, while humus is amorphous and an integral part of the soil. Worms and other organisms carry litter particles into the pores and voids in the soil mass where the litter is transformed. The resulting humus is dark coloured. It cements individual soil particles together into loosely structured aggregates. Humus thus increases the soil's pore volume and water-holding capacity and improves its 'structure', making clay soils softer and sandy soils firmer. Another attractive property of humus is its capacity to adsorb ions such as calcium, magnesium, potassium and ammonium at its surface, releasing these ions later when their concentration in the soil moisture has been lowered by leaching or uptake by plants. In other words: humus maintains a stock of nutrients which buffers variations in element supply and loss. Last, but not least, humus promotes the activity of microorganisms which break the humus down to its elementary components: the aforementioned mineralization. As a rule of thumb, fungi resynthesize approximately one-third of the organic matter attacked within their own bodies. The efficiency of bacteria is on the order of only 1%; their part in mineralization becomes more prominent if temperatures are high. It is a well-known fact that in vitro biochemical processes accelerate two to three times when the temperature rises 10°C. Whether this law holds in nature as well remains to be seen, but there is no doubt that mineralization proceeds much faster in tropical lowlands than at higher elevation or latitude. The quantity of humus in tropical soils soon runs out if litter supply sources are removed by clean felling and predatory cultivation, much to the detriment of soil structure and soil life. Only if a new permanent vegetation can re-establish itself can the humus content of the soil increase and this takes time.

The Mosaic of Species

From the air, one can see clear differences in tone between the canopies of a primary forest on limestone and an adjacent forest on volcanic material. Ground checks confirm the surmised difference in floristic composition: only very few species occur on both soil types. Even a minor change in soil conditions is reflected in the forest. This relation between the properties of the soil and the composition of the forest vegetation was studied by Ashton and discussed in his *Ecological Studies in the Mixed Dipterocarp Forests of Brunei State* (1964), an almost inexhaustible source of information. Ashton analyzed a geologically old and stable area in northwest Borneo, very rich in species and with considerable soil differentiation. He made forest inventories of 50 one-acre (=0.4 ha each) plots on two different geological formations, viz. in Andulau (mainly sandy soils) and Belalong (mainly clay soils), and identified and measured not less than 30,000 trees with a diameter of 10 cm or more. All in all, he worked his way through 2×20 ha and recorded a total of 760 different species. He found 472 different species in Andulau and 420 in Belalong. That means that only 132 species, or 17.4%, were common to both areas. Many plots had not a single species in common with other plots, even within the same land unit. Only 7 species occurred on four different soils or more, 12 on three, 65 on two and 60 species were confined to only one soil type: so fastidious are dipterocarps! I am not quite certain whether the situation is the same in younger forests, but it is noteworthy that Ashton (1964, p. 64) reported that a secondary vegetation in Brunei, which established itself after the primary forest was cleared, gave the impression of being different on different soils. On clay soils he noticed *Anthocephalus cadamba* (Rubiaceae),

Leucosyke (Urticaceae), *Macaranga* (Euphorbiaceae) and *Pterospermum stapfianum* (Sterculiaceae), all pioneer crops with big, thin leaves, intermingled with representatives of the primary forest: *Shorea ovalis* and *S.parvifolia* (Dipterocarpaceae). On sandy soils, he found more species with small and hard leaves: *Adinandra* (Theaceae), *Gaertnera brevistylis* (Rubiaceae), *Ploiarium* (Theaceae) and *Timonius flavescens* (Rubiaceae), a shrub vegetation with few signs of further development.

In another publication (1976b, p.195), Ashton mentions that the floristic diversity of the rain forest, again in northwest Borneo, is greatest where the soil phosphorus content is between 40 and 150 ppm. Both below and above this range, the number of recorded species decreases. The phosphorus content of soils derived from young volcanic material and sandstone weathering is commonly below 100 ppm; soils on shales and granite contain between 80 and 300 ppm phosphorus, but in basalt weathering, values of 3000 ppm and higher are occasionally found.

5 The Trees

General Remarks

The elements of a vegetation comprise many life forms. Each plant species manifests itself in a life form, and sometimes in more than one by transmuting from one form into another. Most lianas, for instance, start life as a treelet or shrublet; the strangler tree starts life as an epiphyte, and epiphytes may be terrestrial at higher altitudes. No unambiguous classification can therefore be given for life forms, but a number of categories can be discussed.

Schimper (1898) was the first to survey the life forms in the rain forest thoroughly. More recently, Holttum's excellent *Plant Life of Malaya* (1954) examines the non-trees, while Hallé's et al. *Tropical Trees and Forests* (1978) is an essential book for understanding the true tree forms.

Trunk

A tree has one distinct, woody clear bole or trunk. A shrub branches close to or even below ground level. Real shrubs are, however, not common in lowland primary forest. Often, the occurrence of shrubs is a sign of secundarization or site degradation, or of both.

Mini- or spindle treelets, on the contrary, are quite common. Even the smallest of these has a distinct stem with a few leafy twigs. By producing flowers and fruits they prove to be mature, and indeed they do not grow much further afterwards. One example is *Ardisia crispa* (Myrsinaceae), a treelet some 50 cm high, often sold in Europe as a house plant which does well in living rooms at moderate temperatures. The leaves are simple, scattered, some 5–7 cm long, and have crenate margins. The flowers are white, the berries globular, 1 cm in diameter and red; the star-shaped calyx is persistent at its base.

Most mini-treelets grow to between 2 and 4 m in height. A rich assortment is found in the Annonaceae, Euphorbiaceae, Myrsinaceae, Rubiaceae and in several other families; they are found in every virgin forest.

Those life forms that possess a thick, unbranched trunk bearing on top a tuft of large leaves must be considered trees as well. The squat *Cycas* (Cycadaceae) in the rain forest of Malesia belongs here as well as the coconut palm, *Cocos nucifera* (Palmae), of tropical shores. A similar habit is displayed by tree ferns, of which there are hundreds of species, especially in the mountains, though some also occur in the lowlands. The diameter of the bole of such trees measures from 1.2 to 20 cm or more, depending on the species; the length is about a 100 times the diameter, just as in ordinary trees. Such life forms are known as *Schopfbäume* in German, as well as in the English language. The many sparsely branched screwpalms, *Pandanus* (Pandanaceae), with some 600 species in the Old World, belong to this group as well (Fig. 5.1).

All thick-boled forms are called *pachycaul*; they are regarded as primitive. There is more about them in Chapter 13 (Evolution).

Slender-boled, much-branched trees are called *leptocaul*, and all tall primary forest trees belong to this category. The average height of the giants of the forest is the same in all three

Fig. 5.1 Life forms of Pandan, *Pandanus* (Pandanaceae). (Stone, Mal. Nat. J. 19: 293, 1966)

large areas of primary forest: 35–42 m, sometimes up to 46 m, with a crown diameter of 13–22 m. Still higher trees exist in the rain forest. The highest trees in Malacca recorded by Whitmore (1975, p. 17) are a *Dryobalanops aromatica* (Dipterocarpaceae) of 67.1 m, a *Koompassia excelsa* (Caesalpiniaceae) of 80.72 m, and, in Sarawak, an individual of this same species of 83.82 m. In order to raise water from the ground to that height, a pressure of 8 atm is needed; our own waterworks make do with 3 atm. The tallest tree in Fig. 2.6 *Cotylelobium melanoxylon* (Dipterocarpaceae), measured 64 m; the clear bole was 40 m, and the diameter was 1.30 m at breast height. As in all tall, tropical, primary forest trees, the trunk is long and slender in relation to the crown. If the same tree grows in a solitary position, the trunk remains short and the lower branches are not shed, which gives the tree its 'park tree' habit. In forest giants the crown itself is usually cauliflower-shaped or takes the form of a giant umbrella; in the smaller leptocaul trees it is more egg-shaped (Fig. 5.2a and c).

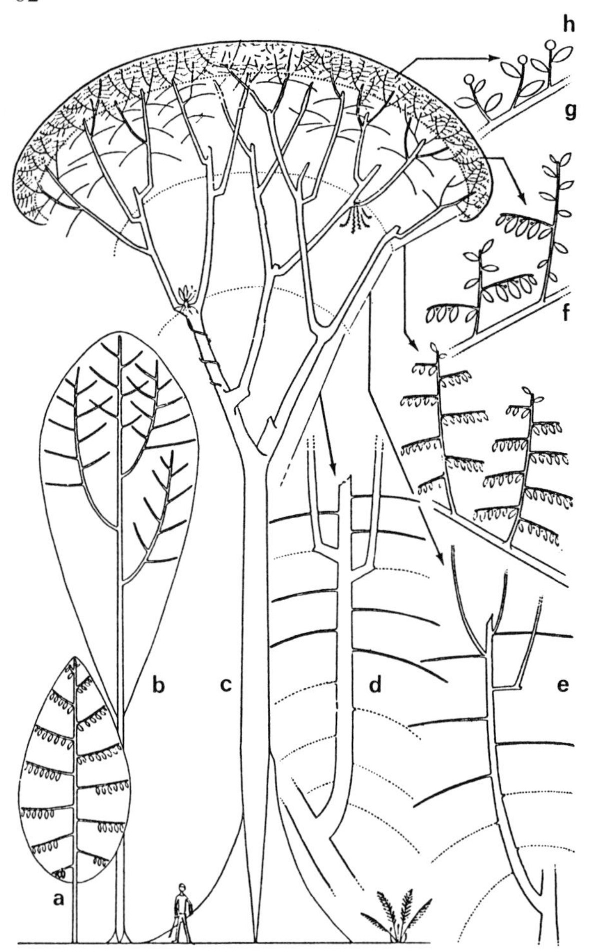

Fig. 5.2 Reiteration in rain forest tree. (Hallé et al. 1978)

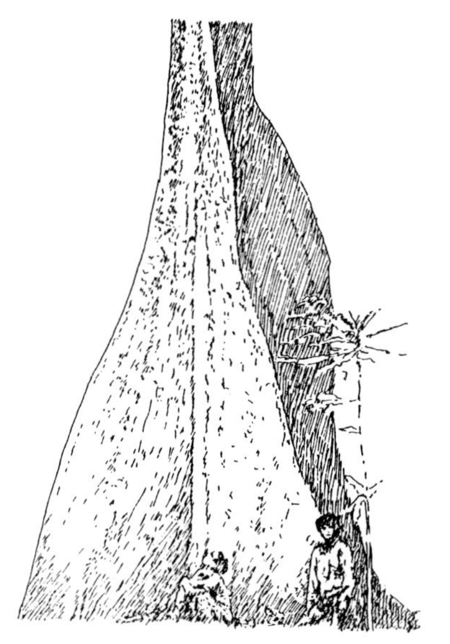

Fig. 5.3 Buttresses of *Kostermansia* (Bombacaceae) (After Corner 1978)

At the base of larger tree trunks *buttresses* are often present (Fig. 5.3). Such tree bases resemble in cross-section an irregular, 2–, 5–, or sometimes up to 10-armed star. Viewed from the side, buttresses are more or less triangular, one or two hand's breadth thick, 1–2 m wide, and 2–3 m but sometimes up to 9 m high. They act like the guy ropes on a tent; if you hit them with the blunt side of a jungle knife you often hear a clear sound, as if they are under tension. To fell such a tree, local people build a small scaffold around the tree, made from thin poles of young trees which they tie together with rattan or other lianas. Two men, balancing on this, can then cut the tree down above buttress height. Other tree species have fluted trunks: some deep grooves extend high up the trunk, up to the lower branches. The resulting, sectionally rounded, longitudinal bulges never extend so far out at the base as buttresses do.

Such trunk features are almost exclusively found in rain forest trees. Richards, who gives them a great deal of attention (1952, pp. 62–74), discusses four theoretical explanations, whose tenor is the extension of footing in case of superficial rooting, but this does not account for their obvious correlation with the tropical rainy climate. Hallé et al. (1978, pp. 286–289) do remark on this connection, pointing out that the balance between the cambium activity and the amount of assimilates produced, which has to be transported by the tree, changes both with the age of the tree and with climate factors. In transpiration-prone climates an

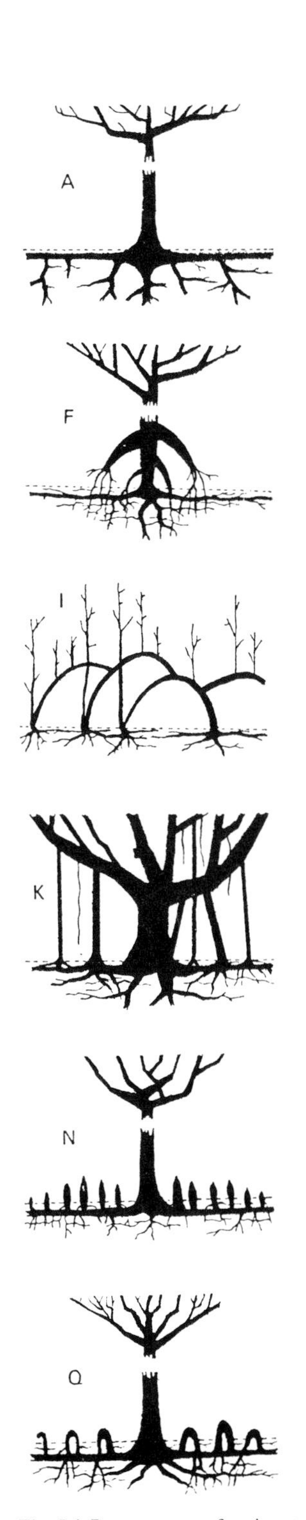

Fig. 5.4 Root systems of various tropical trees (Jenik 1978; also includes a list of species)

enlargement of the cambium surface is advantageous for the purpose of transporting more sap through an increased amount of wood.

Roots

Many peculiarities can also be observed in root systems, a long-neglected subject. In an enlightening article, *Roots and Root Systems in Tropical Trees* (Jenık 1978), 25 models are recognized (Fig. 5.4). The diversity tends towards enlarging the purchase area of trees which, in general, root more superficially than those in the temperate regions, as well as enlarging breathing possibilities for trees whose roots are immersed, and inherent to that, horizontal enlargement.

Leaves

Leaf size adds to the impression of uniformity made by a primary forest; most leaves are 10–20 cm long and one-third to one-half as wide, and in case of compound leaves, their leaflets have about that same size and proportions. Leaf tips are often elongated by a few centimetres, narrowing into a so-called *drip-tip* (Schimper 1903). Such leaves do indeed dry more quickly than those without this feature because the water covering them is shed as liquid drops rather than evaporated; accordingly, after a shower they recommence assimilating and transpiring more quickly.

Prickly leaves are rarely encountered in the rain forest. Rain forest leaves either have entire margins, or are obscurely dentate or serrate. A strikingly large number possess a thickening at the base and/or top of the periole. It is known that this feature permits a degree of 'articulation', enabling a leaf to some extent to change its position.

Flowers

Another typical feature in the primary forest is *cauliflory* or flowering from the trunk. Adventitious buds develop on the trunk and/or the major branches, of which now and then several develop into flowers or inflorescences. Certain species which flower and fruit in this way produce the largest fruits known: the durian, *Durio zibethinus* (Bombacaceae), measures 25 cm or more, and the compound fruit of the nangka or jackfruit, *Artocarpus heterophyllus* (Moraceae) measures 1 m × 50 cm. The low position of such fruit, attached to the trunk by a sturdy stalk, leads one to assume that they attract and are meant to attract large animals. Cauliflory as a trait merges into *ramiflory*: flowering on the older branches or even occasionally underground. A species of the large, inventive genus *Ficus* (Moraceae) in Malesia produces a special kind of inflorescence. Its reddish fruit, lying amidst the litter of the forest floor, turns out, when picked up, to be attached to a 1–1.5-m-long runner sprouting from the base of a small tree which bears only leaves. These so-called ground figs, only produced at the soil surface, are probably dispersed by pigs. A transitional stage between *geocarpy* and cauliflory is exemplified by *Ficus* species in which inflorescences 1-m-long sprout from the trunk near the ground. Similar geocarpy and near-the-ground cauliflory is also present in the Gesneriaceae, Annonaceae and Actinidiaceae families (Van Balgooy and Tantra, 1986).

Architecture

Now something about the architecture of trees. The difference between a papaya tree (*Carica papaya*, Caricaceae) with its heavy, small-branched build, an araucaria (*Araucaria*, Araucariaceae) with its massive central axis and spreading storeys of branches and an *Acacia* (Mimosaceae) with the many slender, strong individual branches, is recognized at a glance, even when these trees are cultivated as house plants. Likewise, there are many differences in

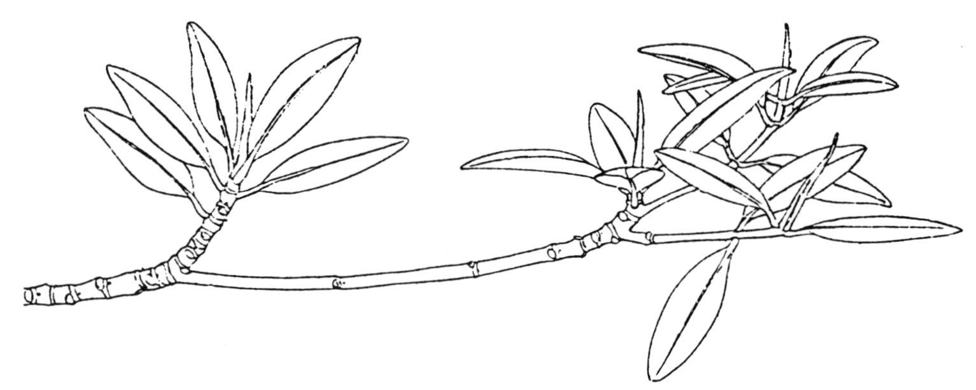

Fig. 5.5 Sympodial branching system: elongation is by side shoots. (P. B. Tomlinson and M. H. Zimmermann ed., *Tropical trees as living systems* (1978)

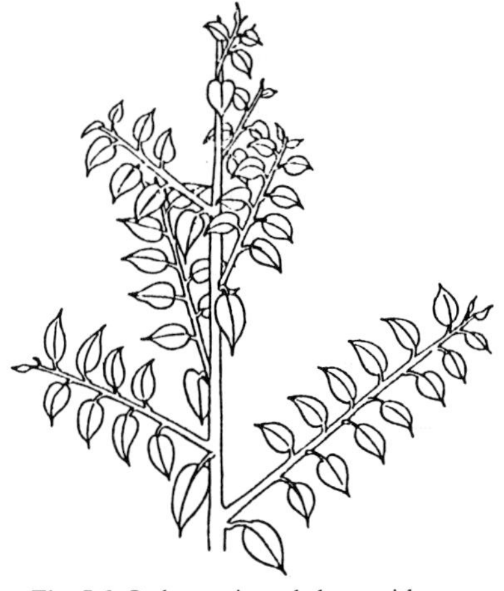

Fig. 5.6 Orthotropic end shoot with plagiotropic side shoots. (P. B. Tomlinson in Meggers et al. 1973)

shape among the broad-leaved species which nearly exclusively compose the tropical rain forest, although it might seem that these differences can be seen less easily than in temperate regions, for tropical rain forest trees rarely, if ever, lose all their leaves as if it were winter. On the other hand, no winter disturbs the growth patterns fundamental to the creation of form diversity. Such forms as the short shoots on apple trees which bear the flowers and fruits, are almost absent in the rain forest. If they occur, as they do in the bougainvillea (Nyctagynaceae), that splendid tropical garden plant, it is a sign of forest secundarization: the primary forest has gone or is damaged.

The position of flowering is important. If this occurs at the side of the axis, as in coconut palms (*Cocos nucifera*), the axis can grow unhampered. This is called *monopodial* growth. But if flowers and fruits are formed at the top of the axis, as in mango trees (*Mangifera indica*, Anacardiaceae) or, as in temperate regions, the horse chestnut, the growth of the axis is taken over by a bud under the top, forming a new axis: such branches are never straight and are said to grow in a *sympodial* way. The trunk of cocoa (*Theobroma cacao*) does the same, but here growth is not stopped by flowering, but by other processes.

In this respect the sago palm (*Metroxylon sago*, Palmae) is very striking since the main trunk is the terminal part of the sympodium and the whole plant dies off after flowering.

Apart from its monopodial or sympodial growth pattern (Fig. 5.5), an axis may or may not grow rhythmically. In the temperate regions, rhythms are predominant because they allow trees to stop growing in winter. But in tropical trees, such as rubber (*Hevea brasiliensis*, Euphorbiaceae), rhythm appears to be independent of climate: six growing periods per year alternate with six leaf elongation periods and later with flowering periods, when growth is arrested. Many tropical trees, however, show no rhythm at all. The axis of coffee (*Coffea*, Rubiaceae), for example, grows continuously. Rhythmic growth most often is accompanied by rhythmic branching and results in branch storeys, separated by clear pieces of trunk, such as in *Araucaria* or the temperate climate Christmas tree species. Continuous growth means continuous branching, and in this case branches cover the trunk from the top to bottom. Nutmegs (the Myristicaceae family in general) and ebonies (*Diospyros* species, Ebenaceae) grow in a rhythmic way, and therefore form branchy storeys like European firs.

Fir trees illustrate another difference in growth form (Fig. 5.6). Some axes, for example the trunks of Christmas trees, are vertical and their needles or leaves grow spirally. Such a form is 'invented' to attain height and is called *orthotropic*. Other axes, such as the branches of these firs and all axes of many Leguminosae (e.g. *Tamarindus indicus*, the tamarind) grow horizontally and their leaves are arranged in a more or less perfect horizontal plane. Such axes are apt to function as light interceptors and are called *plagiotropic*; they often also bear the inflorescences which then are available in dense quantities to insects and other pollinators.

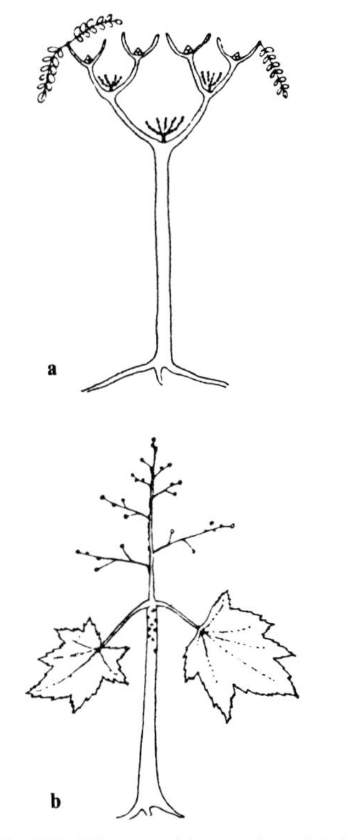

Fig. 5.7a-f Some architectural models in the family Araliaceae (Philipson 1978)

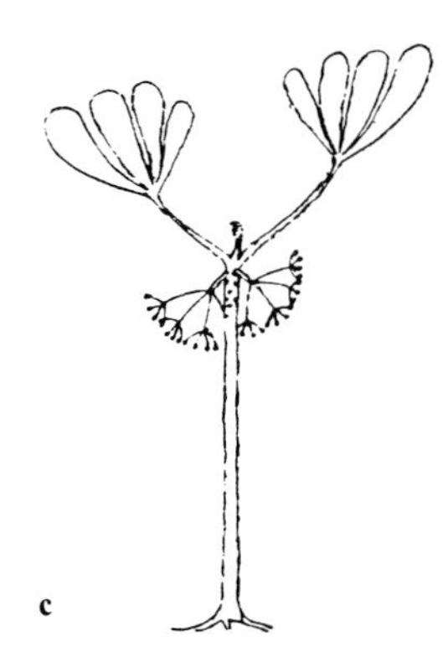

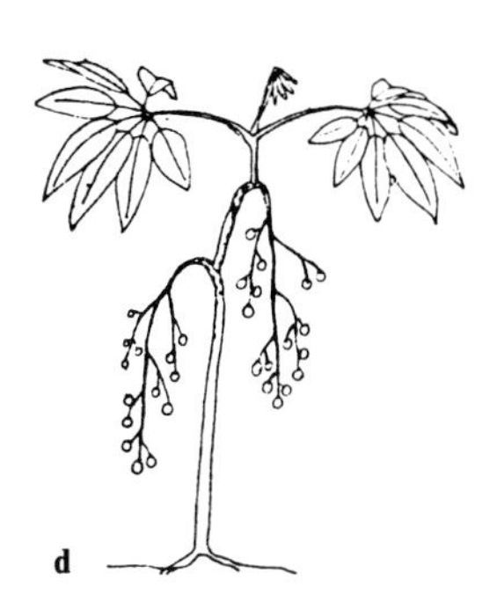

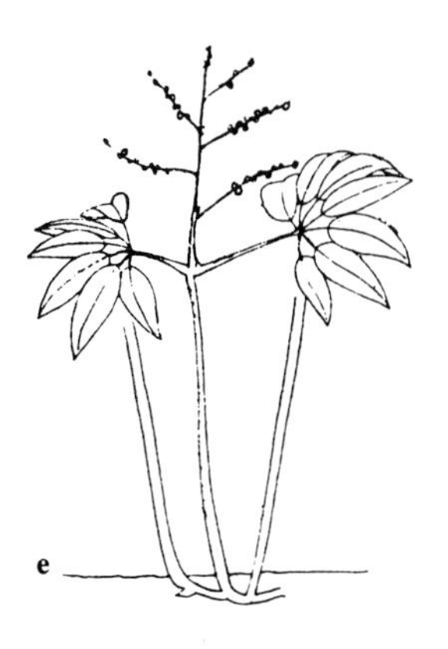

Very briefly, these are a few of the elements important to the architecture of trees. This subject is extensively treated in the book by Hallé et al., *Tropical Trees and Forests* (1978). Hallé and Oldeman, the founders of the study of tree architecture, discovered that each tree species retains the combination of growth characteristics which can be called the architectural model of the tree (Fig. 5.7). This model is not restricted by the tree's dimensions. So far 23 models have been distinguished by these authors and each has been named after a distinguished botanist.

That the architecture is fixed in the inherited plan of the tree can be deduced from the fact that new parts developing after damage or during crown expansion repeat the original model. This is well illustrated by the *water shoots* which develop especially on twists of branches and trunks. They grow fast, straight upwards and have extra long nodes; the large leaves they bear usually deviate from the normal ones, often resembling the tree's juvenile leaves. If unobstructed, such a water shoot may develop in a number of years into a complete, almost independent treelet, growing out of the original tree and imitating the original tree's architecture. This phenomenon is called *reiteration*. Reiteration permits a tree to make the most of its possibilities, as far as the model permits. A spruce, for instance, grows high but not wide. Most broad-leaved trees, however, grow opportunistically. Where they find a bright opening they develop a partial crownlet. This is most striking along riverbanks where tree trunks bend out over the water, and, unshaded, they reiterate. The same process occurs at the top of the canopy, into which the trees grow during the process of becoming adult. On the way up, the crown formed does not just grow upwards, but renews itself many times, filling the empty space present in the area above it, while retreating from the space it formerly filled. Shaded out, major branches die, and are later shed altogether, while time and again the crown splits up into partial crownlets, arranging them as economically as possible in the new situation (Fig. 5.2). Only leptocaul trees have this ability because they invest relatively little material in their slender branches and the leaves on those thin branches are also easily replaced. It is therefore only the leptocaul type of tree that can quickly penetrate the canopy, as in the life cycle depicted in Fig. 3.14.

The life cycle can be divided into several definite stages. First, there is the *seedling*, which grows up to become a *sapling*; when it attains a man's height it is named *pole tree*. When the trunk is 10 cm in diameter, the pole starts counting as a tree. It is then, in the case of a species which will reach the canopy, a *coming tree*. It can remain in this stage for over 60 years. When it grows into the canopy, enlarging its crown in the process of reiteration, we name it a *tree of the present*. It can stand like this for a few centuries; we do not know how long because if growth rings are present in the trunk, they are usually not year rings, but reflect some non-regular check in growth. At the end of a tree's life, its branches start to die. Damage by fungi and insects increases, and the tree becomes a *tree of the past*. When a major branch breaks off, the tree becomes unbalanced and starts to lean over, becoming much more vulnerable to storms. Finally, the tree falls, causing a wound in the canopy, a *chablis*. New, coming trees hurry to grow and fill up the hole (see Figs. 8.2 and 8.3).

6 Other Life Forms

Lianas

In the development of a forest, climbers play a role in all stages of succession. As with trees, each stage is more or less represented by its own species. The herbaceous pioneers are soon followed, in the young secondary stage, by fast-growing woody species of Convolvulaceae, Cucurbitaceae, Vitaceae, and others; only woody climbers are considered true lianas.

Climbers are leptocaul par excellence. In western Europe the virgin's bower, *Clematis* (Ranunculaceae), the ivy, *Hedera* (Araliaceae) and the woodbine, *Lonicera* (Caprifoliaceae) are representatives of the type since in wintertime their stems do not die off. In the tropics, where over 90% of liana species live, they grow to much greater size; a stem diameter of 15 cm and a length of 70 m is by no means exceptional. In a primary forest one can see the liana stems, often twined around each other, disappear into the foliage, or hang down in mighty loops because long ago a supporting branch broke off. Many a tree is entwined by three, sometimes four species; 13% of the standing crop of harvestable trees in natural forests in the tropics is for this reason rendered useless. No wonder that foresters try to get rid of them! Lianas also suffer from 'normal' forest destruction, and for these reasons have become one of the most threatened categories of plants. Few people realize this. It is a fact that after a forest is logged, lianas seem to grow up abundantly; even my expectations of finding many species were high when I investigated a logged forest near Lae in Papua New Guinea. I was badly disappointed: the areas of untouched forest rendered a greater diversity of liana species in flower or fruit than the disturbed forest.

Our knowledge of lianas has increased little since the appearance of the standard work by Schenck, *Beiträge zur Biologie und Anatomie der Lianen/im Besonderen der in Brasilien einheimischen Arten* (see Jacobs 1976b for an outline). Research on lianas is difficult: most profile diagrams do not show them; their 'slash characters' are not known; they are rather variable in their vegetative parts, and often I have searched in vain on lianas 'that had made it' for the characteristic climbing devices present in the younger stages. To obtain good specimens of a liana (and to preserve the wood of all stages in alcohol), the supporting tree must be cut down, and sometimes the neighbouring trees as well if the liana has bound them together. The fast-growing shoots, with their long nodes and small leaves, which saves on weight that would otherwise obstruct their spiral movements in search of support, are very tender, that is, difficult to collect.

The physiology of these tender shoots is interesting; Darwin did some splendid research on this subject in 1905. The anatomy, too, is of much interest. A tree trunk is built like a tube, but the stem of a liana is like a cable in which the elements lie separate and are slightly movable in relation to each other, making the stems highly resistant against pulling, twisting or snapping. Many liana stems are band-shaped in section; others are star-shaped. Lianas are also peculiar in their morphology because of their diverse climbing devices: winding twigs, sensitive shoots which harden around their support, roots, thorns, tendrils and hooks. An exhaustive survey cannot be given here; Troll's survey (1937, 1939, 1942) is 135 pages long.

More briefly, Vareschi (1980, pp. 33–39) surveys types of lianas, especially those found in South America, and Cremers (1973, 1974) discusses the architecture of lianas in tropical Africa. Some lianas can apparently do without climbing equipment, except for the horizontally spreading branches, which prevent the stems from sagging down, and which is in any case a feature of many species. Putz (1980) is not right in assuming that lianas will not usually colonize a tree with either buttresses or smooth bark, an idea also rejected by Boom and Mori (1982).

The architecture of lianas is still insufficiently known. Several species cultivated for this purpose have been studied from the seedling stage to adult plant; several models unknown to Hallé et al. (1978, pp. 251–258) have been discovered. The situation becomes more complicated when species which start their lives as young treelets and later change their stem anatomy to become lianas are taken into account.

The role of lianas in the forest is certainly important, but little documented. That they help to close the canopy can easily be seen along riverbanks; in this way they stabilize the microclimate. They tie tree crowns together so that in a light storm everything remains upright, but in a heavy one everything is blown down. Some data are known about the share of lianas in the production of littler: while trees produce 59%, lianas with a wood mass only 5% of that of the trees, produce 36% of the litter! (Hladik 1974, in Gabon).

Lianas can also offer protection to animals as Montgomery and Sunquist (1978) tell us. "Two-toed sloths may have chosen trees with masses of lianas primarily to gain protection from predation. Two-toed sloths were inactive during daylight hours, and when we climbed to recapture them were often found asleep within a mass of lianas. While they slept in such a location, it was very difficult for us to approach the animal without alerting it to our presence, because the small interlaced branches of the lianas transmitted motions caused by our presence to the sloth. The movements of any potential arboreal predator large enough to kill a two-toed sloth would surely have alerted the sloth as well, giving it the opportunity to attack the predator or to flee through the masses of lianas and out of the tree". Sloths were also observed to make use of lianas as pathways through the canopy (p. 331).

Lianas make up some 8% of the species in a tropical rain forest. In Sabah (Borneo), which measures 78,500 km^2, an estimated number of 150 genera of lianas are present: 13 Asclepiadaceae, 12 Menispermaceae, 10 Rubiaceae, 9 Apocynaceae, 9 Leguminosae and 8 Annonaceae. Baur (1968, p. 60) mentions a plot in the Okumu Forest Reserve in Nigeria containing 65 species of climbers out of a total of 250 species, which included herbs, or 26% of the flora. Montgomery and Sunquist (1978) tell us that in Barro Colorado Island, Panama, of 39 trees sampled 28 had lianas in their crowns, totalling 55 species. The number of lianas found in ten plots of 0.4 ha by Fox (1969) varied between 472 and 1146 (average 849), between 28 and 91 of these lianas were thicker than 5 cm (average 56).

Rattans are climbing palms. Easy to recognize, they inhabit primary forests from West Africa to the islands of Formosa and Fiji. The centre is Malesia, with 479 species (Dransfield 1974), and the largest genera are *Calamus, Daemonorops, Korthalsia*. The stems are solitary or grow in clumps. They are equally thick over their entire length, 1–8 cm thick, depending on the species, and often contain clear, delicious water for the thirsty traveller: if a piece 1 m long is cut off and put to the mouth, the water runs out in a small squirt. Other species of lianas also have this property, but knowledge of the species is required: some species are poisonous. The long rattan stems are crowned by some 10 to 15 leaves. Each leaf ends in, or is accompanied by a long, strong whip covered with sharp, hard, backwards-pointing hooks; with these the plant anchors itself in the vegetation on its way up. The sheaths of leaf and inflorescence bear spines too, arranged in handsome and characteristic patterns. Some species of *Korthalsia* have a small hole in each (somewhat bulging) leaf sheath. When the stem is touched one hears: rrrrrrrt! This is caused by ants which inhabit the sheaths: they clack their heels as they jump to attention; if the disturbance continues they swarm out and bite fiercely. Fruits of rattans are about the size of marbles or walnuts, they are scaly and contain one seed.

Rattan canes (the longest recorded stem measured 165 m) are harvested by pulling them down; the upper part, on which the sheaths are not yet decayed, is cut off. Once processed, the various uses of the canes are so numerous that in Malesia one can speak of a rattan civilization. Corner, in *The Natural History of Palms* devotes a nice chapter to the rattans. The rain forest is the main reservoir of this richness; only a handful of species occurs in secondary forest and very few show promise for cultivation. Rattan is the best known of the 'minor products' of the rain forest in the Malay Archipelago; as long as this forest type remains, its resource potential is almost inexhaustible.

Epiphytes

These are plants growing in trees without contact with the soil, but which are not clearly parasitic. Many orchids and ferns are epiphytic as well as the Bromeliaceae, a family almost completely restricted to the Neotropics. They can best be seen in ever-wet cloud forests in the mountains, where they grow within easy reach, or in deciduous forests where the host trees

Fig. 6.1 *Taeniophyllum* (Orchidaceae), on the trunk of a palm. The green roots grow in all directions, giving a star-like appearance to the plant. The inflorescence is a few cm long, the entire plant ca. 12 cm in diameter. Bogor, Botanical Garden (Photograph MJ 1980)

appear to be loaded with plants varying in size from diminutive to more than 1 m. A total of 45 to 65 species on one tree has been recorded; a heavily invested tree only 14 m high carried 416 individual epiphytes (Johansson 1974, p. 64). The same author found that 40 to 62% of the trees over 10 m in a Liberian rain forest carried epiphytes, contrasting with only 15% of the trees (10 to 30 m) in a secondary forest. In a savanna in Tanzania, however, not a single tree was found without them.

In Liberia, Johansson (1974, p. 21) investigated 153 species of epiphytes on 47 species of host trees (called phorophytes) in primary forest and on 11 species growing outside the forest. There were 101 orchids and 39 ferns (the ratio varied with the altitude: 3:1 at 500 to 700 m, 1:1 at 1000 to 1300 m) and 13 species belonging to other families: Araceae, Begoniaceae, Cactaceae, Melastomataceae and Piperaceae. If not producing berries, probably dispersed by birds, they spread by means of tiny seeds or spores. Such plants have to be well rooted so as not to be washed away by rain. *Tillandsia usneoides* (Bromeliaceae) with the appearance of a vegetable beard is used by some birds as nesting material, the plant may continue to grow afterwards. Parts of the plant may be torn off by strong winds and be carried away. They may even grow on telephone wires.

From where do epiphytes obtain their food? The old idea that epiphytes do not take food from their host has been refuted by Ruinen (1953). She showed that the roots are associated with a fungus that penetrates bark and conductive tissue of the tree. Heavily invested branches gradually decline. If the host tree dies, the epiphytes at first show increased vigour because they receive more light and utilize the decaying bark. But later on, when the substrate is exhausted, the epiphytes perish. Ruinen coined the term "epiphytosis" for this form of parasitism. As to the epiphytes on telephone wires: rich epiphyte growth along roads where inorganic dust is blown about is a well-known phenomenon. Usually, however, epiphytes grow beneath leafy branches, where falling litter can be caught by them. Well-known house plants are the bird's nest fern, *Asplenium nidus*, and the staghorn fern, *Platycerium*. Their leaves stand out to form a kind of funnel in which litter is collected. This turns into humus which at the same time acts as a kind of sponge to retain water. The trees are always teeming with ants, often actually living in the epiphytes, such as in the corms of *Hydnophytum* and *Myrmecodia* (Rubiaceae) or in the hollow rootstock of the fern *Lecanopteris*. Ants come home with their prey: insects containing nitrogen. Termites build corridors up a tree consisting of particles from the forest glued together, in this way they carry inorganic matter upwards which may be utilized by epiphytes.

The epiphytes' greatest enemy is drought, always worse high up in the crowns where the humidity is subject to such rapid changes. Nearly all epiphytes therefore have contraptions to retain water: either fleshy parts such as the characteristic bulbs of orchids to be seen in any nursery, or an arrangement of the leaves as in Bromeliads, whose leaves form a cylindric funnel in which water is collected – this in turn carrying a special fauna. A spectacular adaptation is that found in *Dischidia rafflesiana* (Asclepiadaceae): some of the leaves are about 10 cm long, closed but for a small opening in the top. This hole collects water, which partly fills the cavity, and the plant sends a rootlet into the pitcher to absorb the water. In many orchids (e.g. *Vanda*) the roots are covered by a thick layer of whitish dead cells, the velamen, which readily takes up water and minerals but which limits evaporation. Nearly all epiphytes also have a thick cuticle covering their leaves preventing evaporation.

Went (1940) found that epiphytes grow in associations, varying with the host tree, probably as a result of differences in chemical composition in the bark. Johansson (1974), moreover, showed distribution patterns in epiphytes depending on the place in the tree, as illustrated in Fig. 6.2. Most epiphytes grow on old trees.

Taeniophyllum (Orchidaceae) is a leafless epiphyte (Fig. 6.1). From the centre, which carries a tiny inflorescence, the roots a few inches long spread in all directions. The roots contain chlorophyll.

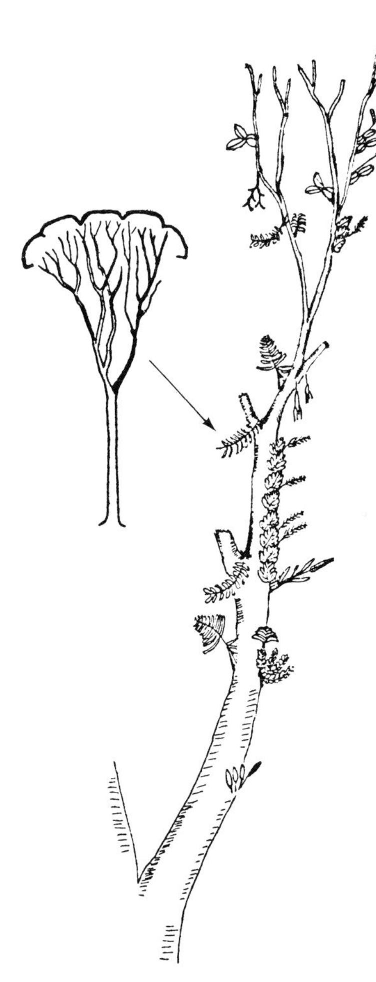

Fig. 6.2 Epiphyte communities. West Africa (Johansson 1974; also includes a list of species)

Herbaceous plants anchored to trunks by their roots are not to be considered as lianas, since these are always woody. The herbaceous ferns of the *Lomariopsis* group, for instance, have rootstocks climbing for several metres in this way, and the stems of a number of Araceae behave similarly.

One peculiar example is *Carludovica* (Cyclanthaceae) of tropical America. The oldest part of the stem dies as fast as the tip grows, so that the plant appears to be creeping up the trunk very slowly like a lizard (Fig. 6.3).

Nearly all Bryophytes in rain forests are epiphytic. The stems of *Spiridens* (Spiridentaceae) may be as much as 1 ft (30 cm) long. Moss growth is an indicator of high humidity. In swamp forests mosses may grow close to the ground, and in mountain forests, especially those often in the clouds, and in valleys, the trunks and branches of the trees may be covered with cushions of moss several centimetres thick, absorbing all sound just as in a studio. But in mountains of Luzon, I found to my surprise that after a few dry days the whole cover could burn like paper (Jacobs 1972).

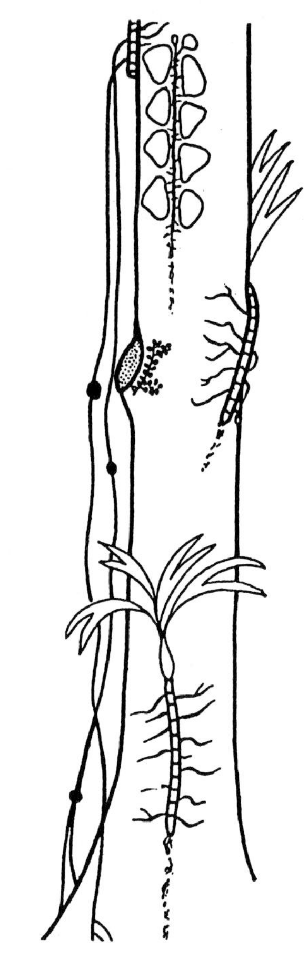

Fig. 6.3 Tree creeper, *Carludovica* (Oldeman 1974)

Epiphylls: A World Apart

In damp places in tropical rain forests old leaves are often covered with a vegetation of minute plants: lichens, liverworts and even mosses. They occupy an ephemeral habitat, which Ruinen called the phyllosphere, analogous to the root habitat, the rhizosphere. The leaf surface, often damp and shaded, is colonized in a fixed sequence. First comes the nitrogen-binding bacteria *Beijerinckia* with *Azotobacter* in its wake; then come fungi, yeasts, algae and lichens, followed by unicellular organisms such as flagellates, amoeba, ciliates and slime molds; after these, small Arthropods and sometimes mosses. The leaf thus enriched with (in)organic matter finally drops and contributes to the cyclic process of the rain forest (Ruinen 1961, 1963, 1965) (Fig. 6.4).

Ruth Kiew (1982) watched epiphyll growth on a low palm (*Iguanura*), of which the leaves persist on the trunk until they decay. She estimates the 'working life' of a leaf as 3 years; in an ever-wet climate, the first lichens appear on the leaf surface after 7 months. The scarce light is intercepted gradually and after 5 years much of the leaf is covered, especially along the veins where water collects after rain. Kiew thinks that epiphyll growth imposes a limit on the functioning of the leaf.

Bamboos

This is the name for woody grasses, the stems of which are solid at the nodes and hollow between the nodes; the largest species may attain a length of 30 m with a stem diameter of 20 cm; climbing species in the rain forest are often the same length, but with stems only 1.5 cm across. The erect species are light-loving; in the forest they establish themselves in places which have been damaged. They tend to proliferate by underground rootstocks. In dry areas they can maintain themselves for centuries and thus bear testimony to ancient patterns of human interference. The centre of diversity of bamboos is Southeast Asia but they have spread throughout the rain forests of Malesia.

There are scores of species and their uses are endless. Houses can be built and furnished using almost exclusively bamboo material; this is commonplace in Indonesia. The material culture of Indo-Malesia is characterized by the use of rigid bamboo material from outside the forest and flexible rattan from within it.

A good introductory chapter on bamboos is in F.A. McClure, *The Bamboos/A fresh Perspective* (1966).

Fig. 6.4 Epiphylls on old leaf with short 'drip-tip'. Malaya (Photograph MJ)

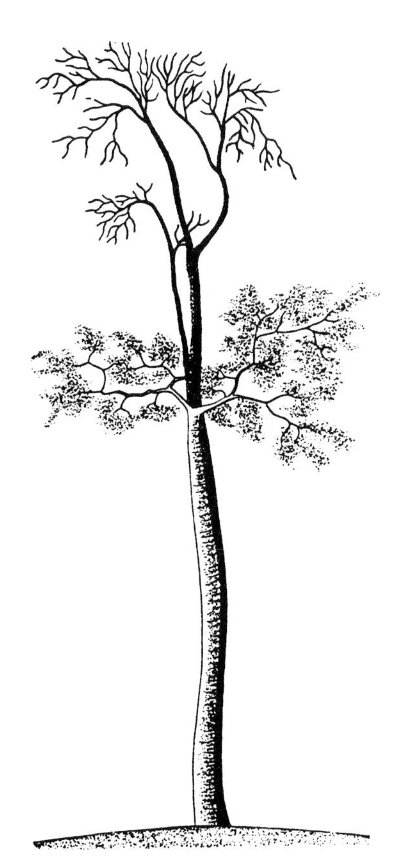

Fig. 6.5 Hemi-epiphyte: *Wightia* (Van Steenis 1949)

Hemi-Epiphytes

At an altitude of about 100 m in West Malesia the sombre green of the forest is sometimes enlivened by pink patches. The plant responsible, *Wightia* (Scrophulariaceae) may grow to a height of 30 m and is all the more conspicuous when in bloom because it is then almost leafless. Its habit is remarkable: too weak for a tree, too bulky, especially at the top, for a liana, the plant leans against a stronger tree and holds on to it with its roots (Fig. 6.5). The flowers are short, tubular and a deep purple, the seeds small and winged.

Wightia starts life as an epiphyte high up in the tree crowns because it needs light to germinate. For a long time it remains inconspicuous, sending down a root with side roots which clasp the trunk. Real growth only takes place after the vertical root reaches the ground.

The root increases in size and begins to look like a trunk. In the meantime the real stem continues growing above the root until in the end it is impossible to distinguish between the two, and a crown is formed. If the host tree dies *Wightia* remains standing like an ordinary tree.

The paradoxical data on herbarium labels reflect this metamorphosis. *Wightia* can also occur in the open, for instance on old lava streams, where it may grow into a small tree 15 m tall (Van Steenis 1949, 1972, p. 38). The hemi-epiphytic habit is often a sign of great versatility, and can be found in several unrelated groups: *Fagraea* (Loganiaceae), *Schefflera* (Araliaceae) and particularly in the strangler figs, *Ficus* (Moraceae).

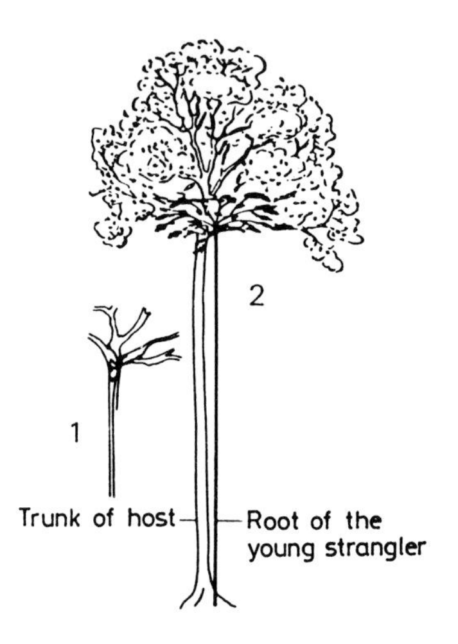

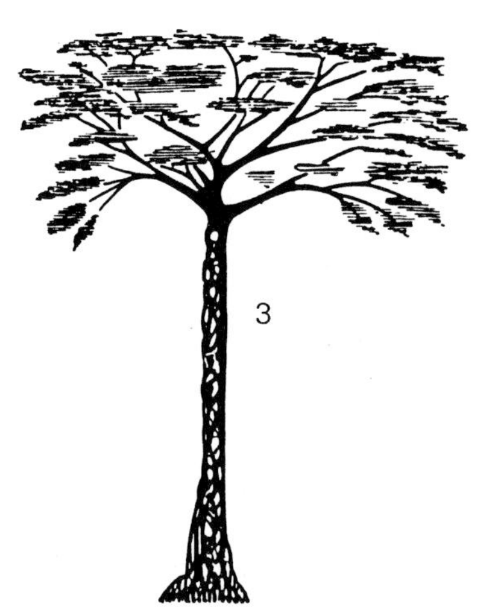

Fig. 6.6 Development of a strangler, three stages (After Corner 1949)

Tree Stranglers and Banyans

These are two life forms of the large genus *Ficus*, which also includes trees and lianas and in which all forms of cauliflory are found. All stranglers and banyans start life as epiphytic shrubs on the branches where their fleshy fruits have been dropped by birds and bats. The first root, which may be as long as 40 m, is thin and unbranched. As soon as it reaches the ground, it expands enormously, growing in time a complete network around the trunk of the host tree, made up of roots 10 cm or more thick. Some species completely strangle their host, the remains of which may be seen for a long time inside the mighty web of roots (Fig. 6.6). Other species do not, and the range of life forms includes hemi-epiphytes (see Corner 1940, p. 658 ff).

In another type of *Ficus*, roots are sent down from horizontally spreading branches. These roots grow into massive columns up to 30 cm across after reaching the ground. In this way a miniature 'forest', more or less circular in outline, is formed consisting of one species, or, rather, of a single individual. The large banyan, *Ficus bengalensis* in the botanical garden in Calcutta (India) dating from 1782 (Corner, p. 675) measured 100 m across and had 666 'trunks' in 1937. In 1920 the central part was damaged by a storm, followed by an infection of a giant fungus, *Fomes* (Polyporaceae). Measures were taken to save the life of the tree (Biswas 1942, p. 9) and in 1973 it was doing well.

This is the species that astonished the army of Alexander the Great when it reached northwest India in 327 BC; the Greeks recognized the tree as a relative of the edible fig they knew. It was on this expedition that Theophrastos described the botanical findings; the very first beginnings of tropical botany (Stearn 1958).

I am not convinced that the banyan life form occurs in a true rain forest, but a transitional form between strangler and banyan is the *Ficus benjamina* (waringin in Indonesian). The larger-leaved *F. elastica* is a strangler, however, and buildings in the ever-wet tropics are often damaged by the expanding and wedging root system of stranglers. The tropical American *Clusia* (Guttiferae) is a strangler as well.

Mistletoes and Other Hemi-Parasites

The only West European member of the Loranthaceae is the mistletoe, *Viscum album*, a compact evergreen shrub with sticky white berries. If these germinate on the branch of a suitable host, preferably a poplar, their sucker roots penetrate the bark and absorb water and minerals from the wood vessels. The plant is capable of making its own organic matter, however, and is therefore called a hemi-parasite.

The family is represented in the tropics by some 1100 species in about 30 genera. In the rain forest they usually grow high up in the crowns. The fallen tubular flowers up to 10 cm long on the floor of the forest give them away (most species belong to a different subfamily than the one to which *Viscum* belongs). The flowers, in various shades of red, green and yellow, are suitable for pollination by birds. These shrubs, with their opposite branches up to 2 m long, are often cauliflorous. Often they send a root along the branch of the host, up to 1 m long and bearing several sucker roots that penetrate the host. Some Loranthaceae are hyperparasitic on other members of the same family.

The exterior of the tree crowns is often inhabited by the related parasitic family Santalaceae, with long, slender, woody stems that extend from one twig to the other.

Holoparasites

These plants are devoid of chlorophyll and are completely parasitic. Well-known European examples are the broomrape *Orobanche* (Orobanchaceae) and the dodder *Cuscuta* (Convolvulaceae). They belong to unrelated families. Parasitism evolved independently in several groups of plants.
The most spectacular parasite of the tropics is *Rafflesia* (Rafflesiaceae), a genus with some 10 to 15 species, whose centre of distribution is West Malesia, though the family, with some eight genera, is of pantropical distribution. A flower bud of *Rafflesia* looks like a cabbage. The open flower of *R. arnoldii* in Sumatra may be 80 cm or more across, and is purplish brown in colour and smells of carrion, which attracts flies and beetles. The plants are either male or female, the fruit of the female being hidden in the decaying flower parts. *Rafflesias* grow on a genus of large lianas, *Tetrastigma* (Vitaceae); the vegetative part of the plant consists of thread-like tissue, comparable to the mycelium of a fungus (see Harms 1935). The flower buds emerge from the stem of the liana not more than 4 m above ground. J.E. Teijsmann, the famous Dutch conservator of the Botanical Garden in Bogor managed to infect a *Tetrastigma* in the garden by rubbing seeds of *Rafflesia* into cuts made in the bark. This and the fact that *Rafflesias* always grow near the ground gives an indication of how dispersal takes place in nature. Probably large ungulates (in Sumatra: elephant, deer, rhinoceros, tapir and wild boar) tread on the fruits and damage the bark of a liana if they step on it. The eastern boundary of *Rafflesia* is formed by Wallace's line to the east of Bali, Borneo and Mindanao. This also marks the eastern limit of the large West Malesian mammals.

'Nature monuments' of a few hectares were once created for *Rafflesia*, but to a person of ecological insight it will be clear that the protection of large ungulates is essential to the safeguarding of *Rafflesia*, which in turn means that there should be enough room for these animals to roam freely.

Saprophytes

Like holoparasites, saprophytes are devoid of chlorophyll; they obtain food from litter and are rare in Europe. The yellow bird's nest *Monotropa hypopithys* (Pyrolaceae), and the bird's nest orchid, *Neottia nidus-avis* (Orchidaceae) are the best-known examples. Saprophytes are not common in the rain forest, and their colour matches the litter in which they live. More often than not the rewards I promised for finding saprophytes for me were not collected, and I found that the best chance to find a saprophyte was when I sat down at leisure to eat my lunch or to satisfy some other physical urge! Saprophytes such as *Epirixanthes* (Polygalaceae), *Petrosavia* (Liliaceae) and *Sciaphila* (Triuridaceae) produce tiny stems 10 to 20 cm long and a very few small pale flowers. Their life cycle is still a mystery, although the Triuridaceae, consisting of three genera and some 40 species, occur in all three rain forest areas.

A conspicuous saprophyte is *Galeola* (Orchidaceae) with pale brown stems 10 m or more long creeping over the decaying logs in which it roots.

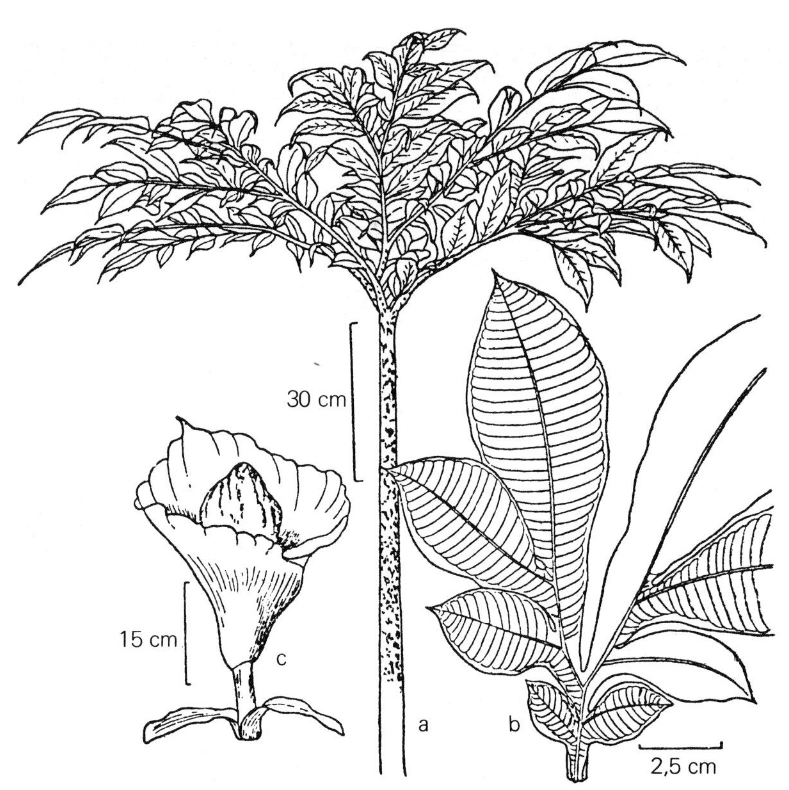

Fig. 6.7 *Amorphophallus campanulatus* (Araceae). *a* Complete leaf; *b* part of leaf; *c* inflorescence. The starchy tuber is used in West Malesia (Henderson 1954)

Terrestrial Herbs

The largest herbs all belong to the monocots. The tallest on record attains 9 m. Best known and to be seen in greenhouses in *Musa* (Musaceae), the banana. The thick 'trunk' consists of closely fitting leaf sheaths and can be felled with a single blow of a knife or machete. The leaf blades (which remain intact in a greenhouse) in nature are usually shredded along the secondary veins to the midrib, though without any harm to the plant. When eventually the inflorescence pushes its way up inside the pseudostem, it emerges and bends down to produce the well-known bunch of fruits, whereupon the plant dies and stools out at the base. Banana plants always grow gregariously, preferably in somewhat open places; they don't thrive in the forest.

Musaceae are closely related to the Zingiberaceae, a pantropical and taxonomically difficult family of some 1300 species in 40 genera. The family name is derived from the ginger, *Zingiber officinale*. Here, too, there is a pseudostem, and the leaves, usually less than 1 m long, are arranged at regular distances along it while the inflorescence is very often (but not necessarily) produced on an underground rootstock, so that from time to time a large, brightly coloured asymmetrical flower opens. Like *Musa*, gingers prefer light, and large groups of them often line riverbanks; the darker forest is inhabited by much smaller species, some of them epiphytic (see Fig. 14.6).

The Araceae are another large family; in Europe, lords and ladies (*Arum*), bog arum (*Calla*) and the aromatic sweet flag (*Acorus*), all grow naturally. *Anthurium* with its fiery red spatha, *Monstera* and *Philodendron* are popular house plants. They all have a stalk densely set with flowers subtended and often covered by a bract, but the family as a whole is very polymorphic. *Monstera*, native in Mexico, may climb up the trees to a height of 15 m and bears leaves 1 m across on a stem as thick as an arm, while the taro, *Colocasia*, bears a tuft of

Fig. 6.8 *Colocasia* (Araceae). North Borneo (Corner 1964)

triangular leaves also on a firm stem (Fig. 6.8). The edible tuber is an important source of starch in Malesia and the Pacific. A similar, but larger, herb is *Alocasia*. Both may be seen in the rain forest, but neither attain the dimensions of some species of *Amorphophallus* (Fig. 6.7). An amorphophallus is usually a solitary plant of the forests of Malesia. The huge corm is rich in starch and produces first a leaf and then an inflorescence. The leaf of *A. titanum* is like a parasol. The stem, 10 cm across and 1.5 m long, is blotched; the leaf blade is finely dissected. After the leaf has wilted, a funnel appears of a weird purplish-green hue. Inside the funnel a bludgeon emerges and rises to the height of a man. The plant can easily be grown in cultivation and has even flowered in Europe. Personally, I think *A. decus-sylvae* from central Sumatra is even more spectacular. The stalk of the inflorescence in this species may be up to 2.5 m high.

The Gleicheniaceae are another group of giant herbs. These are ferns growing in open places in the ever-wet tropics, where they may form impenetrable thickets especially between 1000 and 2000 m. The stem, resistant to the machete, is actually the axis of a composite leaf, it may resume growth time and again, and can attain a length of several metres.

Under the closed canopy the undergrowth is scant, though there are always some ferns, small Zingiberaceae and a few genera of Acanthaceae, Balsaminaceae, Begoniaceae, Cyperaceae, Gramineae (often broad-leaved), Gesneriaceae, Melastomataceae, Piperaceae, Rubiaceae and Urticaceae. Nearly all have small seeds. Scaly rootstocks, so common in plants of temperate climates, are rare in the tropics. One finds them in some monocotyledons where such structures are, so to speak, an essential element of the general design. Terrestrial herbs of the rain forest very often have variegated leaves, which makes them desirable for cultivation, and several species of *Selaginella* (Selaginellaceae) and other genera have iridescent leaves. According to Lee (1977), green iridescence is due to the refraction of diffuse light onto specially oriented chloroplasts by lense-shaped cells; blue iridescence is due to the operation of thin interference films in or on the epidermis. The advantage iridescence has for the plant is said to be the more effective absorption of red wavelengths at the expense of the less important blue wavelengths.

Little has been written on small terrestrial herbs. Burtt (1977) and Kiew (1978) provided some new data, but much more remains to be done.

Flora Brasiliensis, tabula XVIII. Banks of the river Itahype, province Bahia, eastern Brazil. Drawn 1817-1820, published 1842. In the foreground *left* are two Araceae, the larger is *Montrichardia linifera*, the lesser, *Pistia stratiotes*, floats. Between the palms, *Euterpe edulis* (Palmae), a decaying trunk. On the *right*, the large grass *Gynerium* (Gramineae) (Martius 1840–1869)

7 Composition

Aspects of the Composition

What are the elements that make up a tropical rain forest? The composition in the first place is a matter of plant and animal species collectively known as the flora and fauna. But a list of species is merely a beginning, notwithstanding its fundamental value. The following aspects are also important:

1. The life forms in which the species manifest themselves;
2. The part each life form plays in the vegetation;
3. The size of the largest trees and climbers;
4. The sizes of species of trees and climbers, classed according to diameter;
5. The numbers of seedlings and juvenile stages of the various species;
6. The density of individuals per hectare or larger unit, per species;
7. The associations between individuals of one species: scattered or in groups;
8. The associations between species.

These aspects are an expression of the biogeographical situation, as well as of the (micro)climate, altitude, slope, soil, drainage and, of course, the history of damage and destruction, natural and anthropogenic.

This chapter is confined to plants. Some notes on animals are given in Chapter 12.

Heterogeneity of Data

As seen in Chapter 2, by no means all plant species are known to science. Some families have been well studied by taxonomists, to the neglect of others. Some territories have been much better inventoried than others. Ecologists may not even realize how much their scientific results are determined by such peaks and gaps in the available knowledge. And there are other obstacles.

One is the use of different units of area for study. On the continent of Europe and within its sphere of influence, the standard unit is the hectare (ha), 100×100 m equalling 10,000 m^2. In English-speaking countries the unit has been, and often still is, the acre, 0.4 ha. Of course, a larger area contains more species, but not in such a proportion that a hectare contains 2.5 times as many species as an acre. It is possible to straighten out the differences by a mathematical procedure, but this may not match the reality in the forest. Two adjoining acres give a different picture than do two remote acres; we remember Ashton's example in Chapter 4.

Nearly all workers select their objects differently. While they look at trees in the first place, badly neglecting the non-trees, the one confines himself to those 40 cm or more in diameter, another takes 30 cm as the lower limit, again another, 20, again another, 15 or 10. The few who bother to look at the other life forms as well, may or may not include the ferns, and fern allies like *Selaginella*. In counts all these groups and life forms are rarely properly distinguished.

At the time Richards published *The Tropical Rain Forest* (1952), not all forest types were clearly defined, notably the *wallaba* or *kerangas*, the peat swamp and the freshwater swamp. The definition of intergrades between evergreen and deciduous forests is still rather arbitrary; so are the distinctions between *varzéa* and *igapó* along the Amazon. The altitude above sea level is often not recorded below 500 m, or even 1000 m. Past damage may be difficult to diagnose, and goes easily unnoticed by naïve workers. All such factors limit the value of comparisons, not to mention the reliance by ill-prepared workers on native plant names which may result in the strangest blunders, or their failure to collect proper specimens for documentation. Part of the expertise of a rain forest scientist consists of the ability to single out the strong and weak sides of a piece of work presented to him, and to make comparisons in their proper perspective.

Species Richness: A Few Figures

The only analysis I know about which comprised all life forms in rich, ever-wet undamaged forest (Meijer 1959) was made in the Cibodas Reserve in West Java at 1450–1500 m, i.e. in the mountain forest. One hectare was inventoried with the exception of the epiphytes, which due to an intensive study by Went (1940), were estimated at 100 species. The list of life forms reads thus:

	Woody species	Herbaceous species
Epiphytes		100
Trees	78	
Low herbs		39
Shrubs (including, presumably, some treelets)	31	
Woody climbers	20	
Terrestrial ferns including fern allies)		20
Herbaceous climbers		9
Tall herbs		8
Treelets	6	
Woody creepers	5	
Herbaceous creepers		5
(Hemi-)stranglers	3	
Tree ferns	2	
Palms	2	
Pandans	1	
Bananas		1
Mistletoes	1	
	149	82

The total is 331. To this should be added 13 species found only in the juvenile state, and 6 pioneers in a clearing. There were 228 trees on that one hectare. The biggest, *rasamala, Altingia excelsa* (Hamamelidaceae) was 62 m tall and 81 m in diameter. Next came *Castanopsis javanica* (Fagaceae) at 58 m, *puspa, Schima wallichi* (Theaceae) and various other Fagaceae at 45 m. Of the 78 tree species occurring in that plot, half were represented by a single individual only. Of the commonest tree, *Villebrunnea* (Urticaceae), there were 33 individuals, 11% of the total.

It is a pity that we don't have details about the epiphytes. They abound in this montane forest with its ever-high humidity, and it seems doubtful if in the lowlands their percentage would often score as high as the present 30%. As it is, the 78 larger trees made up 25% of the whole (phanerogam plus fern) flora, all trees 39%, all woody plants 45%.

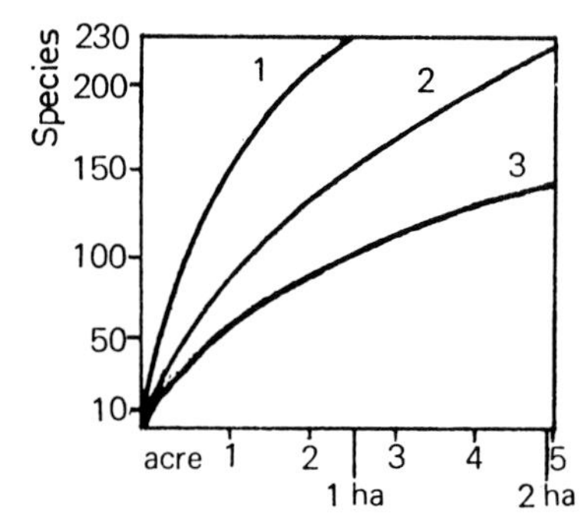

Fig. 7.1 Species number per unit area, for small areas. *1* Rengam Forest Reserve, Malaya, 227 species on 1 ha (trees over 10 cm thick only); *2* Andulau Forest Reserve, Brunei, 220 species on 2 ha (same); *3* Badas, Brunei, 144 species on 2 ha (Ashton 1964)

Here is a calculation from Brunei, in northwest Borneo. Brunei is 5000 km^2 in area, one-seventh the size of The Netherlands. Ashton (1964, p. 4) estimated the number of tree species in the area which were 10 cm or thicker at 2000. The entire flora of The Netherlands consists of 1200 species.

Now some calculations from the New World. In Surinam, Schulz (1960, p. 167) analyzed the tree flora of 2 cm or thicker at Mapane, 40–100 m above sea level. In 1 ha, he mentions 32 common species 35 cm in diameter or more, in 26 genera in 14 families, plus 12 thinner species, which rarely exceed 25 cm. He counted the less common species with less than one tree in 1 ha, and arrived at a total of 206 for 5.6 ha.

In Brazil, Prance et al. (1977) inventoried 1 ha dry land forest near Manaus, in the centre of Amazonia. On these poor soils, the forest does not attain great height: the canopy reaches ca. 30 m, the emergent trees ca. 40 m. The trees are slender, only four attaining a diameter between 65 and 100 cm. In that 1 ha there were 350 trees 15 cm or thicker, belonging to 179 species, plus 56 trees of 5–15 cm. These are very high numbers.

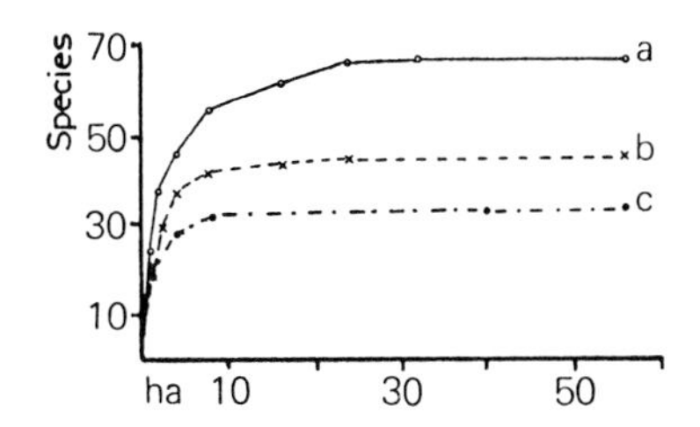

Fig. 7.2 Species number per unit area, large area. *Curve a* Species with more than 10 trees in the area. *Curve b* Regularly distributed as well as frequent species. *Curve c* Species with more than 20 trees in the area. Jengka, Malaya. (Poore 1968)

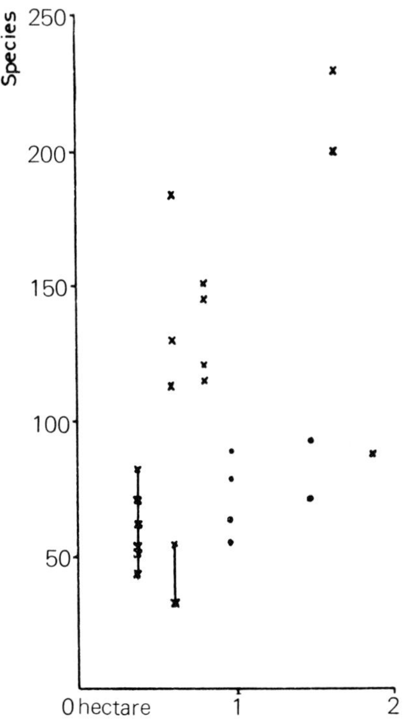

Fig. 7.3 Species per unit area, comparison between various forest areas (After Whitmore 1975)

Species per Area

Although data are not very profuse, they all reveal the same picture, if put in diagrammatic form. If on the left, the numbers of species are recorded, against the area units on the base, the curve is steep initially, then rises more gradually, to level off to some extent but never completely. Every added hectare contains species not yet counted. If the slenderest trees, too, are incorporated, the curve shows no sign of flattening at 2 ha (Fig. 7.1). This result was obtained by Ashton (1964), who worked with numerous small plots of different composition. Only Poore (1968), who analyzed one large plot which was the most homogeneous that could be found, succeeded in obtaining a flattish curve (Fig. 7.2); but he limited his choice to thick trees. Thus, Poore thought a plot of 2–5 ha sufficiently representative, although his upper curve levelled off at as much as 30 ha.

The high numbers of rare species that were found, by Schulz as well as by Poore, strongly influence the picture. Poore's curve does not rise above 70 species because he confined himself to those of which he found ten or more individuals. He who drops this limit can, in places in Malaya, find 200 tree species in 1 ha (Fig. 7.3)! Prance, in Amazonia, found equally high values, as we have just seen.

These are records; the 100–150 species per ha in New Guinea is also a high figure; but 60–80 species per ha is usual in most places, and the curve continues to go up: in Pasoh, 460 tree species were found on 11 ha which is two to three times as high as the average elsewhere in Malaya.

In Cibodas Meijer found that trees represent 25% of the total flora. Does this mean that we may multiply the number of tree species by four to arrive at the total number of species? Meijer's high percentage of epiphytes makes this risky. If, rather arbitrarily, we halve the number of epiphytes, trees can be said to make up 35% of the flora. This estimate tallies better with the customary one of one-third of all species being trees at least 10 cm in diameter or more. Boldly extrapolating, we then arrive at an approximation of 200 plant species in 1 ha as being rather low, 400 as fairly high, 600 as exceptional. But one must always remember that many species are exceedingly rare.

Woody Plants Predominate

Meijer's species were classified into woody (149) and herbaceous (182), amounting to 45% and 55% of the flora respectively. The latter figure, however, was augmented by a number of epiphytes that would be unusual in the lowlands. If we, boldly, I admit, halve this number, which means reducing it by 50, the figures become 149 woody species and 132 herbaceous, with percentages of 53 and 47. The latter includes ferns, which seems reasonable since their stature equals that of many phanerogam families.

This percentage of woody plants of ca. 45–53% is very high in comparison with any temperate flora. In The Netherlands, in a flora of ca. 1200 species, we have 30 indigenous trees (of which 10 are willows, *Salix* (Salicaceae), plus 35 shrubs and woody climbers, that is, if we count the subsections of the brambles, *Rubus* (Rosaceae), which in tropical botany probably would be given the rank of species. Our entire tree flora, 2.5% of the total, is ten times poorer than the 25% found by Meijer in that one hectare in Cibodas. Adding the 35 shrubs and woody climbers, 3% of the total, to the 5.5% woody plants, does not alter the proportions. The comparison of The Netherlands with Brunei is even more telling. The 30 tree species native to Holland could well be found in a Dutch area the size of Brunei; but in Brunei's relatively small land area there are 2000 species. The difference is very great indeed.

This predominantly woody character of the rain forest flora has great consequences for the dynamics of the vegetation. A woody plant needs a much longer time to come into flower than a herb, and when mature will usually stay in its place for many years.

We do not know the age of many rain forest trees, as has been said, but it has been estimated that the large dipterocarps take 50–60 years to flower for the first time, and this may also be true for other emergent species. As for the climbers, in a plot of 33-year-old secondary forest in East Borneo, the thickest stem I found was 10–12 cm. Would a liana of 20 cm diameter – nothing unusual – be as old as a first-flowering emergent?

How long, after reaching maturity, a large tree will remain in place will no doubt differ with the circumstances. Estimates agree on two to three centuries at least, and five centuries may not be exceptional. Smaller trees may not grow that old, but when considering replacement it is the trees, the skeleton of the forest, that count. But even if we take one century as the rough average life span of a fairly large tree, it will be evident that the processes of forest regeneration and succession are much slower than in areas in which the vegetation consists mostly of herbs, such as a marsh.

Another factor which slows down dynamics is the mixed character of the rain forest. In a species-rich system, many different populations have to be built up; this takes more time than the establishment of a few species. As we shall see, a forest functions through many plant-animal relationships which must be established and integrated into a locally developing 'web of life'.

Species Richness and Diameter Classes

In 1968, D. Poore made an analysis of the trees in the richest known rain forest, that of Jengka, in the central part of Malaya, at 45–75 m altitude. Unfortunately, the government of Malaysia later allowed this forest to be cut. Poore's entire plot was 23.04 ha in area, but there were some narrow creeks and other forest types in the plot – a large homogeneous plot is rare in Malesia – so that a total of 22 ha of true mixed lowland dipterocarp forest remained. Poore inventoried the trees which formed the canopy, most of them 30 cm or thicker. The 2633 canopy trees that were measured and identified (an average of 115 ha^{-1}) belonged to 357 species, 139 genera in 52 families. The thickest, a *Dipterocarpus costulatus* (Dipterocarpaceae) was 2 m in diameter, a quarter of the trees were 40 cm in diameter, while nearly half the trees came in the slender 30 cm diameter group.

With the aid of tables one can calculate the 'basal area' of a tree from its diameter. This is the surface area of the trunk near the ground, in section. In the Jengka forest the basal area of the trees 30 cm or more in diameter was 24.2 m^2 ha^{-1}; the joint basal area of the thinner trees was estimated at one-tenth of that.

Here is Poore's table of the best-represented families of the larger trees:

	Number of trees	Number of species	Basal area, % of total	Trees per species
Dipterocarpaceae	771	30	42.8	25.70
Burseraceae	370	23	10.1	16.10
Leguminosae	174	15	8.0	11.60
Euphorbiaceae	155	28	3.3	5.70
Sterculiaceae	132	6	4.3	22.00
Myrtaceae	113	24	3.0	4.70
Annonaceae	89	16	1.6	5.65
Myristicaceae	84	21	2.2	4.00
Moraceae	82	11	2.6	7.45

Throughout the forest, 12% or 44 species occurred so often (if sometimes widely scattered) that they are to be regarded as 'constant species'. An associative relation between the commonest species, however, was not found.

In the 'hall of the forest', of course, all species of trees and lianas are present in all their diameter classes. Schulz (1960) counted 32 common tree species in his Mapane plot of 1 ha in Surinam. These were divided as follows: 866 of 2–5 cm diameter, 459 of 5–25 cm, 85 of 25–35 cm, 52 of 35 cm and thicker. He also found 12 species of low stature, 700 of 2–5 cm, 311 of 5–25 cm. Together, this totalled in 1 ha 1566 trees of 2–5 cm, 770 of 5–25 cm, 89 of 25–35 cm (including four tall trees belonging to smallish species), and the 52 thick ones, i.e. 2477 in all.

A graph of the measurements of all the trees of one species shows a characteristic pattern, from many slender ones to a few with large diameter. This means that the majority dies long before adulthood. And, indeed, when one looks, one occasionally sees a dead sapling standing. Very little is, however, known about death and decay in the tropical forest, and systematic studies would reveal much of interest to botanical science as well as to forestry.

One line of recent enquiry is being pursued by students of Prof. Oldeman of Wageningen. They suspect that there must be some or other distribution pattern for the biggest trees, and found that the lines connecting these form a network of rather equal-sided triangles. A sapling or pole has quite different chances of growing up, depending on its place in a triangle: if too near a standing emergent of the present, it has a very small chance of success, and will die sooner or later (see Van Bodegom 1981, p. 15 seq.). Such knowledge is, of course, of great importance for forest management.

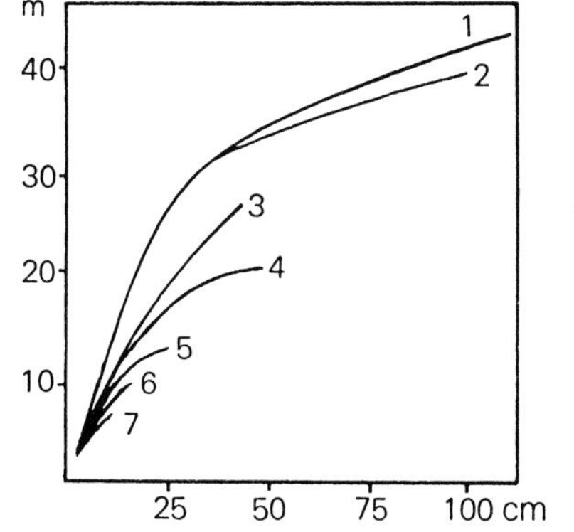

Fig. 7.4 Correlation between trunk diameter and tree height for seven species from Surinam (After Schulz 1960; also includes a list of species)

The diameter is often measured because this is such an easy way of data collecting. The diameter can give an indication of height (Fig. 7.4); a rule of the thumb is: $h = 100 \times d$, but age is hard to determine from it. Under unequal conditions, the trees of one year grow up very differently. Damage may have similar or worse results, which in the latter case will show up in the graphs, through either the abundance of or absence of certain classes. If only thick trees of a species are present (as in Fig. 8.5), this is an indication of some serious disruption in the past: either a long period of grazing which destroyed the seedlings, or clearing followed by a secondary growth of trees which cannot reproduce under their own cover.

Few Individuals per Species

As there is a limit to the number of trees that can be packed into 1 ha, it is obvious that the more species there are, the fewer individuals of each species can be accomodated. The three maps (Figs. 7.5, 7.6, 7.7) illustrate this. They were made in one plot of rich forest in western Sarawak, 12 by 6 chains in size, i.e. 241 × 121 m, which is nearly 3 ha. All trees 20 cm or more in diameter were mapped, and those belonging to the Apocynaceae, Moraceae-Artocarpeae and Sapotaceae families were identified as to species; the three families all have white latex. The Apocynaceae are dispersed by wind, the others by animals. Explanatory notes are given in the captions. The total number of trees was 140, in 26 (sub)species, divided as follows:

Numbers per species	50	15	12	7	6	5	4	3	2	1
Species per number	1	1	1	1	2	1	3	2	8	7

An analysis reveals that 40 m is a small average distance between trees of one species, 70–80 m is common, and a maximum distance cannot be indicated; 185 m happens to be the maximum here, but the plot was not large enough to draw any firm conclusions. We can also extrapolate that the densities vary greatly even within one species: if the map is divided into three parts of nearly 1 ha each, 26 trees of *Pouteria malaccensis* (Sapotaceae) occur in the upper hectare, 5 in the bottom one. This is obviously an exception – without this species, Sapotaceae and Artocarpeae would be much more comparable – but nonetheless we see that averages of 0.3–0.6 trees ha^{-1} are usual, and that 4–5 trees ha^{-1} is definitely a high density.

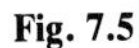

Fig. 7.5

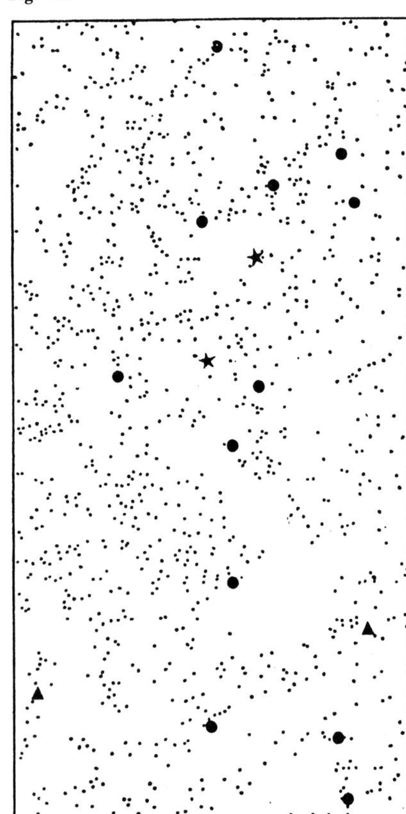

Fig. 7.6

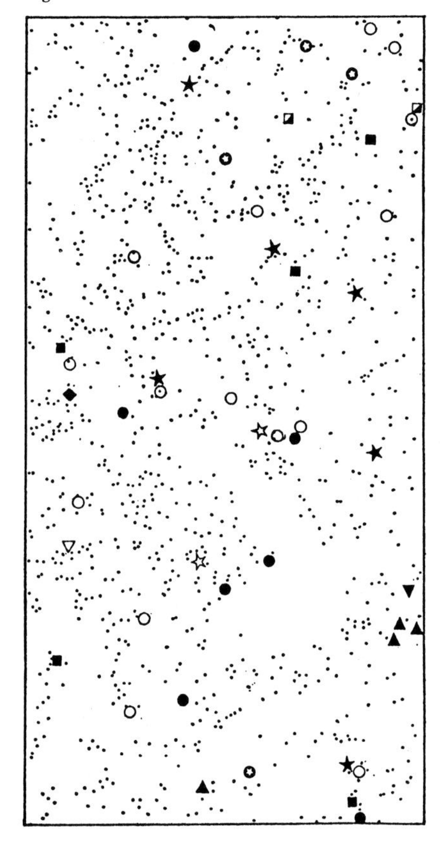

Fig. 7.7

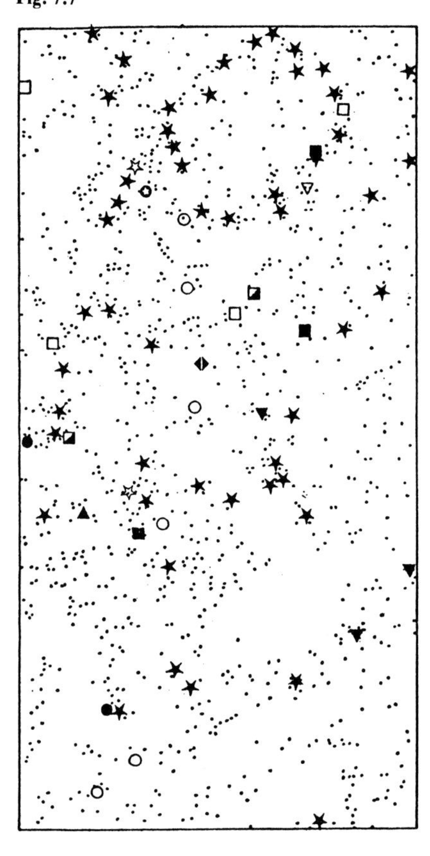

Fig. 7.5 Distances between trees. Specimens over 20 cm thick from the family Apocynaceae, over 2.9 ha of forest in Sarawak (Borneo). All in all 16 trees from three species, all are wind-distributed. *Dyera costulata* has 12 trees, 20–55 m apart; *Alstonia angustifolia* with 2 trees, 35 m apart; *A. angustiloba* also 2, 108 m apart

Fig. 7.6 Distances between trees. Specimens over 20 cm thick from the family Moraceae, tribe Artocarpeae, over 2.9 ha forest in Sarawak. All in all 48 trees from 11 species or subspecies, distributed by animals. *Artocarpus integer* has 15 trees, a small cluster in the upper right corner, one left of the middle and one 70 m distant in the lower right corner. *Prainea frutescens* has 7 trees, 6 in the lower half and 1 at a distance of 95 m. A similar pattern is shown by *Artocarpus odoratissimus* with 6 trees, 5 in the upper half and 1 at a distance of 95 m; *A. anisophyllus* with 4 trees, 3 of which are close together and one 72 m distant; *A. nitidus* ssp *humilis* with 3 trees in the upper right corner and 1 at a distance of 185 m. *A. kemando* does not form a group, with 5 trees at distances of 45–100 m; but *A. nitidus* ssp. *borneensis* may be considered as forming a group with 2 trees 40 m distant. Three (sub)species, 27% of the total number, have only 1 tree on almost 3 ha

Fig. 7.7 Distances between trees. Specimens over 20 cm thick from the family Sapotaceae, over 2.9 ha forest in Sarawak, Borneo. All in all 76 trees in 12 species, distributed by animals. *Pouteria malaccensis* has 50 trees, of which 26 are in the upper third part, 19 in the middle third and 5 in the lower. Particularly in the upper part 18 trees form a conspicuous circle with a diameter of 50–60 m, and 9 trees more can be added to it if the diameter is doubled. *Palaquium leiocarpum* has 6 trees, on the average 40 m apart. *P. rufolanigerum* has 4 trees in the upper half, 55–100 m apart, yet forming a group. *Ganua kingiana* has 3 trees in the lower half, 28 m apart (they might be considered as forming a group), and 1 at a distance of 70 m. *Madhuca erythrophylla* has 3 trees, 55 and 83 m apart, these cannot be considered as forming a group, neither can those of *Isonandra lanceolata*, 2 trees 75 m distant, nor those of *Madhuca beccariana*, 2 trees, 95 m apart. Four species (33% of the total number) have only 1 tree on almost 3 ha

When we divide the map of Moraceae-Artocarpeae (Fig. 7.6) lengthwise into strips 241 m long and 10 m wide, we shall come across an arbitrarily selected species with the following frequency:

1 1 0 1 1 1 1 1 0 1 3 2.

This gives a further impression of the prevailing scarcity of individuals of one species.

Poore's work in Jengka (1968) provides similar figures to Ashton's from Sarawak. Of the most common tree, *Shorea acuminata* (Dipterocarpaceae), there were slightly more than 4 trees ha^{-1} on average; of the least common tree of regular occurrence, *Xerospermum intermedium* (Sapindaceae), there was one individual per 3 ha. But in his entire plot of 22 ha of dry-land forest, no less than 140 species, 37% of the total, occurred only once! (We are talking of trees of reproductive age, thicker than 30 cm)

About the other life forms we know nothing. We can, however, presume that the high canopy, composed of trees of very different heights, offers many extra 'niches' in the five ecological zones visualized at the tree of Longman and Jeník (Fig. 3.8). The smaller trees and climbers are probably less thinly spread. Yet we may judge that in the tropical rain forests, scarcity is common since data have actually been determined for the larger trees.

That there is a large average distance between trees in species-rich forest, as a direct consequence of species richness, is an overwhelming fact. About one-third of the tree species in the Jengka plot may only have occurred once in a square measuring about 500 × 600 m. Those unfamiliar with the rain forest are unable to fathom the consequences of this fact, both for the pollinating and foraging animals and for the plants, among which it increases the phenomenon of dioecism, to be discussed under Speciation (see Chap. 14).

But the consequences of such wide spacing between individuals are important for man as well. In a species-poor system, such as an orchard or a man-made forest, individual trees touch one another. Pests or diseases can spread without interruption. Among insects or fungi, population growth is slow in the beginning, but proceeds faster exponentially, like any other population growth. In a species-rich ecosystem, however, the large distances between individuals are not easily bridged, and the stage of explosive growth is therefore never reached.

Tribal agriculturists know this well, and plant all their crops in combinations, actually imitating the species-rich systems. Whereas in a continuum a pest or disease moves on, the gaps between the plants in mixed cultures provide an initial, often effective, barrier. Pests or diseases are therefore never a problem in mixed rain forests, while large tracts of uniform forests both within and outside the tropics are often devastated. Gray (1978) discussed these matters extensively in his paper.

Various groups of people have to reckon directly with the low species densities in rain forests: the collector of 'minor forest products', the biologist, the forester and the conservationist. Moreover, the pattern is often very complex, notably where trees are grouped together in clusters. This will now be discussed.

Dominance No, Mosaic Yes

Poore's and Ashton's aforementioned figures leave no doubt about the mixed character of the primary rain forest. Even the best represented species makes up but a low percentage of the total, a maximum is about 15%. In the mixed dipterocarp forests, those in which 70% or more of the trees 1 m in diameter and thicker belong to the Dipterocarpaceae, it is the family which dominates, not any species in it.

If forests, for the sake of convenience, are called after a single commercial tree in them, this does, of course, not imply dominance in the botanical sense. Nonetheless, there do exist forests in the tropical belt which really are dominated by a single species. This often indicates past damage, and to suspect it is always appropriate. There are, however, undisturbed forests where dominance occurs, as mentioned above, to a maximum of 15, or possibly 20%. Richards (1952, pp. 254–263) discussed several examples, and concluded that all of them are related to adverse soil conditions, often poor drainage. Although he expressed reservations, he concluded that dominance in a tropical rain forest finds its cause in limiting or damaging factors, mixed composition being the norm.

Two interpretations of 'mixed' are, however, possible, as shown in Figs. 7.5, 7.6 and 7.7. The trees of *Dyera* (Apocynaceae) occur more or less evenly spaced, while among the *Pouteria* (Sapotaceae) and *Artocarpus anisophyllus* (Moraceae) some grouping can be discerned. *Pouteria* even forms a nice circle. Between these two patterns, an even distribution and groups, all sorts of intermediary forms of spacing can be found. The smallest groups are called *consociations*; they vary in the number of individuals, in their diameter classes and in the surface area they occupy. The smallest consociation consists of two. A maximum is hard to define; for *Pouteria* (on the map) it seems to be 30. The group of *Shorea curtisii* (Dipterocarpaceae), of which Richards (1952, pl. ix) shows a photograph, may contain a similar number.

This phenomenon of grouping was described for Africa by Aubréville, and discussed by Richards (1952, pp. 251–254); the latter concluded that an entire forest is to be regarded as one single *association*, albeit one of ever-fluctuating composition. We have noted already that the species composition varies from one hectare to another, at least the details do. Now we are confronted with this phenomenon of highly irregular, small consociations leading to great differences in numbers of individuals of one species per hectare. This makes it even more difficult, actually impossible, to determine a representative, so-called minimum area for plant-sociological work.

This mosaic pattern much complicates the phenomenon of scarcity of individuals per species explained in the previous section. The few conspecific trees per area unit often occur in clusters. Within the clusters, distances are much smaller than the average (see legend of Fig. 7.7); between the clusters, they are much larger. Obviously, the chance that pollen for cross-fertilization is carried from one tree to another is far greater within a cluster than from one cluster to another, and this will directly affect the flow of genetic materials in the forest. We will discuss this further under Speciation (see Chap. 14).

What causes the clustering of trees? The patterns we find suggest a relationship with seed dispersal. *Dyera* (Fig. 7.5) which can attain 60 m in height, produces long narrow capsules with cardboard-like valves. The 6 × 2 cm seeds are flat and very thin, dry and light, which makes wind a likely carrier. *Dyera* seeds are thus distributed at random across considerable distances, with correspondingly few seeds falling on each square metre. Other cases are not so simple. *Pouteria* has rather fleshy fruits the size of a small plum, with a couple of seeds 2.5 × 1 cm. Such fruits travel only with the aid of animals. Thus, a biological factor enters, on which dispersal depends.

As the seeds of *Pouteria* have to be transported by animals, they will be concentrated in a pattern determined by the behaviour of the animal, which will probably result in clustering. But dispersal is only the beginning of a long story. Close to the trunk, where many seeds fall, seed-eating animals, e.g. rodents, but also boring insects, will gather. Seeds deposited at some distance from the tree therefore have a better chance (Richards 1973a, p. 63). Besides, every species has its own range of preference and tolerance with regard to soil and other basic conditions.

One should also realize that, in all biological systems, differences tend to accumulate in the course of time. An emergent tree stands in its place for 2 or 3 perhaps 5 centuries, continuously producing seed. Even in a perfectly homogeneous situation, concentrations must develop, and gaps form in other places. Finally, chance plays a role, not only in dispersal and germination, but in the natural dynamism of the canopy, with its randomly occurring centres of growth, due to the dismantling, natural death and fall of post-mature trees, which determine the possibilities for saplings.

Altitudes Above Sea Level

At higher altitudes, the forest becomes poorer in both species and stature, and assumes a different character. The changes in structure are well illustrated by a series of diagrams made in Malaya (Fig. 7.8). Between 150 and 780 m, the average height diminished by one-third, between 150 and 1500 by one-half, between 150 and 1800 by two-thirds. The lowest limit between two zones is at 1000 ft or 300 m. In the lowermost zone, the zone of 'Lowland Dipterocarp Forest', Robbins and Wyatt-Smith, in their paper *Dryland Forest Formations and Forest Types in the Malayan Peninsula* (1964), distinguished nine types:

1. Red meranti-keruing (*Shorea-Dipterocarpus*) forest;
2. Balau (*Shorea*) heavy hardwood forest;
3. Kapur (*Dryobalanops aromatica*) forest;
4. Kempas-kedongdong (*Koompassia*-Burseraceae) forest;
5. Merbau-kekatong (*Intsia-Cynometra*) forest;
6. Keruing (*Dipterocarpus*) forest;
7. Chengal (*Neobalanocarpus heimii*) forest;
8. Nemesu (*Shorea pauciflora*) forest;
9. Damar laut merah (*Shorea kunstleri*) forest.

In the next higher zone, 'Hill Dipterocarp Forest' between 300 and 750 m (although in steep sites this zone occurs at lower levels), there are six types:

10. Seraya (*Shorea curtisii*) forest;
11. Balau kumus-damar hitam (*S.laevis, S.multiflora*) forest;
12. Balau laut (*Shorea glauca*) hill forest;
13. Balau keruing (*Shorea-Dipterocarpus*) forest;
14. Merpauh (*Swintonia spicifera*) forest;
15. Keruing-resak-mengkulang (*Dipterocarpus-Vatica cuspidata - Tarrietia simplicifolia*) forest.

In the zone of 'Upper Hill Dipterocarp Forest', between 750 and 1050 m, two more types were found:

16. Meranti bukit (*Shorea platyclados*) forest;
17. Meranti sarang punai bukit (*Shorea ovata*) forest.

All these forests are truly mixed. The indicative species name does not imply a dominance. The above classification has a special value in that it resulted from over 3 decades of work by outstanding forest botanists in a country where the flora is both rich and well studied. The occurrence of 17 dry land rain forest types in an area of 131,000 km^2 also demonstrates the incredible variety possible among this sort of vegetation.

While the profiles in Fig. 7.8 show impoverishment in stature and in stratification with increasing altitude, the study by Robbins and Wyatt-Smith does not reveal anything about the vertical distribution of the species. This, indeed, is an extremely difficult subject to study, and data are very scarce. The reasons are the following:

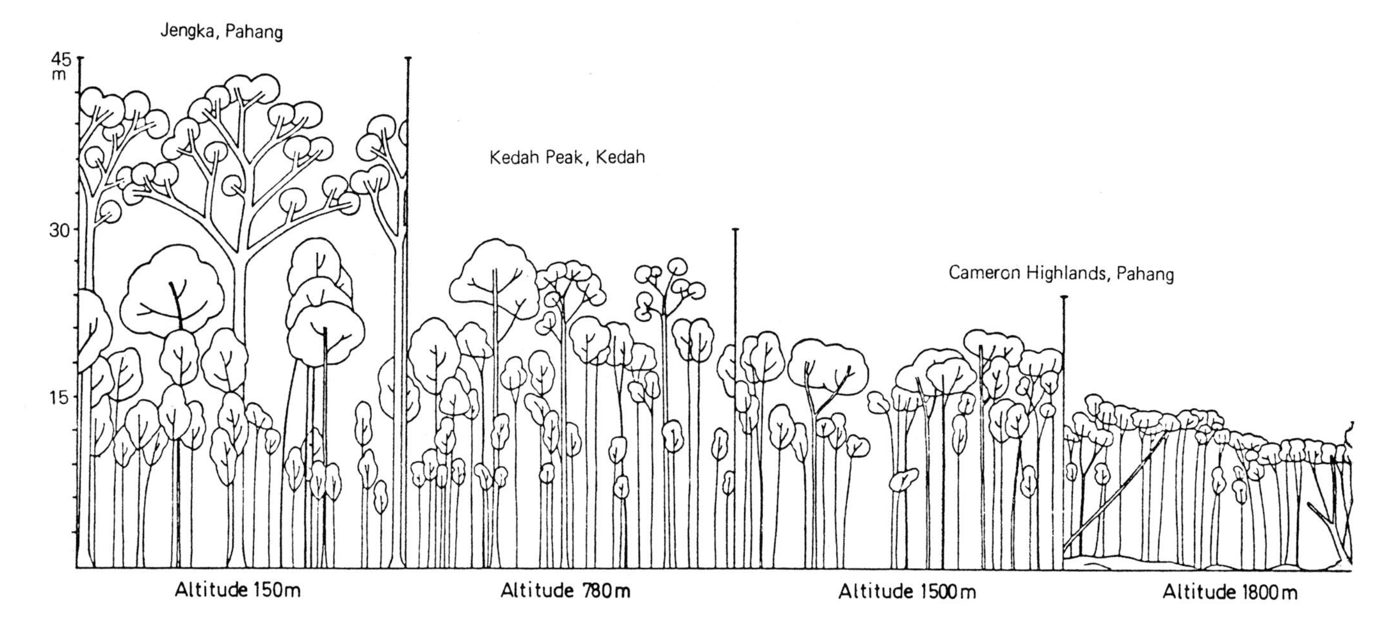

Fig. 7.8 Four forest types in Malaya, at different altitudes. (Robbins and Wyatt-Smith 1964)

1. The only real rain forest region in which mountains occur is Malesia. Neither in Africa nor in tropical America with their extensive lowland basins is vertical distribution a focus of interest.
2. Data must be gathered in more or less adjacent, preferably continuous, tracts of forest, from sea level (where virgin forest is scarce) up to well above 1000 m.
3. Only those data have value which have been collected during field work on large plant families of which the species are taxonomically well known.
4. A transect must contain many individuals of many species.
5. For all the selected species, the complete vertical range must be determined in the forest, amidst many other species.
6. Local conditions must be accounted for, particularly telescope effects (upwards on high mountains, downwards on lower ones), before generalizations can be made.

The only valid set of figures I know about is to be found in the table prepared by Ashton (1964, Table 14), for all Dipterocarpaceae in Brunei (147 species). The data, presented in Fig. 7.9, clearly show the decline in numbers with increasing altitude.

A further analysis of Ashton's data reveals that if the forests in Brunei below 300 m were destroyed, 39 species of Dipterocarpaceae would be wiped out, or 26.5% of the total in that country. Destruction up to 400 m would result in a loss of 65 species, or 44%, of Brunei's dipterocarps.

In Indonesia, 500 m has for a long time been accepted as the upper limit of the lowland forests, due to experience rather than as a result of specific study. This does not need to contradict the 300 m in Malaya because of the so-called telescope effect: on high mountains all zones are broader, and vice versa (Fig. 15.2). Sumatra, Java, Celebes and New Guinea have many peaks of 2500 m and higher; while the highest summit of Malaya is 2100 m, most mountains in Borneo are not as high, except in the far north.

In the *Flora of Java* (Vol. 2, p. 49. 1965) the totals of species have been tabulated per 100 m altitude. These numbers, too, decrease going upward, but this evidence does not help us here, because Java has a heterogeneous flora and is poor in rain forest. The

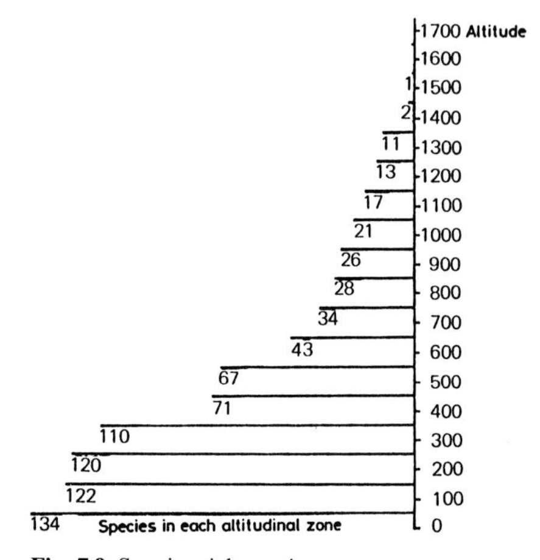

Fig. 7.9 Species richness in Dipterocarpaceae in Brunei, per 100 m increase in altitude (Ashton 1964)

subject of altitudinal zonation was intensively studied by Van Steenis, but almost exclusively above 1000 m. Other authors, like Brown in The Philippines, found no intact forest in the lowlands and worked above 400–500 m out of necessity. The study by Martin (1977) of Gunung Mulu in northwest Borneo hardly mentions the lowland forest. I do not know about other studies. Richards is silent on the subject. Herbarium labels too often give 'lowland', with the meaning any altitude below 1000 m, to be of any use in a detailed compilation.

The pressure of exploitation on the rain forests is heaviest in the lowlands, because of easy accessibility, the size of the trees, and usually lesser need for protection to prevent erosion and to safeguard water catchment. Yet, among biologists and conservationists the subject of zonation below 1000 m is hardly ever considered, even though the closer to sea level, the more species are present, and all evidence points to impoverishment with increasing altitude. Work in tall, species-rich forest is difficult, however, whereas above 1000 m, where the trees are lower, and the forest is simpler in structure and composition – and the climate cooler – everything is so much more enjoyable. This may have contributed to the neglect of the cardinal importance of the lowest lowland forests in the past.

Summary: The Implication of Species Richness

Those who have no intimate knowledge of the tropical rain forest and the large numbers of species composing it cannot realize all the implications of this richness. Species richness is a biological fact, however, and at the same time, the key to an understanding of many of the forest's features, and it would be a fatal mistake to omit composition and complexity from any considerations regarding the tropical rain forest.

We therefore list below the main facts and implications. In doing so, we look both back at this chapter by way of summary, and forward to subsequent chapters where some points will be elaborated.

1. The number of tree species per hectare in tropical rain forests amounts to ca. 80–200. The total number of vascular plant species on a hectare is approximately three times as high.
2. These largenumbers of species and the predominantly woody character of about half of them imply a long tenure in one place. For this reason, both the regeneration and the extension of the rain forest proceed very slowly.
3. No two hectares have exactly the same species composition. A high percentage of species is only found once even in a large plot.
4. No plant-sociological minimum area can be indicated. Consequently, plant sociology as it has been developed in Europe is inapplicable to the rain forest.
5. There is no dominance of tree species. The most common one seldom makes up 15% of the total. Higher percentages than this indicate the presence of limiting conditions or of damage.
6. It is axiomatic that the more species there are per hectare, the fewer individuals of one species can be accomodated in it. The order of density is usually one tree of each species per ha, varying between ca. 25 trees in 1 ha and one tree in 25 ha (or more).
7. The fewer individuals of a species there are in a hectare, the greater the distance between them. A general average is 100 m; a maximum is unknown.
8. A mixed composition with long distances between individuals prevents massive outbreaks of pests and diseases.
9. Long distances have to be bridged to make a cross-pollination possible. This promotes the dioecious state (to be explained in Chap. 14).
10. The more species per hectare, the smaller the ecological niche available to each. This makes them less adaptable to changes in conditions (to be discussed in Chap. 14).

11. As we shall see in Chapter 12, all rain forest trees are virtually dependent on animals for pollination, and for dispersal if they have large seeds. These relationships depend on composition, in quality and quantity. The functioning of the forest depends on these relationships.
12. Dispersal by wind and by animals results in different patterns of seeding. Difficulty in dispersal leads to concentrations of trees in clusters.
13. The longevity of many species and a tendency to clustering result in a mosaic-like composition. This precludes the delimitation of minimum areas, with consequences for the species' protection.
14. The higher the altitude above sea level of a tract of rain forest, the fewer the number of species occurring in it. Most species are confined to altitudes below 300–500 m. This also has consequences for the protection of the rain forest.
15. Damage and destruction of the rain forest will endanger more species per area unit than the damage and destruction of other vegetation types.
16. Because of the small numbers of individuals per species, the elimination of a few individuals has a strong effect on the size of a population and on the various biological balances in which the species is involved.
17. The minimum population with sufficient genetic diversity for survival ad infinitum is found only in very large areas.
18. Presuming that the percentage of plant species useful to man is the same in all vegetation types, the tropical rain forest is the richest pool of such species, known and unknown (see Chap. 16).
19. As the known commercially attractive species are scarce, efforts at exploitation involve searching. Silvicultural practices preclude successful exploration by changing the forest composition in order to favour known commercial species. Silviculture, therefore, becomes one aspect of damage and destruction.

Throughout, we have had in mind the primary rain forests at low altitude on dry (not swamp) land. In such forests the greater the diversity, i.e. the number of species per square kilometre, the more valid the above 19 points become.

8 Primary and Secondary Forest

Germination in Shade or Light

The vertical gash on a slope visible from afar (see Chap. 4 and Fig. 15.1) is the beginning of a long process which commences with bare soil, in this case a narrow landslip, and should eventually result in a forest indistinguishable from the surrounding forest. As regeneration proceeds, the gash changes colour: first the grey or ochre of the soil becomes pale green with young plants, then bright green, then an ever darker green as one stage in the development of flora and vegetation gives way to another; collectively, these stages form a succession.

On bare soil and in hot bright sunshine, the seeds which will germinate belong to species of the first stage of succession, the pioneers. In the shade of the surrounding forest, these species do not survive, and their nearest station may be kilometres away. Yet the seeds are there. Either they were already present in the soil, or they have travelled far. A full-grown pioneer plant produces large quantities of small seeds which are often equipped with tufts of hair or wings, and they are easily carried over long distances by the wind.

The true rain forest tree germinates in an entirely different way. Take for instance the avocado, *Persea americana* (Lauraceae). Its seed is the size of a small egg, weighing 60 g or more, and an avocado seedling can attain a height of 1–2 ft just on the nourishment it contains, with the result that its leaves are unfolded above a fair number of its competitors. The seed of ulin, *Eusideroxylon zwageri* (another Lauracea), the ironwood of Sumatra and Borneo, the largest of all dicotyledon seeds, is about 14 cm long, weighing about 230 g. It often produces a sprout over 1 m tall before the leaves develop.

To make such a start possible, the food in the seed must be in a form suitable for quick utilization. This requires a high moisture content which renders the seed liable to attack by fungi. Most large seeds will therefore need high humidity to remain viable, but they will only retain this viability briefly; for many primary rain forest species, 3–4 weeks is the maximum. This severely limits the collecting transport, and nursing of rain forest seeds. Not all rain forest seeds are as large, of course, but large-seeded species are in general typical.

One consequence of the difference between the two modes of germination is that species which need light from the very beginning will not germinate in the shade of their parents or, as foresters say, will not regenerate under their own cover. A true pioneer thus makes life for its progeny and for other pioneers impossible. Where it has settled, it cannot stay on. As a species, it is doomed to wander, from one spot of bare soil to another. Van Steenis calls them biological nomads. Their small, dry, hard-shelled, long-viable seeds with their contraptions for easy transportation prepare them for this kind of life.

On the other hand, a seed which needs shade to germinate, can germinate under the tree from which it fell, or under any part of the forest canopy. The larger its store of food, the better equipped it is for germination in the shade, but the more it weighs, the more difficult it is to disperse. Large animals can transport such seeds, but they will only do so in order to eat the nutritious part of the fruit.

The two modes of germination, light-demanding and shade-demanding, also reflect two different life-styles. Each life-style consists of a combination of properties, the answer to one of two diametrically opposed conditions: in the one, the light is 100%, in the other, 1% or less. We will now examine the combinations of properties resulting in such different tolerances more closely.

The Life-Styles 'r' and 'K'

Light in large quantity makes high bioproductivity possible; pioneers therefore produce much vegetable matter in a short time. The quality of this matter (understood as durability) will depend on the firmness of the plant tissue, which depends on the absorption of inorganic substances from the soil. In Malesia, on rich soils, i.e. on newly formed sandy flats by the sea, and on young volcanic slopes, pioneer species of *Casuarina* (*C. equisetifolia* and *C. junghuhniana* respectively) grow very fast; yet their wood is compact and makes excellent fuel, and is so dense and heavy that it is hard to work using commonly available tools. On poor soils, however, and these are of course more common, although the pioneers also grow fast, they produce only lightweight wood, often weak and pithy. A good example here is balsa, *Ochroma lagopus* (Bombacaceae), the wood of which can be cut with a pocketknife, ideal for model and glider airplanes. It grows in the secondary forest formations in tropical America; Thor Heyerdahl built his raft Kon-Tiki of whole logs of this wood, the lightest in the world.

Richards (1952, p. 383) gives some examples of pioneer growth rates: *Trema amboinense* (Ulmaceae) of Malesia, 7 m in 3 years; *Macaranga tanarius* (Euphorbiaceae) also from Malesia, 8 m in 2 years, 11 m in 3 years; the kapok tree *Ceiba pentandra* (Bombacaceae) of Africa, 12 m in 3 years; *Masanga* (Urticaceae), 24 m (and a diameter of 60–100 cm) in 15–20 years. Their high bioproduction also expresses itself in early abundant flowering and fruiting, whereby they spread themselves extensively. This explains why time and again in the tropics, thousands of kilometres apart, we meet the same pioneer species. There are, of course, differences between continents, but the same handful of species is usually involved. To know them, look at the Flora of any big city, such as Merrill's *A Flora of Manila* (1912).

Wherever soil is exposed pioneers grasp their opportunity. They quickly produce their large leaves borne by weak stems. They cast shade and leave litter on the soil (which is enriched by this humus), and just as quickly die. Other species less extremely light-loving in their early youth, take their place. Such a vegetation, which we call secondary because it grows where the original, primary forest has disappeared, is initially without order or structure, impenetrable, composed of non-durable materials, a 'throw-away' association of plants, easily replaced. This life-style has been called the 'r' strategy, and we may let 'r' be short for 'rat' (although this mnemonic was not originally intended): short-lived, successful, opportunistic.

The life-style of the primary forest has already been sufficiently described to make do here with a summary. Its economy is one of long-term investment. Flowering and fruiting come late: a dipterocarp may not be fertile before the age of 60. Its slow growth is a sign of the length of time needed to convert scanty inorganic matter into chemical compounds which preserve their wood, and make it hard and heavy. The blackish wood of the ironwood tree or ulin, *Eusideroxylon zwageri* (Lauraceae) of Sumatra and Borneo, for instance, is so heavy that it sinks in water, and its durability is proverbial. In eastern Borneo I visited a former pepper plantation which had been abandoned 33 years before. The ironwood stakes which had supported the pepper were now standing in secondary forest, but were still in perfect condition. Young ironwood trees, perhaps 25–30 years old, were present, too, but had trunks no thicker than 10 cm, so slow is their growth.

Eusideroxylon is never an emergent species, but its style of growth is similar: save for the future. With quiet cunning the emergents move up and then expand their crowns, while only sparingly renewing their leaves. They then stand for ages before a successor can hope to take their place. This life-style has been called 'K' strategy, and 'K' could stand for 'King', for kings are (or should be) majestic, tall or big, long-lived, and hard to replace. But the 'rats' are useful to the 'kings': the light-loving, pioneer, nomadic ensemble creates suitable conditions for the shade-loving or shade-tolerant ensemble, whose shade will dominate and kill light-lovers. A light-loving ensemble is only maintained by granting it light; any gardener knows that grasses (light-lovers par excellence) suffer in the shade. If grass is not mown, it also suffers: the self-shaded plants sag and rot, and before long seedlings of trees pop up, which have germinated even under this low cover. Many kinds of grass, even the tough lalang, *Imperata cylindrica*, must be burnt regularly to retain condition, which requires the penetration of light onto the soil.

Light-demanding plants are the natural camp followers of man, who bares the soil by cutting and burning, thus for ever increasing the land open to pioneers.

The Relationship Between Secondary and Primary Forest

On a landslip, the relationship is simple; the primary forest has disappeared completely together with the topsoil. Pioneers occupy the bare terrain, not many species but a dense tangle of stems, almost impossible to penetrate, which provides soil cover, often a very dense one, due to the light-loving climbers (many Convolulaceae, Cucurbitaceae, Dioscoreaceae) which grow over it like a blanket. The colour of this vegetation is light green, due to the speed of its growth. Such a blanket may become too heavy for the weak, secondary tree species which carry it, whereupon they too succumb (Fig. 17.13). Under the blanket, branches die through lack of light; where first there was an impenetrable mass of stems and leaves, there is now a hollow place with an even microclimate of shade and humidity. Litter drops and is again transformed into humus, and other species enter.

If undisturbed, the succession proceeds: young secondary forest gives way to old secondary forest, and the longer the trees live, the slower the changes. Budowski's Table neatly sums up the process (Fig. 8.1).

After some centuries, the old secondary forest comes to resemble the surrounding primary forest so closely that only an expert, by a detailed examination of its species composition, can detect the difference. We know how long this takes by studying forests abandoned in historical times: the forests surrounding Maya and Aztec ruins in Central America, in Angkor in Cambodia, and in the former kingdom of Blambangan in eastern Java. Learning to recognize when a forest is primary or secondary is, of course, vital in the selection of rain forest areas for conservation.

One might say that the secondary forest acts like a scab on a healing wound in the canopy; though on the smallest wounds no scab is formed: Kramer (1933) found that a gap 100 square inches or smaller is filled forthwith by the available stock of the primary species, any bigger gap will first fill with pioneers.

Gaps in the canopy and undergrowth which occur naturally when trees fall down are called chablis[3] (Fig. 8.2). A large crown of 20 m diameter makes a chablis of 400 m^2; if other trees are dragged down at the same time, one of 600 m^2 may result. Some plants in the chablis will suffer damage or death from the tree crash, others suddenly face new opportunities; the selection is one of unpredictable irregularity.

[3] A medieval French term revived by Hallé et al. 1978 p. 282.

	Pioneer	Early secondary	Late secondary	Climax
Age of communities observed, years	1-3	5-15	20-50	more than 100
Height, meters	5-8	12-20	20-30, some reaching 50	30-45, some up to 60
Number of woody species	few, 1-5	few, 1-10	30-60	up to 100 or a little more
Floristic composition of dominants	Euphorbiaceae, *Cecropia, Ochroma, Trema*	*Ochroma, Cepropia, Trema, Heliocarpus* most frequent	mixture, many Meliaceae Bombacaceae, Tiliaceae	mixture, except on edaphic association
Natural distribution of dominants	very wide	very wide	wide, includes drier regions	usually restricted, endemics frequent
Number of strata	1, very dense	2, well differentiated	3, increasingly difficult to discern with age.	4-5, difficult to discern
Upper canopy	homogeneous, dense	verticillate branching, thin horizontal crowns	heterogeneous, includes very wide crowns	many variable shapes of crowns
Lower stratum	dense, tangled	dense, large herbaceous species frequent	relatively scarce, includes tolerant species	scarce, with tolerant species
Growth	very fast	very fast	dominants fast, others slow	slow or very slow
Life span, dominants	very short, less than 10 years	short, 10-25 years	usually 40-100 years, some more	very long, 100-1000, some probably more
Tolerance to shade, dominants	very intolerant	very intolerant	tolerant to juvenile stage, later intolerant	tolerant, except in adult stage
Regeneration of dominants	very scarce	practically absent	absent or abundant with large mortality in early years	fairly abundant
Dissemination of seeds of dominants	birds, bats, wind	winds, birds, bats	wind principally	gravity, mammals, rodents birds
Wood and stem, dominants	very light, small diameters	very light, diameters below 60 cm.	light to medium hard, some very large stems	hard and heavy, includes large stems
Size of seed, or fruits dispersed	small	small	small to medium	large
Viability of seeds	long, latent in soil	long, latent in soil	short to medium	short
Leaves of dominants	evergreen	evergreen	many deciduous	evergreen
Epiphytes	absent	few	many in number, but few species	many species and life forms
Vines	abundant, herbaceous,but few species	abundant, herbaceous,but few species	abundant, but few of them large	abundant, includes very large woody species
Shrubs	many, but few species	relatively abundant but few species	few	few in number but many species
Grasses	abundant	abundant or scarce	scarce	scarce

Fig. 8.1 Budowski's Table. Comparison between four successional stages, from pioneer to mature primary forest, with special reference to the American tropics (Budowski 1965). *Cecropia* is from the Urticaceae, *Heliocarpus* from the Tiliaceae, *Trema* from the Ulmaceae

The filling up of a chablis also proceeds in a number of stages (shown in Fig. 8.3); in any forest, many such sites of decline and renewal can be seen, and together, they make up a considerable percentage of the forest area; between 3 and 10% is usually in a state of regeneration, though Poore (1968, p. 179) prefers an average of 8%, calculating that in any particular part of the forest the canopy will fall once in 250–375 years. For further discussion, see Hallé et al. (1978, pp. 366–368).

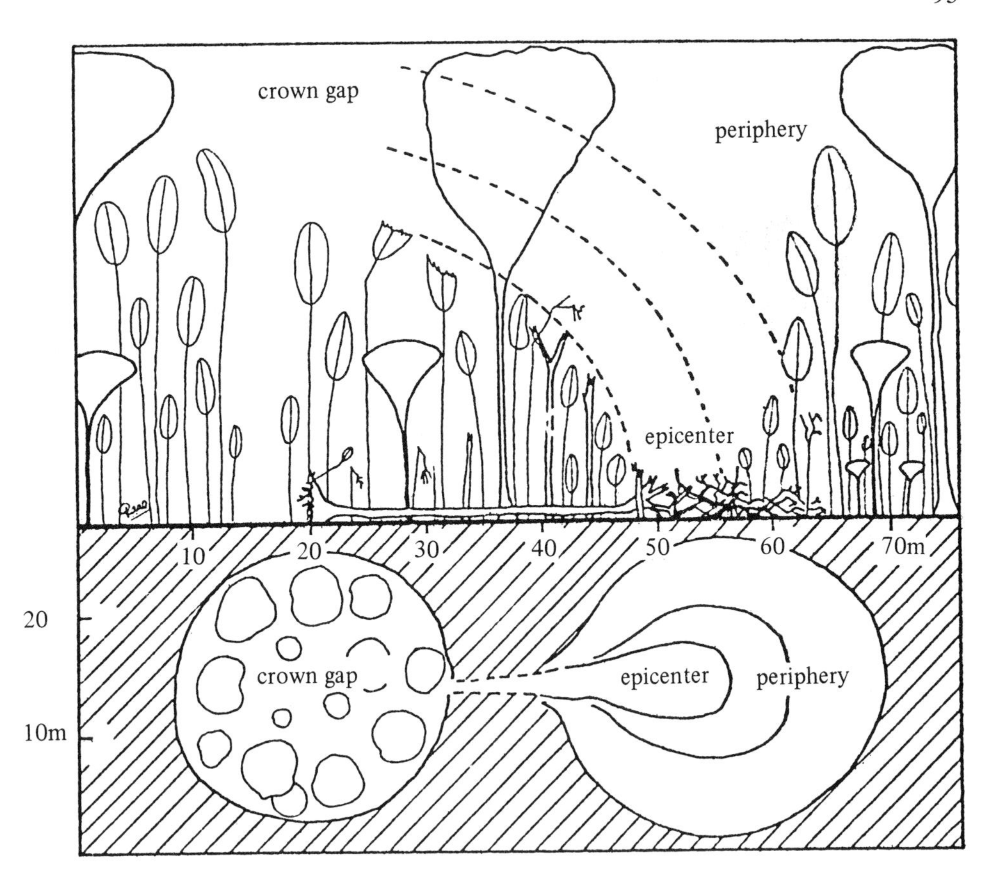

Fig. 8.2 A gap in the canopy caused by a falling tree; in fact, there are two gaps: one in the canopy only, another on ground level as well. The first gap fills up spontaneously (Oldeman 1978)

The character of 'mosaic', discussed in Chapter 7 is partly derived from constellations of chablis in each of which the trees grow up simultaneously (see Fig. 8.4 centre). In this way, each fallen giant makes way for its strongest descendents and its competitors.

Accessibility

A small gap in the forest cover caused by a chablis closes in the same manner as a wound heals, from the perimeter inward. On a landslip site this is also true, except that in this case the regeneration front will be straight. The composition of the flora in the damaged area will, of course, be determined by the seeds that germinate there in all its various stages of recovery. The degree of ease with which seeds and animals can reach a given site is called the 'accessibility' of that site. On this depends the completeness of the regeneration, from pioneer through young and old secondary to primary or climax forest (Fig. 8.5). If the accessibility factor is low, the later stages are not reached, and the primary forest does not return.

Accessibility is not a simple concept. Seed dispersal and animal migration are highly complex subjects in themselves. Dispersal depends on wind, gravity and animal activity; the larger and heavier seeds are, the smaller the distances they are likely to travel. But time is also a factor; after all, the dispersal of a seed is 100 times more likely to occur in a century than in a year.

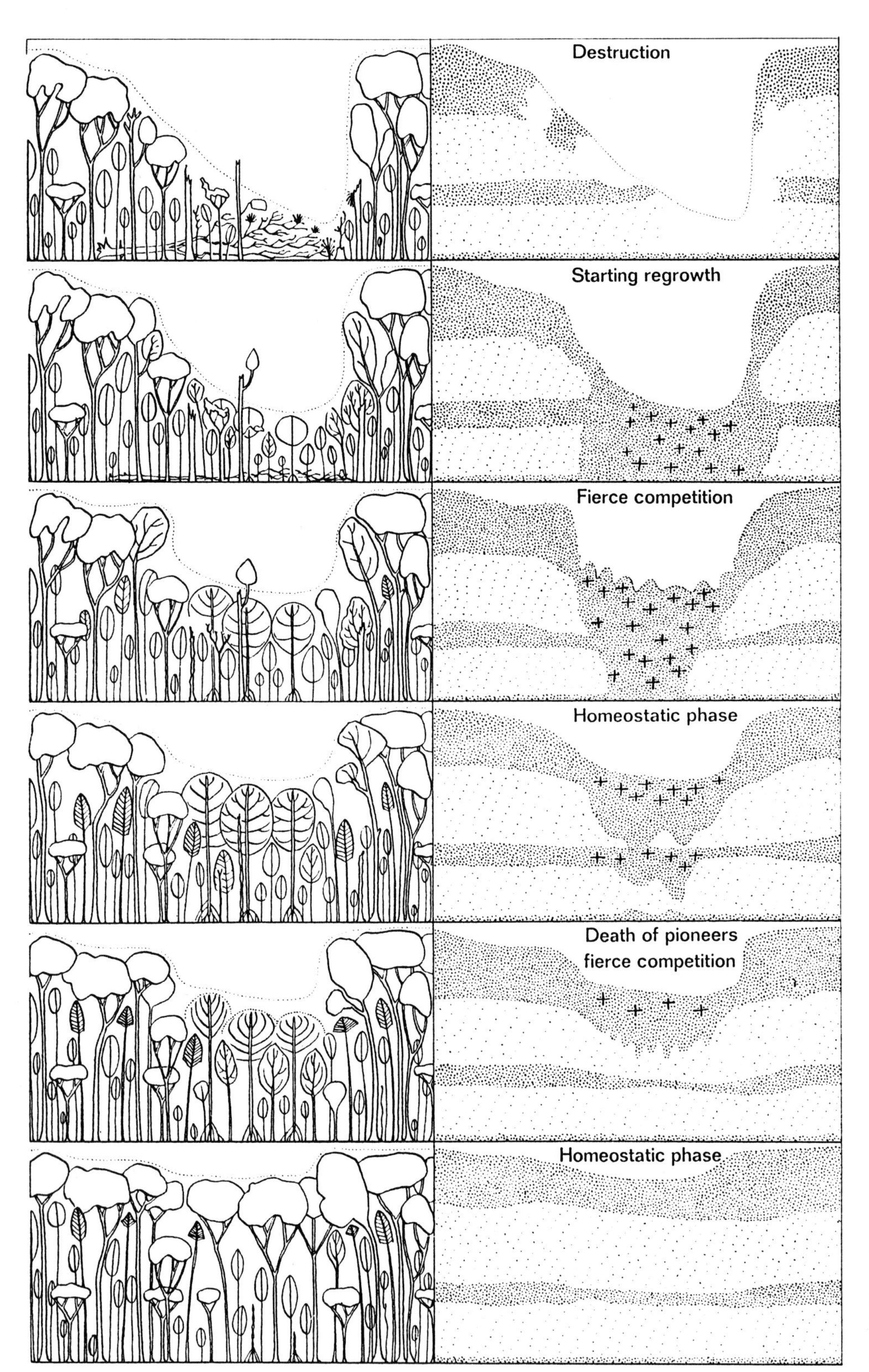

Fig. 8.3 How a chablis fills up on a place where the undergrowth is destructed, in six stages. The *densely shaded* parts indicate the zones where most of the growth takes place; the *crosses* the places where competition is strongest (Hallé et al. 1978)

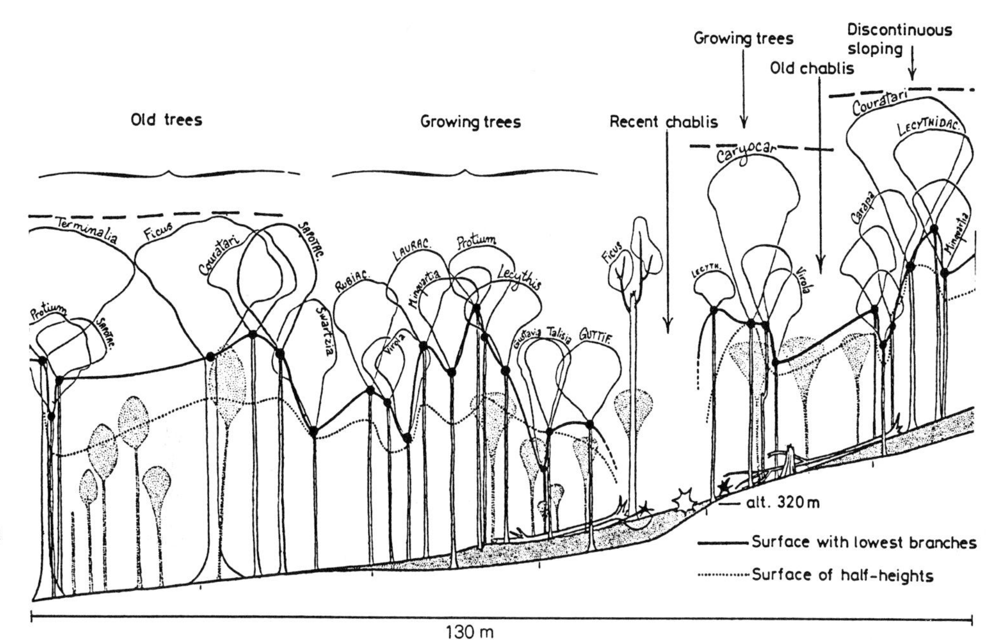

Fig. 8.4 The filling up of a chablis as an aspect of the dynamics of the primary forest. To the *left*, the undisturbed canopy; in the *middle*, a group 'trees of the future/coming trees' on the site of an old chablis; to the *right*, a relatively recent chablis being filled up. On the slope, the canopy acquires a discontinuous character and some similarity to a stairway (Hallé et al. 1978)

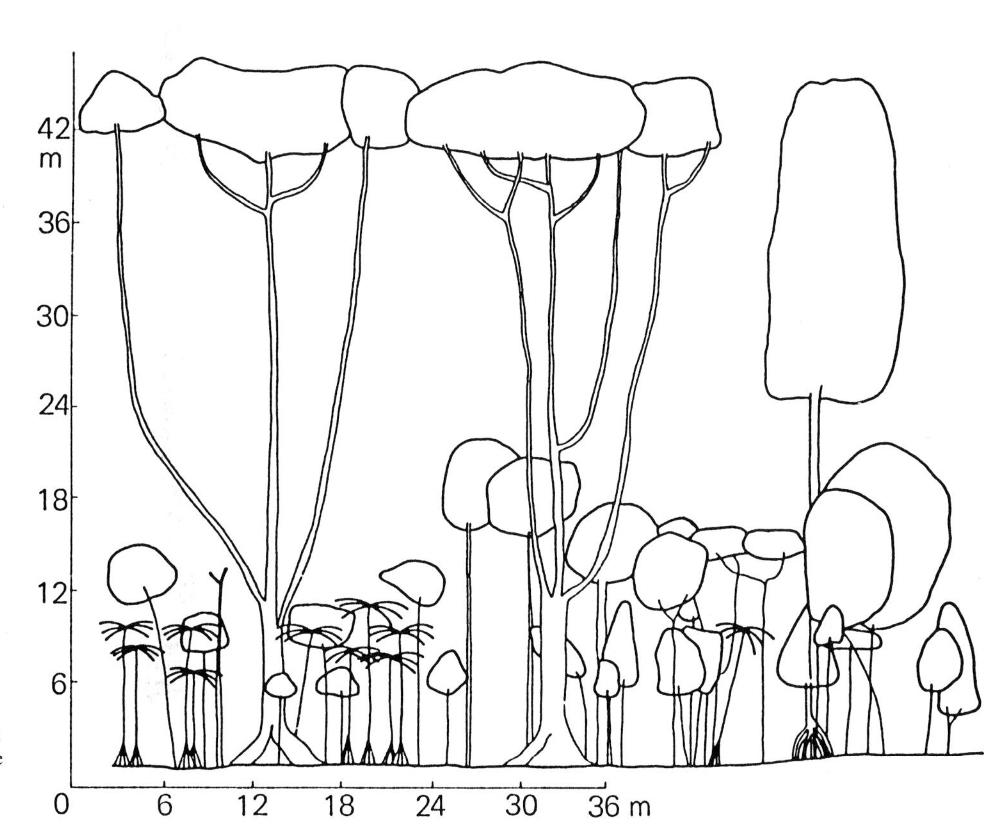

Fig. 8.5 Old pioneers, in this case *Campnosperma* (Anacardiaceae), that cannot rejuvenate under their own crown cover, while obstructing the growth of the trees of the future. Diagram from the Solomon Islands (Whitmore 1975)

Once a seed has reached a site and germinated, complex ecological conditions determine whether or not the seedling will reach the reproductive stage. About these conditions we know virtually nothing, apart from a certain amount of silvicultural experience with a handful of species. For example, some species like jelutong, *Dyera costulata* (Apocynaceae), have defied all attempts at cultivation.

These are the handicaps: (1) the seeds must arrive, and (2) the animals must arrive, and (3) both must do so at a suitable stage of succession; we cannot expect a rain forest to settle in an isolated spot, contrary to pioneer communities which begin with nothing. In other words, the regeneration of a primary forest will only occur where a steady stream of seeds and animals from a mature forest are able to move in.

This expectation is borne out by plant-geographic evidence. One example is the islet of Jarak, 64 km off the Malayan coast. A volcanic eruption in the lake Toba area, the surroundings of which were once under fine forest, covered this island with ash, 34,000 years ago. Only 93 species of seed plants have been found in Jarak, 80 of them animal-dispersed, and there are no Dipterocarpaceae, even though a few members of this family did occur in Rumbia, 22 km from the coast (see Whitmore 1975, p. 63). But rain forest as such never occupied Rumbia. In other words: primary forest never jumps, but migrates only in a continuum, through extension of itself, and only there where secondary forest preceded it.

The Speed of Travel

The profound differences which exist between the floras of the world's three great rain forest areas indicate an ancient separation and an independent evolution. But no one doubts that within each area the rain forest flora must have migrated across land connections. All parts of the western Malay Archipelago, for instance, were connected by dry land (the Sunda shelf), and the New Guinea rain forest migrated into Australia across the Sahul shelf. Later, when the land connection was severed, the rain forest in Australia suffered heavy losses due to a decrease in rainfall, and these losses have not been replaced.

The question of at what speed the forest 'travelled' into Australia is hard to answer. An estimate by Nix and Kalma (1972) of 2000 km in 6000 years does not seem possible. A complete ecosystem must conquer new territory, the many species in it exerting their full influence even if some of them stay behind. After a seed has made a leap in the right direction, its seedling must reach maturity to reproduce: for many rain forest trees this takes several decades, or more.

What do we mean by 'a leap in the right direction'? Here is an example. A certain *Shorea curtisii* (Dipterocarpaceae), with a bole 1.20 m in diameter and a crown 18 m across, and standing alone in a forest on an exposed ridge, produced an estimated 180,000 fruits; 54% of these were dropped within a radius of 20 m, 83% within 40 m, 97% within 60 m; none was found farther than 80 m. They fell in an elliptic area tallying with the prevailing wind direction. And the fruits of *Shorea curtisii* are by no means the largest in the family. The three wings measure 55×10 mm, the seed is 8 mm in size and weighs 0.25 g in the dried state, perhaps 0.5 g when fresh. Whitmore (1975, p. 63), who cites this case, mentions others to the same effect. Even if a variety of winged rain forest tree seeds were to be blown by a storm like snow flakes, few of them would reach a distance of more than 800 m, and the larger seeded ones might travel 500 m at the most. And this is a generous estimate. On this basis, allowing for the 50 years it takes many rain forest species to reproduce, one species would thus travel no more than 1 km a century.

This seems achingly slow, yet it is enough to cover the full length of Sumatra in 160,000 years, and to travel from the mouth of the Amazon along the equator to the foot of the Andes in 300,000 years. In geological terms, 300,000 years is not a long time. The very limited distribution of most rain forest species indicates, however, that in most cases the speed is much lower. Watersheds and rivers pose distinct barriers. Moreover, primary forest species never migrate alone, but do so only in consociations. How fast or slowly a forest as an ecosystem can travel is a matter of conjecture; it is probably much less than 1 km a century. Remember that a completed succession from bare soil to climax forest can take centuries, perhaps as many as 10 centuries. We may presume that a forest's actual speed of spread is rather in the order of 100 m in a century.

This much is certain: that there is no guarantee whatever that a spontaneous, complete return of primary or climax forest to sites where it has been destroyed will take place, even if such sites are protected and adjacent to untouched primary forest. No power on earth has command over the enormous span of time required to complete the necessary succession stages. Strict protection of existing rain forests is far better; nothing else can ensure its survival.

The Primary Forest as Climax Vegetation

The concept of 'climax' plays an important role in tropical botanical science (see Richards 1952). A climax vegetation is one which has reached its maximum development, i.e. it consists of the maximum number of species able to survive under the existing conditions, and one whose composition, if undisturbed, will not change if climate and soil conditions remain the same. Once a vegetation has been properly identified as climax, all other plant associations in the area can be related to it, either as a destruction stage or a regeneration stage, both will be discussed later; in this way the rain forest can be used as the climax or ultimate vegetational sere, a fixed point in our judgment of other tropical plant associations.

However, the limitation imposed by the criterion of 'unchanged conditions' should not be taken lightly. All primary rain forest is a climax vegetation par excellence, but of course not all apparently climax tropical forests are actually primary rain forests. Until the end of the last century the great rain forests were little disturbed. But now there has been so much human pressure everywhere on the forest vegetations, that the identity of an apparently undisturbed forest can no longer be assumed to be primary. Conversely, a formerly untouched natural forest now selectively logged every 80 years will develop into a kind of climax forest, a forest which does consist of its maximum number of species – under this pressure. But what an impoverishment! Therefore, let no one who deals with rain forests unconditionally and uncritically accept the label 'climax'!

Summary: The Dynamics of the Forest

The circular figure from the book by Hallé et al. (1978, p. 377) aptly summarizes the present chapter (Fig. 8.6). On a logarithmic scale of time, the four main stages have been drawn into which the development of a forest from the beginning can be divided. They are comparable to, perhaps even identical to, those in Budowski's Table (Fig. 8.1): I, pioneers; II, young secondary; III, old secondary; IV, primary. The duration of stages III and IV, however, can only be estimated.

In each stage a dynamic phase of growth (d) is followed by a homeostatic phase of equilibrium (h). The arrows inside the circle indicate the relations between all stages and phases. In a chablis the succession is resumed: the larger the gap, the farther the regression. But where a chablis is small, and favourably located, it may accommodate species properly belonging to a subsequent phase; such a spot of forest may even advance a stage before its time.

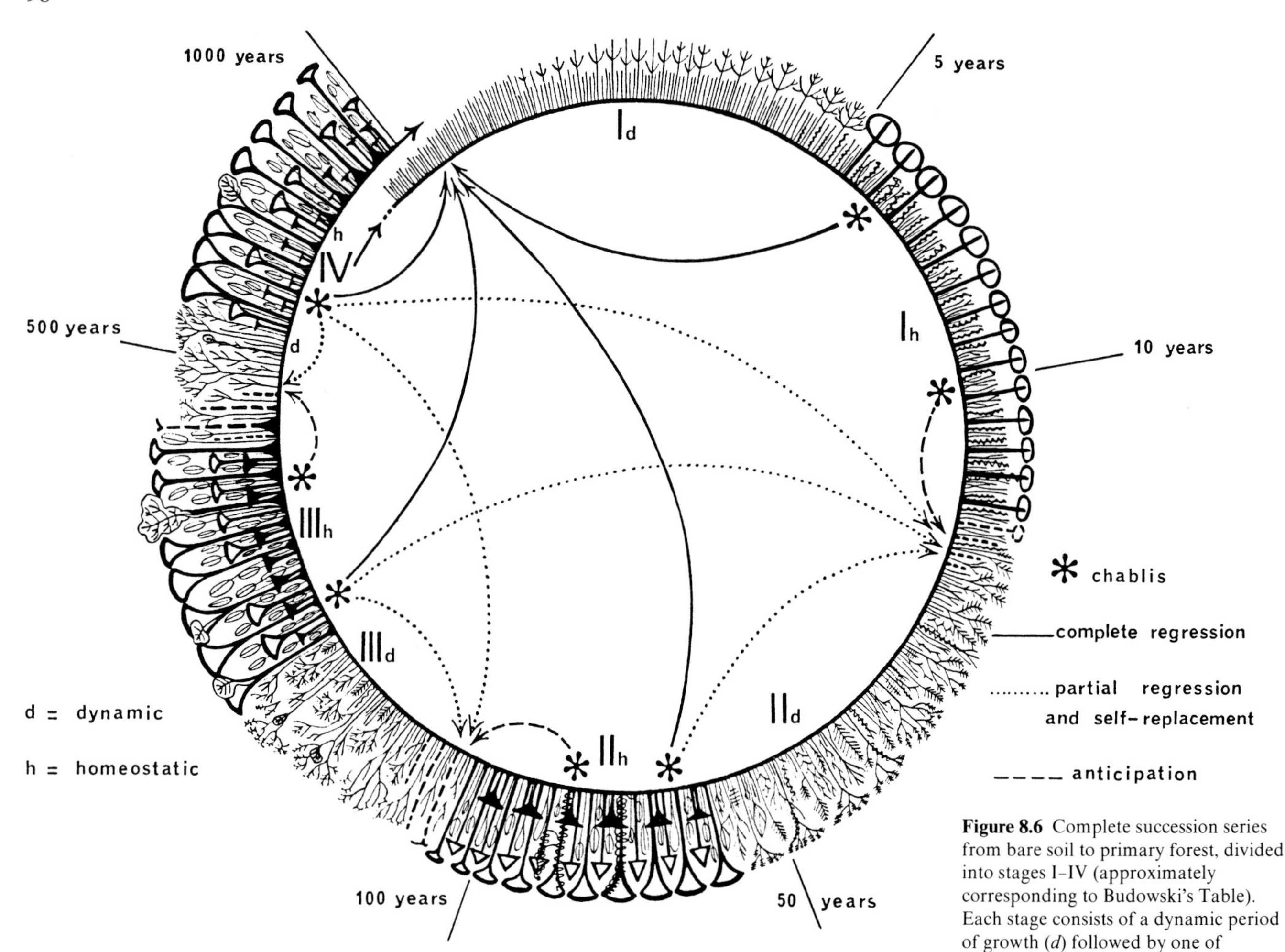

Figure 8.6 Complete succession series from bare soil to primary forest, divided into stages I–IV (approximately corresponding to Budowski's Table). Each stage consists of a dynamic period of growth (*d*) followed by one of homeostasis (*h*). From each stage regression may occur leading to the previous one, if a gap in the canopy occurs. With sufficient accessibility, however, occasionally a later stage may be anticipated (Hallé et al. 1978)

Strata in the Primary Forest?

The question of whether or not the canopy is composed of definite layers or strata has been disputed at least since 1920. Richards (1952, Chap. 2) summarized the various investigations and viewpoints, and concluded that primary forest canopy does indeed consist of three marked layers or strata. Stratum A consists of the emergents, which mostly grow far from one another. The continuous canopy consists of an upper Stratum B and a lower Stratum C. Sharp limits and measurements cannot be indicated, because microclimate, soil and other factors differ between the several plots of rain forest, but B and C strata should be easily discernible, or so it seems.

This three-layered pattern found wide acceptance until a recent large-scale investigation revealed that such a pattern, or any pattern, is in fact non-existent. Hallé et al. (1978, p. 333 ff) extensively explored this theory and pointed out the static nature of Richards' vision of structures which are continuously subject to processes of renovation and senescence.

Figure 3.14 shows us, from left to right, the complete career of an eventual emergent, as it grows through trees of all ages which mature at various heights. That they do not form definite strata was established by Rollet in 1978, even though there is certainly an orderly vertical arrangement of elements in the canopy.

Agricultural and Silvicultural Aspects

In connection with some observations in Chapter 4, we can now also assign tropical crops to the four categories in Budowski's Table, although there are too many crops for a complete treatment.

Crops which belong to the pioneer formation can readily be identified. All grassy ones (Gramineae) are among them: rice, *Oryza sativa*, sugarcane, *Saccharum officinarum*, both Indo-Malesian, and maize, *Zea mays* of the Neotropics. Cassava, *Manihot utilissima* (Euphorbiaceae) and tobacco, *Nicotiana* (Solanaceae), are also neotropical.

Species of secondary formations include the neotropical papaya, *Carica papaya* (Caricaceae), a fast-growing unbranched tree-like plant with soft, almost spongy wood and large, delicious, small-seeded fruits. It is cultivated throughout the tropics, as is the palaeotropical banana, *Musa* (Musaceae). As for trees of secondary formations, *Anthocephalus chinensis* (Rubiaceae), so abundant in the secondary forest of Malesia, is a typical species. It grows quickly to a height of 15–18 m and shows much promise for pulp and light timber uses.

Species of primary formations include virtually all the comestible fruits that originate in the Malesian forests: the various species of *Artocarpus* (Moraceae), citrus, *Citrus* (Rutaceae), durian, *Durio* (Bombacaceae), mangosteen, *Garcinia* (Guttiferae), mango, *Mangifera* (Anacardiaceae), rambutan, *Nephelium* (Sapindaceae) and the nutmeg, *Myristica* (Myristicaceae). Coffee *Coffea* (Rubiaceae), from Africa, and cocoa, *Theobroma* (Sterculiaceae) and rubber, *Hevea* (Euphorbiaceae), both from tropical America, are also primary forest species.

All such trees, like the heavier timber species, must be sown in the shade since only later can some of them tolerate full sunlight. Only species from outside the primary rain forest, either from secondary formations or from deciduous forests or savannas are sufficiently light-loving from the beginning to be used as plantation trees, to create that euphemism known as a man-made forest. Teak, eucalyptus species and pines are good examples here.

9 Tropical America

By R.A.A. Oldeman

Knowledge of the American Rain Forests

Rain forests cover an important part of the Neotropics. The forest complexes covering the Amazon and Orinoco basins can be considered as the central part, surrounded by peripheral rain forests west of the Andes, on the Caribbean Islands, in the Guyanas and in Central America. The Mexican evergreen forests, a small northern part of which still survives in the National Park in Los Tuxtlas (Veracruz State), the southeast Atlantic evergreen forests in Brazil and the Florida mangroves (USA) are not rain forests in the strict sense of this book. They should be mentioned, however, to illustrate the difficulty of defining rain forests. For no South American country do the forest maps made by different authors show the same limits between the rain forest and other vegetation types, or between different types of rain forest. Gradual variations occur not only with an increasing duration of the dry season and with increasing altitude, but also from one soil type to another. The thesis by Kahn (1983), *Architecture comparée de forêts tropicales humides et dynamique de la rhizosphère*, is a spectacular documentation of the many different forests within 'the' Amazonian rain forest (Kahn 1983, p. 255).

Much of what is written on tropical American rain forests is in Spanish or Portuguese, published at Latin American universities or research stations as *informes*, mimeographed reports. These sources often are imperfectly linked to the northern information networks which nowadays are largely computerized. This warning should precede any list of general reading matter, because an enormous amount of direct information remains difficult to obtain, while the texts which can be easily consulted often stem from authors with an outside view. The inside, Latin American, view is supported by more than a century of local university tradition, exemplified by such splendid works as *Flora Brasiliensis* (Martius et al. 1840–1906) or the thousands of plant names due to Ruíz and Pavón. More recently (1976), Gomez-Pompa and Vasquez-Yanes published the standard text on *Regeneración de selvas* in Mexico, while Canadas Cruz (1983), with his *Mapa bioclimático y ecológico del Ecuador* refined the ecological mapping of that country, where Alexander von Humboldt started research in 1802 (see Botting 1973).

Whoever wants to understand the efforts of tropical Americans to analyze their forests, will have to consult books from the school of Holdridge who, since his first publication in 1947, worked at the Tropical Science Centre in San José, Costa Rica. *Forest Environments in Tropical Life Zones*, written in 1971 by Holdridge and four others, gives the most complete overview of his methods. 'Life zones', or 'zonas de vida' are defined by yearly averages of temperature and rainfall, and the amount of evaporation and transpiration calculated for each combination. The famous 'triangle' (Fig. 9.1) summarizes these zones in an elegant way. The use of averages is the basis of all the maps and descriptions made by this group of researchers and their students, even when refining the scale of observation to small regions, adding soil characteristics and human impact, or describing forest patches, from which average 'idealized diagrams' are constructed. The average frequency of species is then used

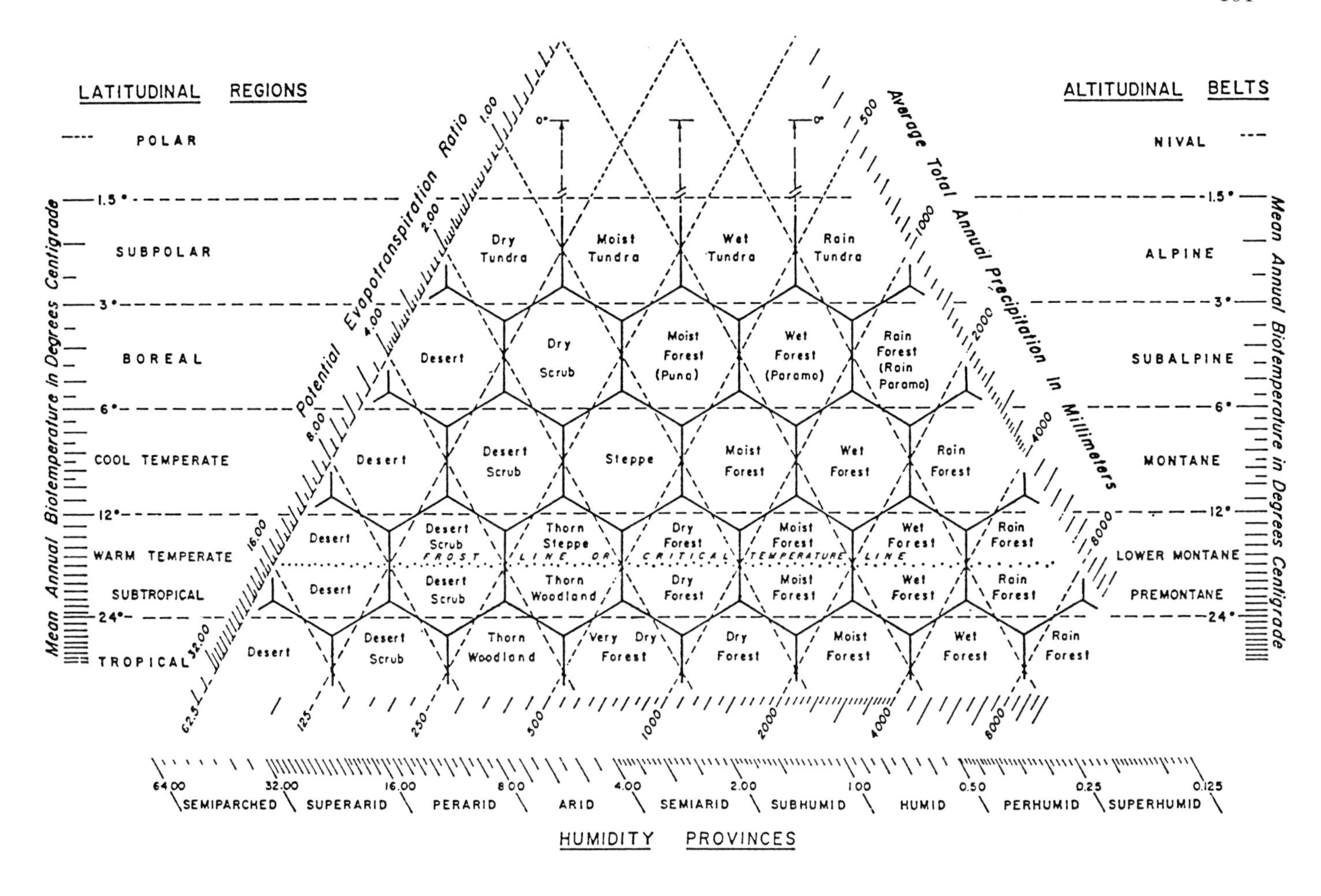

Fig. 9.1 Holdridge's triangle, a climatic diagram used in many tropical American countries to classify life zones, each with their own vegetation type. Note the *three* types of rain forests (in the sense of this book) distinguished by the author at the lower right side of the triangle: tropical moist, wet and rain forest (Holdridge et al. 1971)

to 'correct' the 'standard profile'. Features which are 'averaged out', however, include the age distribution in species, because no young plants are considered, and light tolerance, because plants are drawn at arbitrary light or shadowy places in the 'idealized profile diagram' (Fig. 9.2). The risk of averaging out significant differences, and the dependence on networks of meteorological stations which, at the moment, are still too widespread and too recently established to be completely reliable in rain forest zones, are the weak points of this method. The strong points are the creation of overview documents showing the average properties of whole vegetations and the usefulness of such documents for decision-makers. As can be seen, the increasing use of the term 'tropical moist forest' instead of 'tropical rain forest' is derived from the 'Holdridge triangle' (Fig. 9.1).

As a general introduction to neotropical botany, Verdoorn's *Plants and Plant Sciences in Latin America* (1945) is still a most valuable book. Its subject includes geographic and historical material, agriculture, forestry, natural vegetation and fauna in relation to the organization and possibilities of research. A book often forgotten is *Etudes écologiques des principales formations végétales du Brésil*, written by Aubréville in 1961 after a year's field research. He cites the phytogeographic synthesis for Amazonia by the Brazilian Adolfo Ducke in 1953 as a life work which no mere outsider could pretend to equal. Figure 9.3 shows

Fig. 9.2 Forest at Barranca (Costa Rica), situated in Holdridge's 'tropical moist forest life zone'; see Fig. 9.1. *Above*: 'Standard profile', drawn to scale after measurements in the field. *Below*: 'Idealized profile', idealization consisting of the spreading of adult tree silhouettes from the standard profile over a normalized length, in numbers consistent with statistical species frequency data. This analysis is a visualization of statistics representative for large forest surfaces, but not for small organized forest patches. Young and ageing plants are left out and ecological niches are not considered, e.g. the *Scheelea rostrata* palms (*Sc*) grow in the light under a broken canopy in reality (standard) and in deep shade on the idealized diagram. Neat layering of trees is also an artefact brought about by idealization of this particular forest (Holdridge et al. 1971)

two profile diagrams made by Aubréville, illustrating the enormous variety within the Amazonian rain forest, which is supported by floristic studies made at the same sites. Aubréville's book contains a critical analysis of South American vegetation classifications before 1960. *Die Wälder Südamerikas* by Hueck (1966) and the corresponding *Vegetationskarte von Süd-amerika* by Hueck and Seibert (1972) are successors to Aubréville's work and constitute standard sources on South American forests. Meggers' et al. book *Tropical Forest Ecosystems* (1973) should also be mentioned. Many data have been united in a more or less heterogeneous way by Prance and Elias, *Extinction is forever* (1977). The possible fate of the tropical American forests can barely be understood without reading *Amazon Jungle: from Green Hell to Red Desert* by Goodland and Irwin (1975), and the recent paper by IUCN

Ecological structures and problems of Amazonia (1983). The first text concerns the risks which are being taken, human impact on Amazonia being what it is, whereas the second surveys the ways to develop a non-destructive strategy for Amazonia. The IUCN publication covers a symposium held in Brazil at the Federal University of Sao Carlos.

The Caribbean

Only one peripheral part of the South American rain forests can be compared directly with Malesia, i.e. the Caribbean forests. In their entirety, they constitute a scattering of forests over many islands. From the south to the north, they correspond less to the descriptions of the rain forest given earlier (Chap. 1), because of an increasing summer/winter seasonality. Nevertheless, the geomorphology of these islands, which are the emergent tops of a chain of ancient volcanoes, some of which are still active, creates many climatological nooks and crannies in which conditions prevail for rain forest-like vegetations. No standard work exists on the Carribean forests and knowledge has to be obtained from material written about each individual island. Some examples will be given here.

The island of Trinidad lies closest to the Continent and was separated from it most recently, geologically speaking. Its forests still resemble those in the Orinoco basin and in the Guyanas, and it was here that Beard (1944) made his classical dichotomical key for the identification of South American forest types. Trinidad was also one of the early proving grounds for South American silviculture. Half a century ago, for instance, foresters here produced a variety of teak (*Tectona grandis*, Verbenaceae) well acclimatized for forestry plantations outside its native Burma and Thailand.

Whereas Trinidad has considerable areas of lowland, the Lesser Antilles are most often microcosms of continents with numerous bioclimates. In Martinique, for instance, numerous vegetational types are preserved, among them some rain forest areas, in a network of national parks described by Fiard (1979) in his brochure *La forêt martiniquaise*. In the mountainous half of Guadeloupe, Basse-Terre, there is a large national park which includes the fascinating rain forests moulded by hurricanes, with their 'shorn' canopies: the winds here do not uproot most of the trees but instead trim the tree crowns, so that from a distance the canopy more or less resembles a lawn. Also very curious are the forests covering the mountain peak which surges from the sea as the island of Saba. Its whole forest area, now being enclosed in a national park, seems to have been worked over by human activities, but without too many indigenous species having been lost (Romeijn, pers. comm.).

In the Greater Antilles, the human impact on forests has been heavy. The famous study *A tropical rain forest*, edited by Odum and Pigeon (1970) unites three large volumes of data on an intact, submountainous forest system in Puerto Rico. This was the first large-scale study on energy and nutrient cycling in tropical forests, and it is interesting to know that this research was financed and published by the US Atomic Energy Commission in order to assess the impact of nuclear radiation on tropical forests. Probably the most deforested island of all is Hispaniola; in Haiti, erosion is now the major scourge of all rural areas. Cuba, as the northernmost of the Greater Antilles, has the most seasonal climate; on its mountains there are pine forests. Cuba's forest area has become insufficient to cover the needs of its population and large-scale, internationally supported reafforestation projects have been in existence for 20 years.

As in Malesia, the island geography of the Caribbean has favoured the evolution of many endemic species, occurring on one island only. For instance, whereas there are no snakes on the island of Guadeloupe, a crotaloid species occurs in Martinique. Descriptions of the plants and animals, and floras of these islands have been made since the 16th and 17th centuries, many of the early ones written by Catholic missionaries. The well-known ornamental frangipani tree, now widely cultivated all over the tropics, bears the scientific name *Plumeria*,

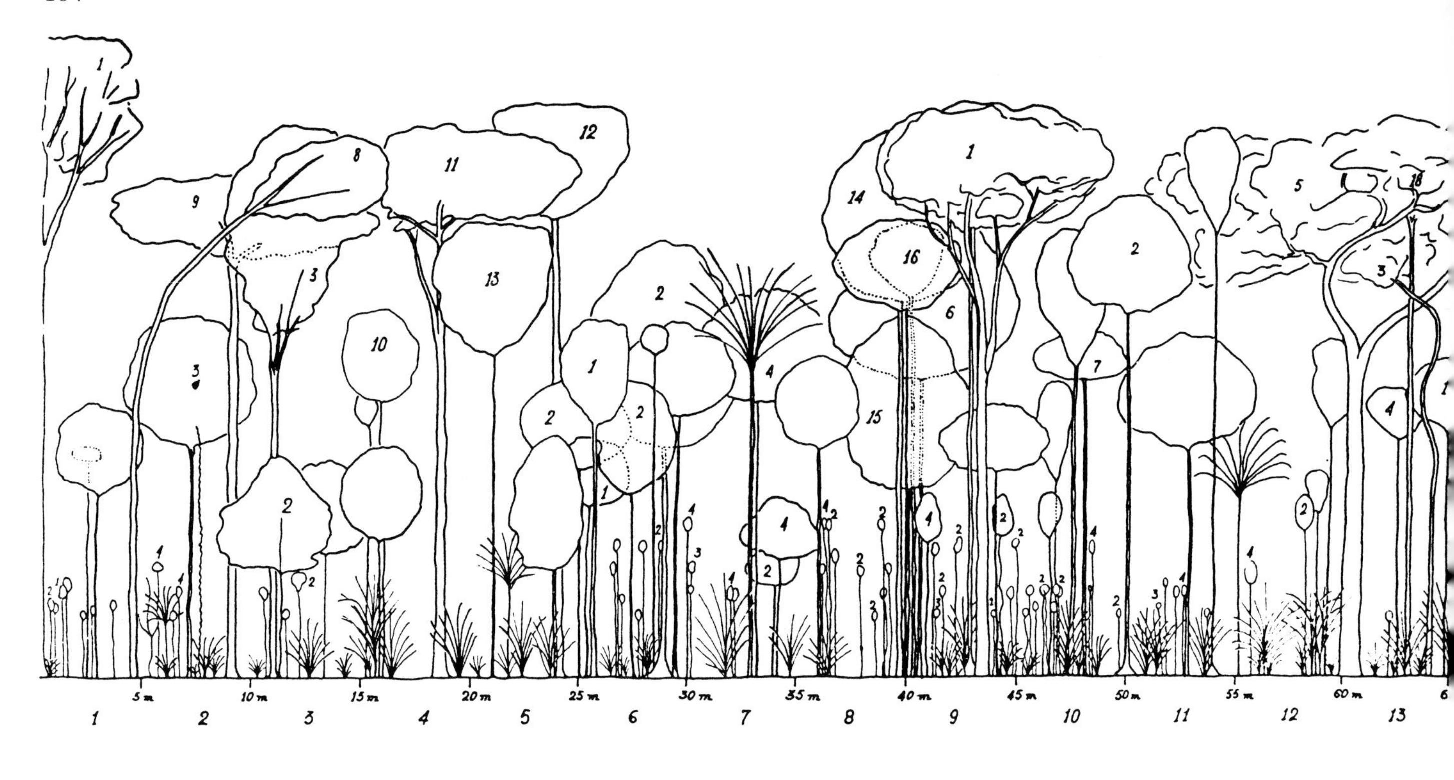
5 m 10 m 15 m 20 m 25 m 30 m 35 m 40 m 45 m 50 m 55 m 60 m
1 2 3 4 5 6 7 8 9 10 11 12 13

0 1 2 3 4 5 6 7 8 9 10 11 12
20 m
40 m

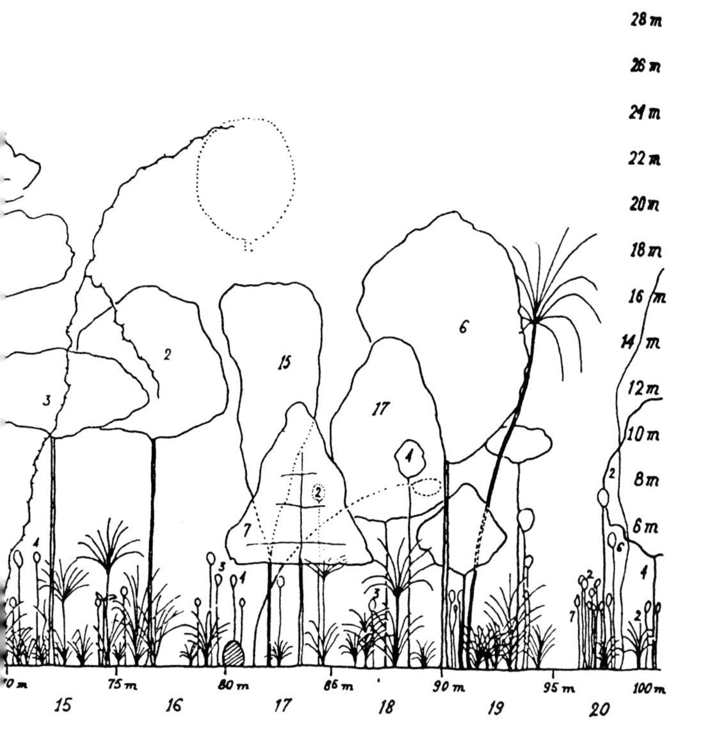

(1) abiurana. *Pouteria* sp.
(2) breu. *Protium* sp.
(3) mata mata preto. *Eschweilera odora* (Poepp.) Miers
(4) louro. lauracée
(5) piquia amarelo. *Caryocar* sp.
(6) caripe. *Licania* sp.
(7) ucuuba. *Virola punctata* Spruce et Benth.
(8) faveira légumineuse
(9) sucupira encarnada. *Andira parviflora* Ducke
(10) envireira surucuru. *Unonopsis* sp.
(11) uchi rana. *Vantanea* sp.
(12) ucuu barana. myristicacée
(13) pintadinho. *Licania* sp.
(14) amapa. *Brosimum* sp.
(15) itauborana. *Guarea* sp.
(16) inga. i. *Inga* sp.
(17) ripéro amarelo. *Eschweilera* sp.
(18) macucu murici. *Saccoglottis guianensis* Aubl.

Fig. 9.3 Two profile diagrams of Amazonian forests drawn and studied by Aubréville (1961), including all palms taller than 1 m and all other plants taller than 3 m. *Upper diagram*: 'Terra firme' forest on clay plateau near Manaus. *Lower diagram*: 'Varzea' forest on inundated river flats of the Rio Guama near Belém do Pará. Each of these forest types can be subdivided into many more, quite different ones

after an anagram of the name of the Reverend Father Plumier, an early botanist in the Antilles. It is to be hoped that many of these dispersed data will find their place in the *Flora Neotropica* now being published by the New York Botanical Gardens which has a long tradition of neotropical studies.

Although the Antilles were the first areas to be colonized by Europeans and consequently have been subject to the longest and most intensive human impact on their forests, it is encouraging to realize that so many remnant forests have been conserved throughout the Caribbean. Many islands, like Barbados, are forestless and all but covered by sugarcane. But the forests that are left warrant an international effort to study, conserve and use them wisely. I have given the Caribbean some space here notwithstanding the fact that only a few examples can be cited. It is too often neglected and this incomplete overview may stimulate interest in this fascinating part of the neotropical forest biome.

The Mountainous Backbone

The northwestern part of the neotropical rain forest area is like another archipelago, but without the sea. From Mexico on southwards, it is dominated by the mountainous backbone of the Americas, the islands being the numerous lowland valleys. Here again, human impact has been increasingly heavy during the last quarter of a century. In most Central American countries and in the coastal plains to the northwest of the Andes, reaching nearly to the Gulf of Guayaquil in Ecuador, no really large forest areas have been left. In Yucatán (Mexico) and Guatemala, deforestation goes back much further. During the Maya civilization (400 to 1400 AD), this whole region was converted to agriculture, agroforestry and silviculture, a fact established by recent Mexican research by Barrera et al. (1977). According to another hypothesis, excessive urbanization and wars led to the breakdown of agriculture and consequently to that of the Maya civilization. The huge ruined cities found back in the jungles of Yucatán in the beginning of the 20th century point to the late secondary character of these forests. They certainly were not 'pristine', and Barrera and his co-authors think that their actual species composition is due to Mayan silviculture.

More to the south, less-influenced forests may remain to the east of the mountains (Atlantic forest formations) and to the west (Pacific forest formations). For the Pacific side, the book by Allen (1956), *The rain forests of Golfo Dulce* remains one of the most valuable documents. Arranged mostly like an encyclopedia, after the more general introductory chapters, the main part of the book consists of data on species, forest products and vernacular names, in alphabetical order. The studies by the school of Holdridge in the same country, Costa Rica, have been mentioned before. The CATIE, *Centro Agrícola Tropical de Investigación y Ensenanza* in Turrialba (Costa Rica) is an international centre which has produced many studies on forests and forestry in the Central American region, such as the *Studies on forest succession in Costa Rica and Panamá* by Budowski (1965) and many relevant articles in the periodical *Turrialba*. Still more to the south, one of the best studied rain forests in the world are the 1500 ha on Barro Colorado Island in Panamá. It became an island in 1914 because the Chagres river was dammed during the building of the canal, and since that time Barro Colorado has been a forest reserve. Its scientific study began in 1916. The book edited by Leigh Jr., Rand and Windsor in 1983 under the title *The ecology of a tropical forest: seasonal rhythms and long-term changes* gives a 468-page account of the most recent information on plants, animals and periodicity. The whole forest regeneration system is examined, both on large and small surfaces and openings after the disappearance of the old forest, and the life cycles of plants and animals are analyzed. It should be mentioned that Janzen's theory on synchronous flowering of tropical forest trees and his model of seed dispersal and periodicity as a function of seed predators (1967, 1970) come both from field research in Panamá and Costa Rica. This theory of biological regulation depends on a feedback mechanism between plant and animal behaviour. During a period of scarcity of pollinating animals, it is inefficient for most trees to flower, so there is a shift in flowering towards periods with many pollinators. And, conversely, pollinators adjust their life cycles to periods with many flowers. The same mechanism comes into play in periodic fruit production, with the difference that the trees fruiting when there are *minimum* numbers of fruit eaters, succeed in establishing a *maximum* number of seedlings.

When considering Central American forests, however, from Yucatán to Panamá, one has to bear in mind the statement by Cook who wrote, as early as 1909: "Truly virgin forests seem not to exist in Central America. Relics of ancient agriculture occupations seem nowhere to be lacking, even in regions now entirely uninhabited, in dense forests as well as in open desert regions".

The very ancient history of the Pacific coast of South America, on which ceramics were found dating from 2000 BC in Valdivia, Ecuador, also raises heavy doubts as to the primary character of the rain forests in that region. In the coastal provinces of Ecuador, where a semi-arid climate now prevails, the ceramics were in the shape of rain forest fruits and animals, as illustrated in Lathrap's catalogue of the 'Ancient Ecuador' exhibition of the Field Museum of Natural History in Chicago (1975, e.g. Figs. 334 to 374). All evidence points to a dense population west of the Andes around the beginning of the Christian era. On the other hand, the land becomes more and more divided by smaller and larger mountain chains as one travels to the north, and the scarce botanical inventories show that the richness of species in the remaining forests is high, whether they are truly primary or not. North of the river Esmeraldas in Ecuador, and in sourthern Colombia, it might still be possible to establish forest reserves of several hundred thousands of hectares, but *programas de colonización* and other development projects with the aim of relieving the population pressure in the Sierra are already leading to deforestation and other conditions which make conservation measures difficult to implement.

An overview of the vegetations on and next to the northern Andes has been produced by Ellenberg, whose *Vegetationsstufen in perhumiden bis perariden Bereichen der tropischen Anden* (1975) summarizes a huge body of information gathered by himself and his team. The *Flora of Ecuador*, in preparation at the University of Aarhus (Denmark) in collaboration with the universities in Quito will increase published knowledge on the species from the Pacific rain forests. Existing documentation includes the book by Little and Dixon, *Arboles comúnes de la provincia de Esmeraldas*, published by FAO in 1969 and publications by Cañadas Cruz, e.g. his article on the swamp forests around San Lorenzo (1965). Another source is the prolific author, Acosta-Solís, whose *División fitogeográfica y formaciones geobotánicas del Ecuador* (1968) is illustrated by drawings.

The Pacific rain forests of tropical America, because of their isolation and possibly because of ancient human impact which, at that time, tended to conserve and enrich the useful species, are a unique and valuable gene source, not unlike Madagascar in the Old World. In a forest near Río Caoní, Ecuador, Arévalo and Oldeman (unpubl.), for instance found 40-m-high canopy trees belonging to the genus *Conostegia* in the pantropical family of Melastomaataceae, which was only known to contain undergrowth trees and shrubs up to 10 m. The balsa tree, *Ochroma lagopus* (Bombacaceae) known for its use in the construction of Thor Heyerdahl's Kon-Tiki raft, probably comes from these forests because they contain other and insufficiently known species of *Ochroma*. The Pacific rain forests might also well be the area of origin of the pantropical cotton tree, *Ceiba pentandra*, as other *Ceiba* species exist in this region, as well as a remarkably rich assortment of other species belonging to the same family, the Bombacaceae. This sacred tree of the Mayas will be mentioned later again.

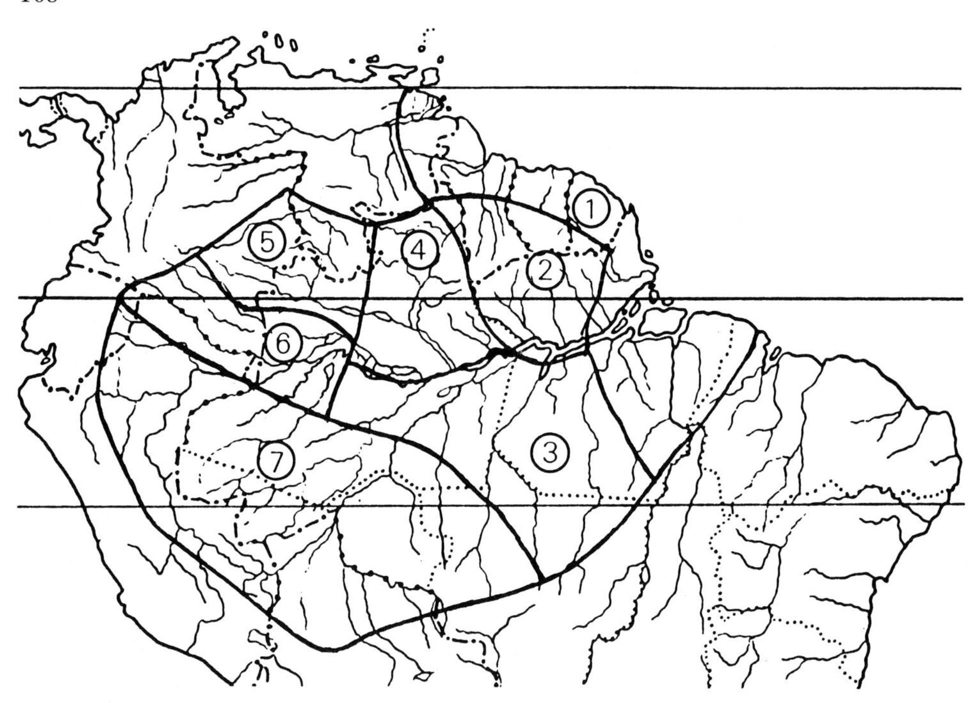

Fig. 9.4 Amazonia, with the seven phytogeographical provinces proposed by Prance (1977). This is one example out of many acceptable phytogeographical and ecological classifications that can be found in the literature

Guyanas

In the eastern and northern part of the South American continent one encounters the Orinoco rain forests and the Guayanese forest belt. These are linked up, at the neck of Central America, to the Pacific and Central American forests and one can follow their decreasing affinity, by using for instance the *Flora of Panamá*. Many species can be identified with it in Colombia and Venezuela, less in the Republics of Guyana and Surinam, and its usefulness ends somewhere past Cayenne in French Guyana. As a very rough border, the fifth parallel North would give an indication of the zone where the Guyanese forests start to blend into Amazonia, between the Pakaraima Mountains and the Atlantic. The Highlands of Guyana, which have been considered as the limit in former times, in reality are quite open to infiltration from the south. At least in meridional Surinam and French Guyana they are low and fragmented and may just be seen as the place where Prance's phytogeographic province 2 becomes completely Amazonian (Fig. 9.4).

The forests along the Orinoco were first researched by Humboldt and Bonpland in the beginning of the 19th century. "In his time, the wild life of these regions was much more abundant than it is today and as his canoe glided along the banks of the river he was treated to a wonderfully close view of birds and animals of every conceivable variety", writes his biographer Botting (1973, p. 105). The 30 volumes of *Voyage de Humboldt et Bonpland 1799–1804*, published between 1805 and 1834, contain descriptions and illustrations of, among other things, the Orinoco forests of 2 centuries ago. Much of these forests still exists. Just as in Humboldt's time, the delta of the Orinoco shows a patchy mosaic of rain forests and savannas in the north, and of rain forests and swamp forests or mangroves in the southern parts, where they blend into the Guyana forests. On the upper Orinoco, the rain forests, separated from the delta by the savannas called *llanos* (plains) make a gradual transition to the Amazonian massif: San Carlos de Río Negro, which was reached by Humboldt on the 7 May 1800, is fully Amazonian and now hosts a research project in the MAB-Unesco network (see Alder et al. 1979).

Fig. 9.5 Frontispice of Aublet's *Histoire des Plantes de la Guyane Françoise* which, in 1775, was the first systematic treatment of the plants then known from French Guyana. Note the useful plants depicted by this *'botaniste du roy'*, whose portrait can be seen in the *inset*. He paid for his travels deep into the tropical rain forest with his life and died soon after having completed his four volumes

In all, the Guyanas form a much-studied and largely intact rain forest regions. The four wonderful volumes by Fusée Aublet (1775) *Histoire des plantes de la Guiane françoise* constitute the first thorough rain forest study of the continent (Fig. 9.5). Numerous plant species described by Aublet have been rediscovered only during recent years. It was Aublet who started the French tradition of latinizing local plant names, such as the genus *Qualea*, from the Indian *kouallí*, which belongs to the tree family Vochysiaceae. In tropical South America, this family seems to be the biological equivalent of the Malesian dipterocarps, described in Chapter 10 (Malesia). The architecture and growth dynamics of the Vochysiaceae, their mast flowering (every 3 to 6 years), their winged seeds and their flower biology all closely parallel those of the dipterocarps. Even their wood is used in the same way, i.e. for plywood fabrication and construction wood. But in tropical America, the family does not dominate the forest: this role is played by the Leguminosae. The presence of kouallis is a sign of some ancient catastrophes such as inundations or fires; they belong to the category of very late secondary species. Where they dominate, something has happened long ago, as for example on the Comté river in French Guyana where the Jesuits left their plantations around 1770, so that it is known that the forests in this region are usually about 2 centuries old. This successful biological strategy under such conditions might be an indication of either the frequency of natural catastrophes or of the existence of former population pressure which has since disappeared, with relevance in Asia as well as in the New World. A more precise comparative study of dipterocarps and Vochysiaceae would therefore be useful.

The names of many famous botanists and forest ecologists are linked to the Guyanas. Aublet, Botaniste du Roy, Sagot and Richard in French Guyana, Pulle and Lanjouw in Surinam, Schomburgk, Jenman and Gleason in the Republic of Guyana are only examples from more than 70 researchers who have worked here. Davis and Richards (1933–1934) studied here *The vegetation of Moraballi Creek* which became a classic for forest transect studies; Schulz published his *Ecological studies on rain forest in Northern Surinam* in 1960, an exemplary floristic and climatological analysis, and Oldeman in 1974 laid the basis for his architectural forest studies, described in the introductory chapters of his *L'architecture de la forêt guyanaise*.

Except for extensive use, the major part of the Guyana forest massif has not been disturbed until now. Such extensive use not only includes shifting cultivation by the indigenous Indians and imported Africans descending from escaped slaves, the bush negro tribes such as Paramacca, Saramacca and Boni with their largely West African cultures. It also includes the production of such 'minor' forest products as rosewood oil distilled from *Aniba rosaeodora*, a canopy tree belonging to the Lauraceae. Both in the Guyanas and in northern Brazil this oil, which still cannot be synthesized, has been produced in mobile distilleries, transported far inland in parts by canoes over the rivers and rapids (Fig. 9.6). From the 1890s on, the Guyanas have also known a veritable gold rush, attracting adventurers from all over the Caribbean and from Europe, which has disrupted the social structure of the agricultural villages only established after the abolition of slavery around 1860. With the economic crises of the 1970s gold prospecting recommenced after a marked decrease in the 1950s and 1960s.

The Guyanas, including Venezuelan Guyana and Brazil's Territorio do Amapá, which until 1902 formed part of French Guyana, and was conquered in a war where 200 French marines opposed 300 Brazilian infantry, are still among the least populated parts of the continent. Only the Republic of Guyana counts more than a million inhabitants. Good land-use planning, including the conservation of vast tracts of rain forest, is still possible here. The Surinam experiments with natural silviculture monitored for 20 years are the only such experiments in this region; data on them were published by De Graaf (1986), publications by Schmidt and others will follow. They represent a first step in solving the problem of non-destructive wood production and harvest in certain planned land-use zones in the Guyanas. As in all silvicultural systems, commercially interesting tree species are favoured at the expense of non-marketable forest components, but the aims of the system include the survival of non-economical species in smaller numbers so that the forest may recover if and when this kind of management ceases and the forest is left to itself once more. The period of recovery in this case would be at least 1 century. More recently, experiments in forestland use have been made in French Guyana under the research program ECEREX (ECology, ERosion, EXperimentation), in which many scientific institutions are collaborating, such as ORSTOM, CTFT, the Forest Service, the Paris Museum and several universities.

Fig. 9.6 Rosewood distillery owned by Mr. Quammi on the upper Approuage river, French Guyana, 1967. Parts of the distillery were transported by canoe, a 3-day trip upriver over the rapids, then on foot, 3 h overland. The wood of the large forest tree *Aniba rosaeodora* (Lauraceae) is reduced to pieces, visible in the centre of the photograph, then distilled to yield rosewood oil, a minor forest product used in perfumes. This small-scale industrial method of forest exploitation is still close to original gathering methods of original neotropical inhabitants (Photograph R.A.A. Oldeman)

The Amazon

The word hylaea is frequently reserved for the large Amazonian forest region, which lies at the centre of all the tropical American forests described earlier. Hylaea is derived from the Greek world *hyle* (which means, among other things, wood or forest), and the term was coined by Humboldt to denote the forests of the Amazon basin.

The Amazonian hylaea, like the Central African or the Malesian forest regions, is a universe in itself. It differs from Malesia, but is like Africa, in its historical interactions with other vegetation types. In Malesia, the forests are surrounded by the sea, but on the continents of Africa and America, the hylaea interacts with drier forest types, mountain forests and savannas. These interactions have changed dramatically with the climate. During the Ice Ages, the tropical humid climates were much drier. Müller and other zoologists studying South American amphibians were the first to find a score of the same species in certain widely spread centres assumed to have remained humid throughout the Pleistocene Ice Ages. Later, other species groups were considered. A recent synthesis of different theories on these regions, called *refugia*, has been made by De Granville in 1978; it is reproduced here in Fig. 9.7. These refugia are considered as natural regions where species have been conserved, and from which species migrated outwards and formed new species when the climate became humid again. In no publication, however, has the question of which region was a refugium during which one of the four most recent Ice Ages yet been considered: the time element remains unclear. From the viewpoint of conservation, however, these refugia deserve special attention, not so much because of their species richness, but because they will probably function as safe houses or speciation centres again, were minor or major climatic shifts to occur. The actual species in a refugium are not necessarily the most numerous, but they include the oldest ones which survived the dry millennia in South America.

Several genera in the Amazonian forest show that speciation is recent or still going on: it is difficult, for instance, to establish clear limits between the Apocynaceous trees of the genus *Aspidosperma*, containing forest trees up to 60 m in height. Over gradients which cover thousands of kilometres, the characters of a species may gradually change so much, that the extreme forms would be, and have been described as, different species as long as the

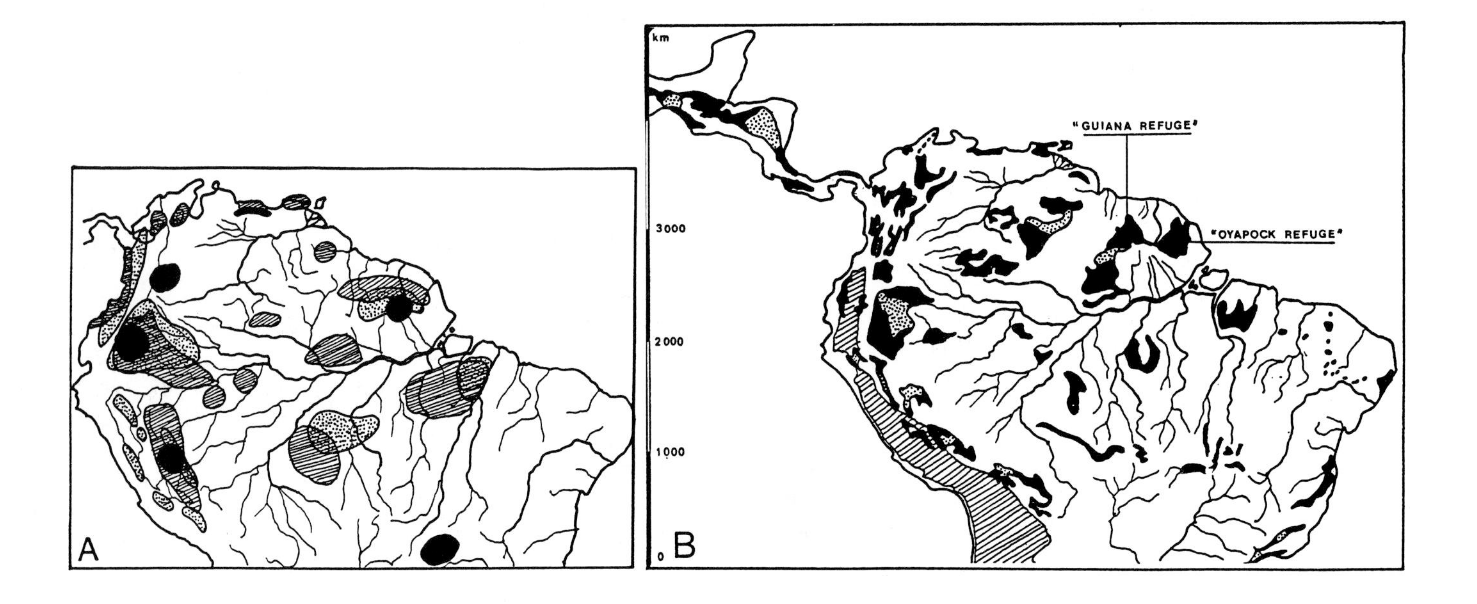

Fig. 9.7 A, B. Forest refugia during the end of the Pleistocene, between 22,000 and 13,000 years ago, as synthesized by De Granville (1978, pp. 137–138) from different authors. **A** Refugia as outlined by Haffer (1969; *dotted*), Vanzolini (1970, 1973; *black*) and Prance (1973; *hatched*). **B** Refugia after Brown (1977) who distinguishes 'subrefugia' (*black*) and 'semi-refugia' (*dotted*), which differ from 'primary refugia' by a reduced number of conserved species. *Hatched*: Andean zones above 1500 m altitude. The authors cited include both botanists and zoologists: much work still has to be done before location, surface and precise period will be clearly known

Fig. 9.8 Due to conditions in the hylaea and its surrounding forests, species and even genera of plants are not always easy to define and may still be speciating. Six fruits taken from the same tree in Saül, French Guyana, certainly belong to the Lecythidaceae, a family which is all but completely neotropical. But according to the revision by Miers, the fruits to the *right* would belong to the genus *Lecythis*, those to the *left* to *Eschweilera*. Except for the size, there is also a marked difference in the number of seed compartments, which according to Miers is a critical characteristic. This can be seen in the two woody 'lids' of the fruits, above and to the right of the matchstick. The family also contains *Bertholletia excelsa*, the Brazil nut, a well-known minor forest product from Brazil (Photograph R.A.A. Oldeman)

intermediate forms were unknown. The century-old discussion on the delimitation of genera within the South American family of Lecythidaceae (one African genus only) falls within the same pattern: in French Guyana, fruits which ought to belong to the genera *Lecythis* and *Eschweilera* (Fig. 9.8) were found on the same tree! This situation profoundly differs from that in Malesia, where in most places there simply is no space on one island or peninsula to reach such migration-cum-speciation distances. It may also be, that climatic conditions in Asia changed less thoroughly, and that old established species survived without much pressure towards new speciation. The whole Amazonian situation seems to have been and to have remained more fluid than elsewhere, for plants and animals alike. The enormous species richness of the pioneer tree genus *Cecropia* (Moraceae) in South America and that of the Asian pioneer genus *Macaranga* (Euphorbiaceae) both point to a rich array of pioneer niches in evolutionary history, but this may well be due to very different causes. No absolute certainty has as yet been reached, and new facts sometimes add to the confusion. It is generally accepted nowadays that once every 1 or 2 centuries a series of abnormally dry years without rainy seasons dramatically increases the fire hazards in the otherwise fireproof tropical rain forests. In 1976, the Guayanese rain forest burned locally, and in 1982 and 1983 large surfaces of rain forest burned in West Africa and Kalimantan. The extent of these fires was certainly very much increased by human clearing practices. But even if only 10% of such surfaces burned naturally every century, this should profoundly influence secundarization and speciation. A preliminary study on the aspects of such factors is included in the ECEREX project in French Guyana.

Roughly, the Amazonian hylaea covers an area of 2000 km from north to south and 4000 km from east to west, where it gradually blends into submountainous and mountainous Andean forest formations. The largest part is Brazilian, but the Venezuelian, Colombian, Ecuadorian and Peruvian parts still may be as much as 20% of the total. Most of the

Amazonian rain forest is situated on land without mountains, but often with a steeply differentiated microrelief in the shape of half-oranges. Hence, the parts most vulnerable to direct deforestation effects such as erosion are the non-Brazilian ones. If *indirect* deforestation of the Amazon basin, due to mistaken forms of land use, were to occur it would originate in the west. In Colombia the transition from the Amazonian lowlands has been mapped and evaluated and the results published in the six large volumes of the PRORADAM project (1979). Notably, this document gives the proportion of Colombian Amazonia which can be used for agriculture as only 6%. The same conclusion, but without exact figures, can be read from Canadas' conclusions for East Ecuador, already cited (1983). The reasons for this conclusion are not only the low soil fertility and its compact structure, but also the perhumid character of the climate, with often more than 4000 mm precipitation per year. The average height of natural forests in perhumid regions is significantly lower than that which occurs in humid regions (2000 to 4000 mm $year^{-1}$), a fact which seems to hold true throughout the Amazon basin. In most of the area there is a dry season of between 1 and 3 months, which also causes forests to be of a different type. The soils are another differentiating factor. The new UNESCO vegetation map for South America (1981) takes into account many of these factors.

The colour of the rivers, the traditional routes for travellers in the Amazon region, reveal soil type and precipitation in the parts they traverse. The well-known white-water rivers (Río Branco) come from regions with lower rainfall and less erodible vegetation than the black-water rivers (Río Negro), said by McCreagh (1961) to contain "... really 'black water', clear thin ink when dipped up in the hand, and black night at the depth of ten feet", which carry a high load of dissolved mineral and organic matter leached from the forest. The coffee-with-milk rivers come from the Andes and bear soil particles dislodged by erosion; in general, such rivers indicate either steep slopes or soil mismanagement upstream. It should be made clear, however, that some erosion occurs everywhere, even under intact forest, when slopes are steep enough. Erosion is a normal landscape-forming process. The current concept of erosion, however, means the erosion accelerated by poor land use. *Extreme* natural erosion can be measured in centimetres per year, *extreme* human-induced erosion may amount to a loss of metres of soil depth per year. This is the reason why heavy erosion will occur even in the plains, where the landscape in half-oranges results in many short, steep slopes, as soon as human intervention becomes more than very slight: one can observe erosion gullies even along trails which are only trodden or used once a week.

Well-known forest types in the Amazon region are *varzéa*, called *pri-pri* in French Guyana, which are inundated by river water in the rainy seaon and hence grow on recently formed, fertile alluvial soils, and the *terra firme* forests in which the classification of all different forests on soils which are never inundated are lumped, although this category has been refined by Figueiredo and Cals (1982). The *wallaba* or *caatinga* are situated on extremely poor sandy soils. This last forest type may be compared to the *kerangas* of Borneo, mentioned earlier. The colour of the river water coming from these regions is also different.

The Amazonian flora is very rich, with tens of thousands of species. As long as the *Flora Neotropica* is still so very incomplete, and because a critical inventory of collections and literature is lacking, one cannot be more precise. Moreover, the collections themselves remain very incomplete. Current collections contain 1% of specimens belonging to species new to science, and 2% are new for the region where the collection was made. Given the fact that many specimens belong to well-known and frequent species, one may assume that 10 to 20% of the higher plants have not been scientifically described. All over the Amazon, however, numerous forests are floristically dominated by leguminous trees, which may comprise up to 60% of the tree species. Their role is less conspicuous than that of the dipterocarps in Asia, because this big family contains many more genera and species and shows a much wider array of forms and functions. Vochysiaceae and Lecythidaceae are typical South American tree families with very few representatives in Africa and Asia. Burseraceae, Lauraceae, Moraceae, Sapotaceae and Chrysobalanaceae (formerly included in Rosaceae) are other frequent tree families. Annonaceae, Rubiaceae and Melastomaceae are very important and species-rich families in the undergrowth, and among the climbers or bush ropes there is a wide array of families, with maybe some preponderance of Leguminosae, Bignoniaceae, Apocynaceae and Malpighiaceae. There are some curious and typical American monocotyledonous families. The Rapateaceae, their spectacular leaf sheaths fitting into each other at the base of the plant, live on the forest floor and in swampy valleys. The colourless Burmanniaceae, the minute, seemingly lifeless straws emerging among dead leaves are saprophytes, living on decaying leaves and branches. But best-known are the Bromeliads, a large family with 2000 species in 50 genera, members of which can be found in many apartments as ornamentals. One of its rare terrestrial representatives is the pineapple, *Annanas comosus*, but most are epiphytes, living high in the trees on the bark of trunk and branches, on suspended soils in places where decayed matter is accumulated. Their leaf bases often form cylindrical sheaths, where rainwater is collected on which they can survive. In this stagnant water, whole mini-ecosystems with a rich fauna of cockroaches, mosquito larvae, other insects and tiny frogs can be found. The monocotyledons have been studied closely by De Granville (1978). Although his book only concerns French Guyana, the ecological conclusions may be extrapolated for a large part of Amazonia.

Palms

The diversity of palms is staggering. Moore (1973, p. 69) gives a total of 837 species in 64 genera for South America and 1147 species in 81 genera for the whole New World. Let us illustrate the importance of palms by quoting from Darwin's contemporary, Wallace's *Palm trees of the Amazon* (1853, pp. 9–11, quoted by Corner 1966, p. 31): "Suppose then we visit an Indian cottage on the banks of the Rio Negro, a great tributary of the river Amazon in South America. The main supports of the building are trunks of some forest tree of heavy and durable wood, but the light rafters overhead are formed by the straight cylindrical and uniform stems of the Jará palm. The roof is thatched with large triangular leaves, neatly arranged in regular alternate rows, and bound to the rafters with *sipós* or forest creepers; the leaves are those of the caraná palm. The door of the house is a framework of thin hard strips of wood neatly thatched over; it is made of the split stems of the Pashiúba palm. In one corner stands a heavy harpoon for catching the cow-fish; it is formed of the black wood of the Pashiúba barriguda. By its side is a blowpipe ten or twelve feet long, and a little quiver full of small poisoned arrows hang up near it . . . it is from the stem and spines of two species of Palms that they are made. His great bassoon-like musical instruments are made of palm stems; the cloth in which he wraps his most valued feather ornaments is a fibrous palm spathe, and the rude chest in which he keeps this treasures is woven from palm leaves. His hammock, his bow-string and his fishing-line are from the fibres of leaves which he obtains from different palm trees, according to the qualities he requires in them – the hammock from Mirití, and the bow-string and fishing-line from the Tucúm. The comb which he wears on his head is ingeniously constructed of the hard bark of a palm, and he makes fish hooks of the spines, or uses them to puncture on his skin the peculiar markings of his tribe. His children are eating the agreeable red and yellow fruit of the Pupunha or peach palm, and from that of the Assaí he has prepared a favourite drink, which he offers you to taste. That carefully suspended gourd contains oil, which he has extracted from the fruit of another species; and that long, elastic, plaited cylinder used for squeezing dry the mandioca pulp to make his bread, is made of the bark of one of the singular climbing palms, which alone can resist for a considerable time the action of the poisonous juice. In each of these cases a species is selected better adapted than the rest for the peculiar purpose to which it is applied, and often having several different uses which no other plant can serve as well, so that some little idea may be formed of how important to the South American Indian must be these noble trees, which supply so many daily wants, giving him his house, his food, and his weapons". If we have counted correctly, 15 palm species are mentioned here.

Animals

The richness of animal life in the Amazonian and other tropical American forests is all but incomprehensible to the human brain. For the higher animals, Goodland and Irwin (1975, p. 87) have made a table of estimates. 1171 species of 77 bird families are known in Amazonia, many of which can be found in Descourtilz' water colours, which were reprinted in 1960 from the original edition in 1855. Others are described and illustrated in Francois Haverschmidt's *Birds of Surinam* (1968), covering 602 species and giving a detailed history of the centuries of ornithological research in that region. Lowe-McConnell (1969, p. 54) mentions 1383 fish species for Brazil and 456 for Central America. By comparison, there are only 192 fish species in the whole of Europe, and 172 in the Great Lakes of North America. Some figures on mammals are given in Chapter 12. A recent and courageous study on the multitudinous Amazonian insect species was made by Akker and Groeneveld (1984) in their *Reconnaissance of Amazonian rain forest architecture and insect habitats*. The enormous number of species and the spectacular forms of neotropical American butterflies are well known: who has not heard, for instance, of the metallic blue *Morpho*, with a wing span of almost 20 cm? New data on reptiles and amphibians continue to accumulate, for instance at the Paris Musée Nationale d'Histoire Naturelle, where researchers like Gasc and Jean Lescure continue to uncover new biological treasures.

Such treasures are all the more vulnerable to habitat destruction, and hence to extinction, because they are small. It is known that insect genera may be represented by different species from one neighbouring valley to the next, or that two related frog species exist, one of which lives exclusively on the underside and the other on the upper surface of the leaves of the same riverside tree. Human land use inevitably will lead to their extinction: in the first case because certain valleys will inevitably be converted to agriculture or other land use, in the second case because the riverbanks are the first human places to be settled. For larger animals, it is the minimum area of an integral forest reserve which is the important factor; for the smallest ones, it is the degree of disturbance.

Useful Plants

The botanical resources of the neotropical rain forests, as exemplified, for instance by the palms, are exceedingly rich. It is better not to limit ourselves here to the Amazon region, because many of the existing crops date from the Maya exploration of the continent, as is clear from the article by Barrera et al. (1977), cited earlier. If Southeast Asia is undoubtedly a centre of selection and improvement of fruit trees, the same can be said for the northern half of South America and Central America. There are close equivalents. What rice (*Oryza sativa*) is to Asia, maize (*Zea mays*) was to the Indian civilizations. When there are bad harvests of rice, tropical Asians fall back on breadfruit, *Artocarpus incisa*, a Moracea with fruits containing proteins and starch. The Mayas started to select another Moracea with the same goal in mind: *Brosimum alicastrum*, which, however, has smaller fruits. When old, both species have also formed valuable wood. The mango (*Mangifera indica*, Anacardiaceae) is a vitamin-rich fruit from Asia; its tropical American counterpart from the same family is *Spondias mombin*, though this too has markedly smaller fruits. *Psidium guajava*, which bears the delicious guava fruit, is the neotropical counterpart in the Myrtaceae family to the numerous *Eugenia* species of which the fruits are eaten in the Old World, although the reciprocal introduction of these fruits has since disguised this distinct origin; there are also native neotropical *Eugenia* species. The cherimoya is a fruit from *Annona cherimolia* which parallels the Old World soursop (*Annona muricata*). The papaya fruit (*Carica papaya*), now pantropical, and without Old World parallels, originated from Central America as did a lesser known fruit from the same Caricaceae family, *Jacaratia mexicana*. Avocado (*Persea americana*, Lauraceae) is a Central American fruit, and *Theobroma cacao* (Sterculiaceae) was introduced and selected in Mexico more than a 1000 years ago. In Brazil, one may buy *doce de Cupuacú*, jam made from another *Theobroma* fruit. Most of these American fruit trees were cultivated in home gardens, the traces of which in Yucatán have similarities with traditional Asian ones.

As in Asia, the array of traditional medicinal plants is enormous. For the Amazon, Nicole Maxwell's *Witch doctor's apprentice* (1960, reprinted 1975) illustrates this. In French Guyana, Francoise Grenand, who lived with her family among the Wayapi Indians for 6 years, described many of their useful plants and the myths attached to them in *Et l'homme devint jaguar* (1982). The Indians are well aware of the artificially selected character of their crops and, according to Jim Yost (pers. comm.), who lived in Ecuador among the Wauwrani (known as Aucas in the Quechua language), they are regularly experimenting with wild species in order to discover their use. Some important and miscellaneous examples are given here. The bark of the *Cinchona* tree (Rubiaceae), yields quinine, the best remedy known against malaria over several centuries. Another product originally medicinal in use is vanilla, the unripe fermented stick-like fruits of a Central American orchid, *Vanilla fragrans*, one species of more than 10 Amazonian. Maté, the dried leaves of *Ilex paraguayensis*, is the equivalent of the Asian tea; many plants we now use as spices or drinks were originally selected for their effect on health. Spiritual health was also a preoccupation of the Indian civilizations and there are records of a large array of psychedelic drugs to be found in South American ethnobotany: coca (*Erythroxylum coca*) in the west of the continent, the peyotl in Mexico (*Psilobius* sp., a fungus) and tobacco (*Nicotiana tabacum*) are some of the better-known species, but there are numerous others, for instance the blood-coloured juice from the bark of certain Myristicaceae, or the milky juice from Euphorbiaceae and Moraceae in the forest. Some milky juices were also known by the Indians to yield latex. Cauchú (Spanish: caucho; French: caoutchouc) is the Indian word for rubber (*Hevea* sp. div.; *Castilloa elastica*); balatá for the latex from *Manilkara bidentata* (Sapotaceae). All these latexes were used by the local inhabitants to waterproof bags, long before this product became known in Europe. The chewing gum made out of the latex from another Sapotacea, *Achras sapota* was probably an ancient ingredient of one or more medical or psycho-medical chewing preparations, in which the latex was loaded with other products, as we load it now with peppermint or other flavours.

A large array of ornamental plants were also selected by such Indian societies as the Mayas and the pre-Incan civilizations which, according to legend, stretched far out into Amazonia from the Andes. The flowering *Jacaranda* tree now planted everywhere in the tropics is one example. One category of crops is not to be found in the traditional Indian civilizations, however. This is the group of special timber trees. Fuel wood and timber probably were minor forest products harvested from the many fruit-bearing and ornamental trees, at the moment in their life cycle when yields of fruit and flower started to decline and the maximum of wood had accumulated in the trunk and branches: this point marks the beginning of the decline in vitality, its senescence. For fuel wood grown as hedge plants, coppicing seems to have been practiced, at least in Yucatán and Central America. One tree still has to be mentioned which, because of the occurrence of related species, seems to be of

American origin. This is the sacred tree of all ancient Indian civilizations: the *Ceiba pentandra* or cotton tree. Here, a curious parallel with the Old World is the sacred bombacaceous tree from Africa, *Adansonia digitata*: the baobab, however, is not a forest tree, but grows in savanna regions.

The Mayas called themselves, and their language, *Quiché*. In this word, *qui* means many, and *ché* means tree. This people hence considered itself as a forest civilization. The fact that the Indian languages in the Andes are called Quechua would indicate that a common culture existed more than a 1000 years ago, before the rise of the Incas, Aztecs and Toltecs, who were the first barbarian inheritors of the resources so carefully selected and managed by their predecessors. Then came the Europeans.

The economist Shanahan, in his *South America, an economic and regional geography with an historical chapter*, written in 1927, shows that up to that time the colonial and post-colonial societies had developed mainly around the Amazon basin, on the coast and, later, also inland in the south. Until this time, forest use had largely remained as it probably was in Indian times, but without the Indian civilization centres and their refined agroforestry and forestry systems. Travel in the huge, inaccessible bulk of the Amazonian forest was along the rivers: to feed themselves, the immigrants practiced the shifting cultivation of the original inhabitants, as well as their hunting and gathering methods. By the beginning of the 20th century, two crops of worldwide importance, rubber and pineapple, had been smuggled out of Brazil and French Guyana respectively, for cultivation in large plantations in Asia and Oceania. Meanwhile, in Brazil there was the balata boom: Manaüs saw multitudes of balata tappers going into the forest, tapping the latex from wild trees, and selling their produce in the city. The great opera house in Manaüs, where companies from Paris came to perform, dates from this rich, but ephemeral period which ended when cultivated Asian rubber took over the markets. The Brazil nut (*Bertholletia excelsa*, Lecythidaceae), which is a delicacy at cocktail parties worldwide, is still gathered in this way. The 18th century plantations of this crop by Jesuits in French Guyana were abandoned, after the Fathers were exiled from *la France équinoxiale* by the King. As for timber exploitation, after the exhaustion of the true mahogany (*Swietenia mahogany*, Meliaceae) forests in the Antilles and in northern South America, only a few precious woods were gathered in small quantities for special purposes such as snakewood and letterwood (species of the Moraceous genera *Brosimum* and *Piratinera*) used to make musical instruments. Large-scale timber exploitation was largely limited to West Africa until the Second World War and Asia followed in the 1960s. But although a relatively modest commerce exists in some scores of species, tropical America only has a 5–10% share of the world market for tropical hardwoods as indicated by the FAO and ECE figures in *Tropical hardwood utilization*, edited by Oldeman (1982).

Estates

Since colonization, tropical America's troubles have derived from the desire to imitate success stories from other continents. The Spanish and Portuguese colonizers brought with them the first ideals: mining and large-scale estates, such as those existing in Iberia, run by warrior-like overlords, to sum up Shanahan's diagnosis. The large-estate approach nowadays is still vividly represented in the whole of Latin America, as is the ideal of manly virtues. In the beginning, neither the means of transportation nor the tools existed which would have been needed to extend this vision to the large forest areas, so the margins of the continent were the first to be transformed. Then, in the beginning of the 20th century, the huge effort to create railways, a success story copied from the US coast to coast and the Russian Trans-Siberian lines some decades earlier, opened up the subtropical south. Cattle raising, an activity par excellence for large estates was hugely successful in the southern part of the continent, and was imitated in the 1940s in Costa Rica and in the 1970s in Brazil, where three successive hamburger connections were to procure beef for the US and Europe, starting with the Argentinian corned beef in the First World War. Machines were created in the meantime to cut and sell the *Araucaria* forests in the south. To understand the present pressure on all the great tropical American rain forest areas, one must take into account both the constant factor of thinking in terms of large estates and the dynamic factor of the pincers of human pressure closing in mainly from the south and west, but also more recently from the east. Combined, these factors have led to ambitious aspirations for the 'opening up' and 'valorization' of the Amazon basin which seemed to have become possible by building highways and through use of heavy machinery for which the roads gave access to the forest.

Not unlike the railway wave in the south at the beginning of the century, it was only some decades after the highway system had become a success story in Europe and the US that the Brazilian President announced in 1970 that 14,000 km of trans-Amazonian highways would be constructed to form the infrastructure for a settlement program towards the west. The same movement was occurring towards the east from such countries as Peru, Ecuador and Colombia, but without the ambition of creating such spectacular roads. The Brazilian program is shown in Fig.9.9. Between 1970 and 1973, 200 million US dollars were invested in these roads, but the impetus has gradually petered out and they were still not finished in 1985, a fact also due to the successive oil crises. Whether the settlements were called 'colonisacao', in Brazil, or 'colonisación', in the Spanish-speaking neighbour countries, their results were deceiving. Aerial photographs of such regions, 4 to 5 years after settlement, show the forests as a moth-eaten carpet. A typical story, from one of Brazil's neighbours, is that of the family which leaves the overpopulated Sierra and for US $10 goes east by bus along some mining road. Here they settle. After clearing and burning 50 ha, they receive a deed of property and a bank loan for cattle raising. Grass is planted, and cattle are bought. Because of lacking knowledge about the carrying capacity, too many cows are established per hectare. After 5 years, the soil is totally exhausted, the meagre cattle are slaughtered and the meat is sold. The family then disappears in the city with the proceeds, not to be found again. Thus, the estate idea lives on in the minds of small farmers just as it does in the minds of large owners in South America. But in this case, the results are loss of money by the agricultural bank, irreparable loss of 50 ha rain forest and soil and no long-term solution to the poverty and misery of so many people. One gets the impression that more colonization takes place spontaneously, by squatters, than in a planned way, in settlement projects; 2000 settler families mean the loss of 100,000 ha forest, and no one benefits. There is no official evaluation of the number of *colónes*, but over the whole Amazon basin there have been hundreds of thousands. The number of supporters for colonization programs of this sort are gradually decreasing in tropical America, although even in 1982 the local press was announcing a plan from a governor to shoot or catch and sell all wild animals in his region, then to exploit the total biomass of the forest, and finally to establish a prosperous agricultural society on the cleared lands. There is, of course, a big time lag between the reformulation of national policies and the penetration of such ideas at provincial or federal state levels.

In the 1970s, the large-scale, estate-inspired projects went on, largely unchecked and even encouraged by tax measures. Goodland and Irwin, in their well-documented book *Amazon Jungle – Green Hell to Red Desert?* (1975) referred to the old name for the Amazon forest (green hell) and the colour of degraded tropical soils (red), when formulating a strong warning against the risks of this kind of development. Prance, in 1977 (p. 211) exposed the part played in forest annihilation by foreign investors: "Two large multinational car factories recently extended their interests by buying enormous terrains in Amazonia for cattle raising. The capital was raised for 30% in the USA, 20% in the Federal Republic of Germany, 10%

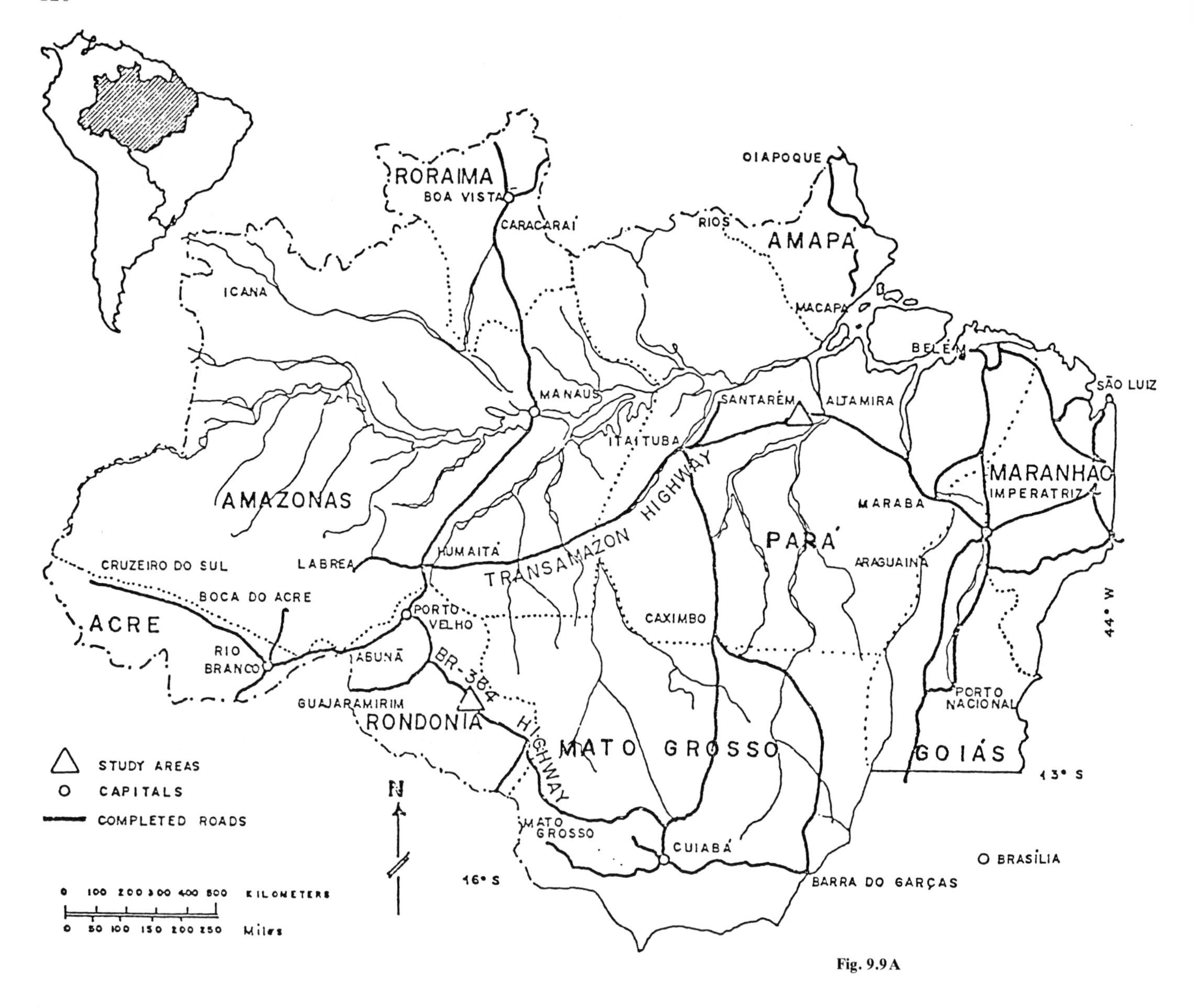

Fig. 9.9 A

in Japan, 10% in other foreign countries, 20% by inhabitants of Sao Paulo and only 10% by the inhabitants of the region itself".

Jarí

One of the most spectacular of these development projects was that of the North American millionaire Daniel K. Ludwig (born 1897), known as the Jarí project. It is the apotheosis of the combination of South American estate thinking and North American industrial ideas. The estate covers 1700 km^2 on the Jarí river, a tributary of the Amazon at 1° S and 53° W. It started in 1967 with an investment of 1 billion US dollars. Ludwig foresaw worldwide paper

Fig. 9.9 A Brazil's highway system, as shown on a map provided by CNPq in 1982. B Road to Caracas, north of Manaus, Brazil, right through the Amazon rain forest; note erosion in progress on the embankments and the settler's village in the distance (Photograph MJ 1982)

9.9B

shortages and wanted to achieve the grand solution. His plantations were to feed a colossal floating paper mill, which had been constructed in Japan and transported by sea, and production was eventually to amount to 700 metric tons of pulp per day. Figure 9.10 represents the organization of the enterprise. This is the figure stated by C.B. Briscoe, up to that time manager of the forest and livestock division at Jarí, in 1981 at an agroforestry meeting in Naïrobi, organized by ICRAF (International Council for Research in Agroforestry). According to a personal communication from Briscoe, the legal Brazilian requirement to conserve 50% of the natural forest in the area was scrupulously respected, although the conserved forests are situated on the less fertile and most hilly soils, as sketched on Fig. 9.10 (upper line). One hundred thousand hectares were cleared first, the more than 400 species of trees being used as lumber, veneer, pulp and fuel wood. Leaves, branches and bark were used to fertilize the plantations and to prevent weed germination by out-shadowing it ('herbicide'). Plantations, according to the soil, included mostly melina, *Gmelina arborea*, an Asiatic Verbenacea yielding harvestable pulp wood in 7 years, and *Pinus caribaea* var. *hondurensis*, a tropical fast-growing pine species. Together they covered 103,000 ha in 1981. A further 2000 ha were covered by experimental plantations with *Eucalyptus* (two species), and *Anthocephalus chinensis* as exotic crops and *Jacaranda copaia*, an indigenous Bignoniacea. Very early, food crops such as rice (Asian origin) and cassava, *Manihot utilissima*, an

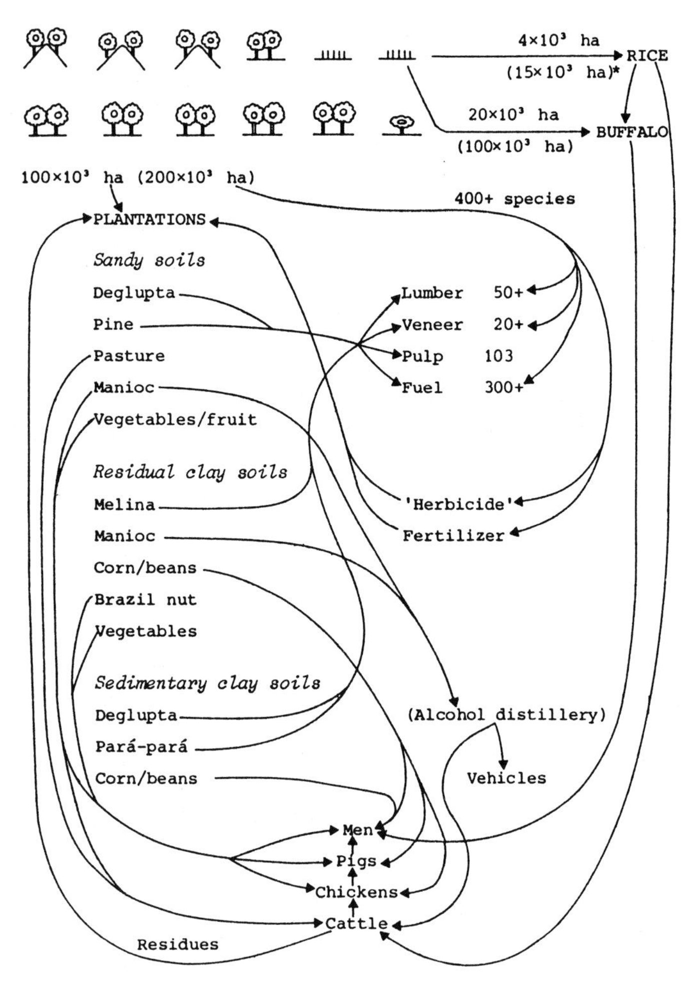

Fig. 9.10 Organization diagram of the Ludwig enterprise on the Jarí river, Brazil, as given by Briscoe (1983). Since 1982 this enterprise has been taken over by a consortium of Brazilian financiers, the diagram dates from 1980

Amazonian Euphorbiacea which had been cultivated since time immemorial, were added. 12,000 head of buffalos on 20,000 ha varzéa and later also under the pine plantations were meant to close this huge, oversimplified agrosystem cycle, which was also self-sufficient in energy by the use of fuel wood and gas oil from the distillation of wood. In 1981, the third generation of melina was growing, and it was claimed that this deep-rooting tree had mobilized so many new nutrients from the subsoil that the plantations could go on along a sustained wood production cycle. Since then, however, rumours have circulated that yields started to drop because of accumulated nutrient losses. The enterprise is now nationalized and seems to be used as a settlement site: there is no way to predict the consequences of this abrupt change in population density and probably also in agricultural methods, except for one. The rain forest is lost forever in this area. One can only hope that this very high price has not been paid in vain and that at least some of the serious problems of the rural population have been more than passingly solved.

Conservation

At the end of the 1970s, some signs became discreetly visible that South Americans realized both the risks of industrial-like development of the rain forests and the deceiving results of such developments. *Tropical Hardwood Utilization*, a book edited by Oldeman in 1982, contains a large section on resources. Chapter 8 gives the guidelines for Brazilian forestry policy from 1979 to 1985 by the Brazilian Institute for Forestry Development (IBDF), in which the "Objectives" contain terms such as "reforestation, compatible with natural resources" and the "Aims" include the institutionalization of conservation units, and their definition and implementation to fill existing gaps in ecosystem representation. The Special Environmental Agency of the Brazilian Ministry of the Interior published a brochure in 1977, entitled *Program of Ecological Stations* and described 19 conservation areas from a few hundreds to 500,000 ha each, 16 of them totalling more than 1.7 million ha. The Colombian PRORADAM project, aiming at land-use evaluation for a part of South Colombia, published *Principales plantas utiles de la Amazonia Colombiana* in 1979, in which wood-yielding trees are not the principal class of plants described. Many species are included which are useful for other purposes, for instance minor forest products. By 1980, however, a total of 129 areas had already been protected in Central America, nearly 5 million ha taken together. The largest of them are Darién in Panamá (575,000 ha), Amistad in Panamá and Costa Rica (450,000 ha) and Cayos Miskitos in Nicaragua and Honduras (110,000 ha), according to an IUCN report (1981).

It is difficult to predict the direction in which the development of tropical American rain forests is moving. In Brazil, between staunch opposition to the present kind of development and technological presumption, a clear-minded group of people is trying to find the very precarious balance between the quality of human life and the conservation of the teeming life of the forest. That the book by Goodland and Irwin (1975), and its serious warnings, has been very well understood and digested by Brazilians is proved by a Brazilian answer, the book *Amazônia; desenvolvimiento, integracao, ecolôgia*, written in 1983 by Eneas Salati et al. and published by Brazil's highest scientific authority, the CNPq (Conselho Nacional de Desenvolvimiento Científico e Technológico). Its first 144 pages give a good overview of the essence of Amazonia's ecology, while the sixth chapter occupies 183 pages that follow, and cover the "ocupacao humana". This chapter includes many data on the Amazonian Indians and their sad history and prospects. The problems, if not solved, are at least clearly stated here. Others, in a more polemical way, are working at revolutionary new designs for ecologically adapted settlements in Amazonia. Morais Pupa Nogueira is a professor at the School of Engineering Sciences of San Carlos, and an architect. According to a Sao Paulo newspaper of November 1983, he wrote a book on the *Urban-rural integration in the project 'Cidade Amazônica'*. "A proposition without utopian connotations", the paper adds in a subtitle. Nor is Brazil the only tropical American country where the consciousness of the forest problems is increasing. The report by Julio Cesar Centeno (1984), from the outstanding Instituto Forestal Latinoamericano in Mérida, Venezuela, bears a title which speaks for itself: *El recurso forestal y el drama económico de la América latina*.

Facing the immense problems of the tropical American forests under increasing human pressure and the necessity to design with the utmost care and respect each forest operation or each conservation decision so that they will last, the last word in this chapter might be given to the desperate statement by Prof. Nogueira (Folha Sao Paulo, Nov. 26, 1983): "In a flagrant and disrespectful attack upon Nature, they transform the forest into a desert after having used it as a ground for beans and fodder in rapid passing. Millions of years to construct the most prodigious factory of self-regulating life, with the highest density of life per cubic meter on earth. And the government, availing of the science of universities, plans, designs and finances its destruction".

Such pronouncements seem to have been heard. In 1984, the tax allowances for large cattle farming have been abolished in the Brazilian Amazon. It is to be hoped that this is a sign indicating a late 20th century in which both the neotropical rain forest and neotropical man find a durable and fruitful way of co-existence.

10 Malesia

Geography

The phytogeographic region known as Malesia lies north and south of the equatorial line for over one-fifth of the world's circumference (Fig. 10.1). The well-known Dutch writer Multatuli called it the 'Emerald Belt'. It consists of ten distinct subregions: Sumatra, Malaya, Java, Lesser Sunda Islands, Borneo, The Philippines, Celebes, the Moluccas, New Guinea and the Solomon Islands. In UNESCO's book *Natural Resources of Humid Tropical Asia* (1974), which contains several good papers, the article I called *Botanical Panorama of the Malesian Archipelago* reviewed all the important literature on Malesia available then.

Whitmore's *Tropical Rain Forests of the Far East*, published in 1975, also brought together and illustrated much information on Malesian forests. Kartawinata (1974) presents supplementary information.

Malesia differs from both Africa and Amazonia in two important geographic aspects: it has many mountains, often in the form of elongated ridges 2000–3000 m high, and it is very insular in form. Both features are responsible for the rather uniform humid climate, and the rain forest is therefore the natural climax vegetation throughout most of the region. But it is not only altitude and coastline that determine the actual extent of this vegetational type. Monsoon forests and savannas cover much of East Java, the Lesser Sunda Islands, Celebes and the southern parts of New Guinea, where the influence of the dry southeastern monsoon winds is most strongly felt. The east and west monsoons in this area are discussed in Chapter 3. The low-lying eastern coasts of Sumatra and the western and southern coasts of Borneo, as well as part of southwestern New Guinea are covered in swamp forest; peat forest and kerangas (heath forest) types are also common, especially in Borneo; scattered throughout the archipelago, limestone outcrops have their own, specialized flora. The effect of man as a limiting factor is seen most strongly on the highly volcanic and therefore mineral-rich land of Java and Sumatra; the rest of the archipelago is only fertile around a single volcano or in the alluvial plains.

There are two large centres of rain forest in Malesia. One, in western Malesia, consists of (in sequence of richness): Borneo, especially west of the central watershed, Malaya, Sumatra, The Philippines and Java. During the Ice Ages when the sea level was low and the shallow parts of the China and Java Seas were above water, this whole area, called the Sunda Shelf, or Sundaland, formed a continuous land mass. The forests of Sundaland are characterized by an enormously rich tropical flora and Java with a mere 4500 species of native flowering plants is by far the poorest of the lot, for which volcanism, greater aridity and man are responsible.

The second centre of rain forest is on the island of New Guinea, with extensions as far as the Solomons and northeast Australia. Interestingly, the flora here is distinctly Asian in character even though the mammalian fauna is clearly of Australian derivation. After the breakup of Gondwanaland, New Guinea was pushed northwards in front of Australia into the tropics, and by way of the geologically unstable area northwest of New Guinea, plants

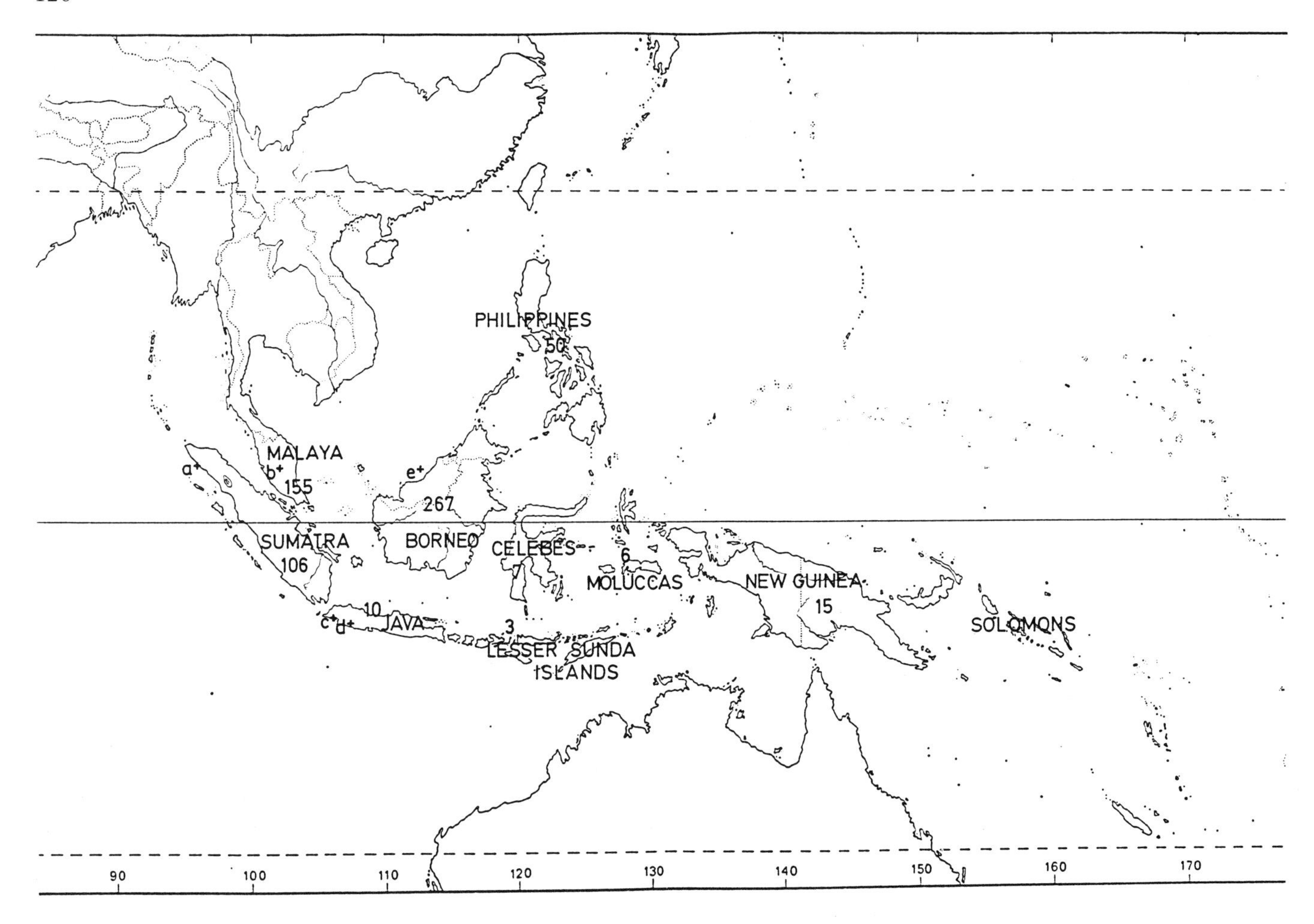

Fig. 10.1 Malesia (inside the circle), surrounding areas and some important localities: *a* Gunung Leuser Reserve in North Sumatra; *b* Pasoh in Malaya; *c* and *d* Bogor and Cibodas in West Java; *e* Andulau in Brunei, northwest Borneo. The *figures* refer to the number of Dipterocarp species in each subdivision of Malesia (After Ashton, *Flora Malesiana* 1982)

were able to migrate from Sundaland via Mindanao to the south and southeast through Celebes and the Moluccas and thence eastwards to New Guinea. There, a relatively poor temperate flora, which had for the most part succumbed in the heat of the tropics, was quickly superseded by the Malesian taxa. This was accompanied by a strong speciation, giving New Guinea its special character.

The search for primitive elements in the Malesian flora has been successful. Van Balgooy (1976) briefly surveys such taxa as Fagaceae, Magnoliaceae and Winteraceae; various thick-stemmed aril-bearing relics will be discussed in Chapter 13 (Evolution), but bottle-shaped trees like the baobab, which one can think of as super-pachycaul specialists, do not occur in Malesia.

Flora

The total number of Malesian species of all vegetational types is, according to Van Steenis (1948, p. X), between 25,000 and 30,000. The flora of Borneo consists of around 10,000 species, of the Malay Peninsula perhaps 8500 species, of Sumatra probably slightly less, of The Philippines around 7000 and of New Guinea 9000. About two-thirds of all these species are found only in the lowland forests. Such species numbers may not look very impressive,

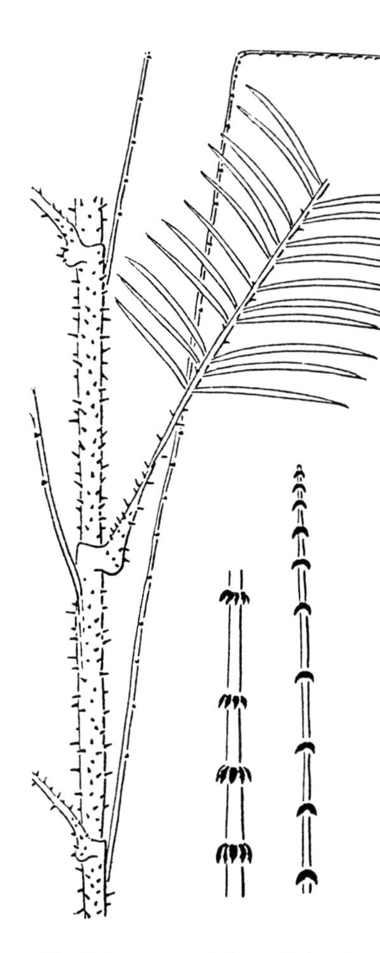

Fig. 10.2 A rattan, *Calamus* (Palmae), part of the stem with leaf and armed whips (flagella). In this case, these represent modified inflorescences, which are implanted on the sheath of a leaf obliquely below the 'knee' or node. The distance between each knee is ca. 30 cm. On the *right*, a central part and tip of a whip, enlarged, showing the recurved barbs by means of which the plant adheres to objects (Corner 1964)

but this is a conservative estimate, based on the 'wide species concept' accepted in *Flora Malesiana.* In fact, Malesian rain forests are the richest in the world in this sense. The palm flora of Malesia and surrounding areas is also the richest in the world with 1385 species in 97 genera (Moore 1973, p. 69) including 560 species of rattans. In Chapter 6 (Other Life Forms), some space has been devoted to rattans (Fig. 10.2); here, we just want to stress their enormous usefulness and value. Rattans are primarily denizens of virgin rain forest; only a few can be grown outside this habitat.

The parasite *Rafflesia*, with the largest flower in the world, is also mentioned in Chapter 6. This botanical curiosity is confined to West Malesia. The well-known pitcher plants, *Nepenthes* (Nepenthaceae), which occur from Madagascar eastwards through Ceylon and Malesia to Queensland and New Caledonia are represented by several scores of species in the archipelago. They prefer more open habitats. Bamboos are best represented in the more arid regions, especially in continental Southeast Asia.

Useful Plants

Sundaland deserves special mention as the homeland of many fruit trees. To mention only the best known: nangka or jackfruit, *Artocarpus heterophyllus* (Moraceae); campedak, *A. integer* plus 45 wild representatives in the forest; a dozen or so jeruk, *Citrus* (Rutaceae) and at least as many wild species; durian, *Durio zibethinus* (Bombacaceae) with about five wild relatives of which the fruits are also edible, and some 20 others; jambu air, *Eugenia aquaea* (Myrtaceae); jambu bol, *Eugenia malaccensis*, and other edible species of a genus of about 700 species; mangosteen, *Garcinia mangostana* (Guttiferae) and some ten wild relatives with edible fruits belonging to a genus of around 200 species; duku or langsat, *Lansium domesticum* (Meliaceae); mangga, *Mangifera indica* (Anacardiaceae) belonging to a genus of 23 species of which many have edible fruits; bananas and plantains, *Musa* (Musaceae), a rather large genus of which a dozen have edible fruits, many wild species occur in New Guinea; rambutan, *Nephelium lappaceum* (Sapindaceae) and kapulasan, *N. mutabile* and some 20 wild relatives; salak, *Salacca edulis* (Palmae) with 12 wild relatives. This is only a small selection; Loh (1975) reports that of the estimated 100 species of fruit trees grown in Malaya on an area of 50,000 ha (= 125,000 acres), 80 species are native, and nearly all come from the lowland forests of Malaya and Borneo. Indeed, the forests of Sundaland are the largest gene pool in the world for the cultivation of tropical fruit. This is 'point 7' in the Table of Values of the Forest in Chapter 16. In the light of the durian theory (Corner 1949), the reason for this is to be found in the very long and undisturbed co-evolution of plants and large mammals in western Malesia. Large mammals never reached Luzon or East Malesia.

I have not yet mentioned the spices. Cinnamon, *Cinnamomum* (Lauraceae) is probably originally a native of South India/Ceylon, but there are hundreds of wild relatives in the forest of Malesia and some of them are medicinally important; cloves, *Eugenia caryophyllus* (Myrtaceae) of the same family as the jambus is a native of the Moluccas, as is the nutmeg, *Myristica fragrans* (Myristicaceae), a genus of around 100 species, very common and characteristic of the Malesian forest; pepper, *Piper nigrum* (Piperaceae) also belongs to a genus of several hundreds of species.

The rain forests of Malesia are indeed exceptionally rich in useful plants, as testified by the monumental works of Heyne, *De nuttige planten van Indonesië* (1950), Ochse and Bakhuizen, *Vegetables of the Dutch East Indies* (1931, 1977) and Burkill, *A Dictionary of the Economic Products of the Malay Peninsula* (1935, 1966).

Furthermore, nowhere else are native plants so universally and extensively exploited as in Malesia. The resources are rich, and numerous generations of inventive people have over the past millennia accumulated and augmented their knowledge of the plants' useful properties. Upon entering the villages one is struck immediately by the varied and clever ways in which plant parts are being used (Fig. 10.3). In Africa plants are distinctly less intensively used, while in the Neotropics the new immigrants seem to have failed almost completely to take over the indigenous plant lore.

Since the climate of Malesia has remained essentially unaltered over a geologically long period of time, the rain forests here have never suffered from prolonged periods of drought, even when during the Ice Ages the altitudinal zones were lowered and arid areas must have been more widespread than now.

Fig. 10.3 Malesia: idealized landscape. In the *background*, a volcano surrounded by clouds. To the *left*, a rain forest tree with buttresses and epiphytes on its branches. Behind the trunk a tree fern, farther right a pandan, both pachycaul trees. The large flower in the *foreground* is a *Rafflesia*, normally growing only in closed forest. The liana on which it grows as a parasite is invisible. *Centre right*: Kapok trees, water buffaloes, wet rice fields. Between these and the volcano a clump of bamboo can be seen and coconut palms along the coast (*Flora Malesiana*, I, i, 4, 1948, frontispice)

Threats

The indigenous people, hunter-gatherers such as the Kubus in Sumatra, the Orang Asli in Malaya, the Punans in Borneo and small tribes in The Philippines, Celebes and New Guinea made little impact on the forests in which they live. The rain forest therefore remains a distinct climax vegetation in areas inhabited by such people. Over the millennia there has never been a hard to recognize mosaic of damage and regeneration. Large-scale destruction is all recent: by the type of shifting agriculture resulting from overpopulation, by mechanical logging and by the government of Indonesia's transmigration policy in which people from Java (where the population increases by 5500 day^{-1}) are resettled on other islands, often in places with poor or already exhausted soils. In The Philippines rapid population growth is also disastrous, and large tracts of rain forest there have fallen victim to illegal occupation. In these ways tens of thousands of square kilometres of rain forest throughout Malesia have been replaced by alang-alang, *Imperata cylindrica* (Gramineae): see Chapter 17 (Damage and Destruction). The situation in Indonesia is discussed more thoroughly in Chapter 19 (Forest and Man).

To conclude I would like to mention the elegant, comprehensive paper by the zoologist Furtado (1979), who discusses both the rain forest and its destruction patterns in West Malesia. This area is dominated by the important family Dipterocarpaceae to such an extent that a separate section must now be devoted to these distinctive trees.

Dipterocarps

The timber of the dipterocarp genus *Dryobalanops* has long been used to line the famous 'camphor chests' of the East, and several other genera yield resins of commercial importance as illuminants and ingredients in high-grade natural varnishes (minyak keruing and damar). But it is not for these uses that the dipterocarps have become in this century the *'enfants chéris'* of the loggers and timber merchants. The trunks of these mighty trees reach high into the canopy, and clean boles of 40 m are not unusual. The wood is hard, but not excessively so, and uniform in structure. They are easy to recognize and apparently plentiful. Because of these qualities about a quarter of all the commercial hardwood cut today consists of dipterocarp species – meranti (*Shorea*) and keruing (*Dipterocarpus*) have become well-known trade names – with the result that now, towards the end of the 20th century, dipterocarp forests have become the most threatened in the world. Nearly all of them, up to 800 m in altitude, have been leased to timber companies for exploitation; if none are reclaimed for conservation, these lowland forests will be gone, at the latest estimate, by the year 2005.

Malesia enjoys a quite extraordinary concentration of this family: of the 16 dipterocarp genera, 10 are Malesian or, at a species level, 386 of the family's 550 species grow in Malesian forests. Most of them are confined to western Malesia, and many to the island of Borneo, where 267 species grow of which 60% are endemic, i.e. they are only found on that island. In Malaya, 155 species are found of which 15% are endemic, Sumatra has 106 species with 10%

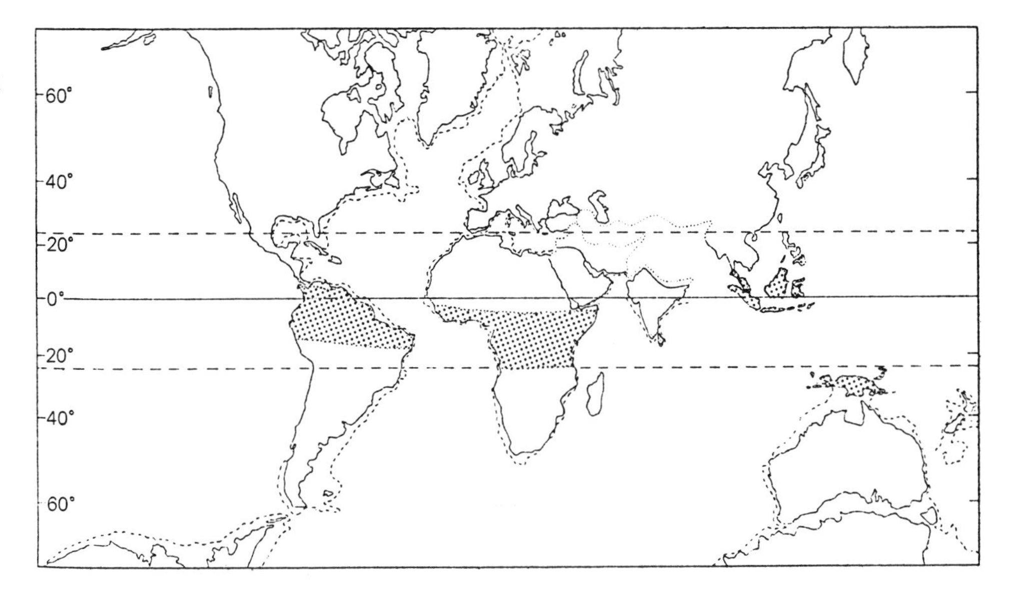

Fig. 10.4 Position of the continents at the beginning of the Tertiary, ca. 50 million years ago. India and New Guinea are still travelling north. The *shaded* areas are now all situated within 10° from the equator. Distribution of land and sea in the Malaysian Archipelago was not the same as now (Flenley 1979)

endemic, and The Philippines 45, of which 50% are, or were endemic: Philippine dipterocarp forests have been almost entirely logged.

East Malesia is much poorer in dipterocarps and they do not dominate the forests. Distribution elsewhere is relatively sparse. There are a few species common to the monsoon forests of India and continental Southeast Asia excluding the Malay peninsula, and Ceylon is relatively rich in them. But in Africa only one (more primitive) subfamily is to be found; the two genera: *Marquesia* and *Monotes* occur in savanna country. Another subfamily of only one genus with a single species occurs in South America.

It was formerly believed that the family, because so many species grow in Sundaland and so few elsewhere, must also have its origin there, but lately opinions have changed. The great dipterocarp specialist is Dr. P.S. Ashton, now at Harvard, who wrote the *Flora Malesiana* account of the family (Ashton 1982). Ashton suggests that the homogeneity of the Malesian dipterocarps points to recent evolution, and considers Africa to be their original homeland. In Cretaceous times India and Ceylon were connected to Madagascar, where there is still a closely allied family, the Chlaenaceae. The ancestral dipterocarps spread over this area. When plate tectonic movements separated India and Ceylon from Madagascar and moved them northeastwards to collide with Asia some 40 million years ago (Fig. 10.4), dipterocarps thus made their entry into Asia and continued their slow migration first into periodically dry forest, but then into the ever-wet forests of West Malesia, which they found much more to their liking, and in which they developed in such a spectacular way (Fig. 10.5). The earliest proof of their occurrence in Malesia is fossil pollen in Northwest Borneo more than 30 million years old (Muller 1970).

In West Malesia's constantly wet climate and on poor soil they have managed to build up the world's most differentiated forest type. More will be said about this in Chapter 14 (How Species are Formed).

Dipterocarp leaves are simple, firm, some 8–20 cm across and often perfectly pinnately veined. Young parts and sometimes the older parts, too, bear characteristic stellate hairs. The stipules, large and caducous, are distinctive of each species, and collectors must therefore look out for them. The flowers are five-merous with calyx and corolla, the latter typically contort (i.e., each lobe partly covers the other on the same side), and often asymmetric. There are

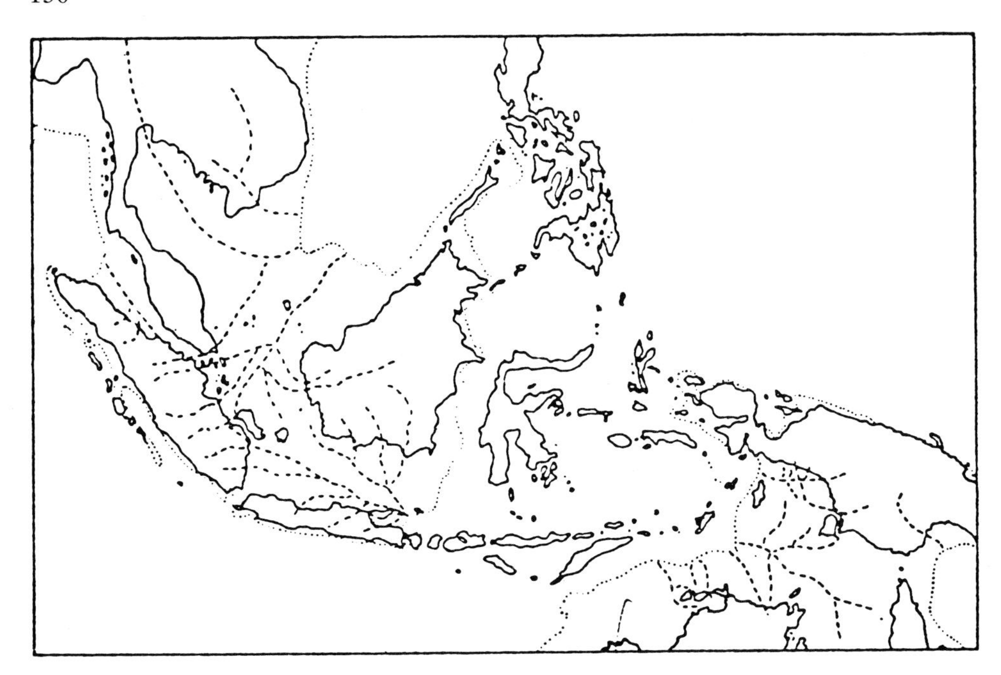

Fig. 10.5 Malesia with Sunda and Sahul shelf; inside the *dotted line* the parts that were dry land at the time of maximum glaciation

usually ten stamens, sometimes many more; they are small and bear a long appendage. In the characteristic dipterocarpaceous fruits the calyx grows out to cover the single-seeded fruit and provides it (with some exceptions) with long wings. Fruits of the genus *Dipterocarpus* have two wings, hence the name, *Shorea* fruits have three wings and *Dryobalanops* five.

All these organs are subject to variations in form while strictly adhering to the same theme: a complete botanical "*wohltemperiertes Klavier*". Complete fugues could be composed from the leaves of *Shorea* (Fig. 10.6), varying in size, thickness, venation and hairs; with countermelodies of fruits and stipules; a single fermata for a particularly large flower; and a tremolo for an aberrant calyx. And from time to time the recurrent melody of the flowers' exquisite scent.

A quite remarkable feature of this wonderful family is the fact that they flower so rarely and irregularly. Flowering, occurring at intervals of 5 to 9 years or sometimes longer, is probably triggered by a period of drought after a non-flowering year. Suddenly an enormous mass of flowers appears on every member of one species closely followed by the trees of another species littering the forest floor with their star-shaped corollas. Some species, belonging to other families, also take part in such dipterocarp 'festival years'.

A large fruit production usually follows in turn. The tengkawan or íllipe nuts are obtained from some Bornean *Shorea* species which grow along streams (Anderson 1975). The locals collect the nuts in the water with baskets and nets. The prices these nuts fetch are quite high because the fat obtained from them is excellently suited to the making of soap and candles. A dipterocarp year may therefore mean a bonanza for the local population, but in view of the irregular seed production there can be no stable market.

One might expect that such large-winged fruits could travel some distance on the wind. This is not true: the majority of the fruits land right under the parent tree. Distances of over 100 m are exceptional and the registered record is 800 m for species with the lightest fruits. It follows therefore that dipterocarps have had to disperse gradually over land, and that all the places where they now grow must at one time have been connected by land.

Fig. 10.6 a-g. *Shorea fallax* from Borneo. **a** Branch with inflorescence, 2/3 nat. size; **b** flower bud; **c** open flower; **d** petals, all × 4; **e** some stamens, shown from outside (*left*) and inside (*right*), × 15; **f** hypanthium with ovary and a few stamens, × 4; **g** fruit with calyx lobes grown out into wings, × 1.5 Herbarium specimen bb 29625, drawn by Amir Hamzah at Bogor for Dr. D. F. van Slooten

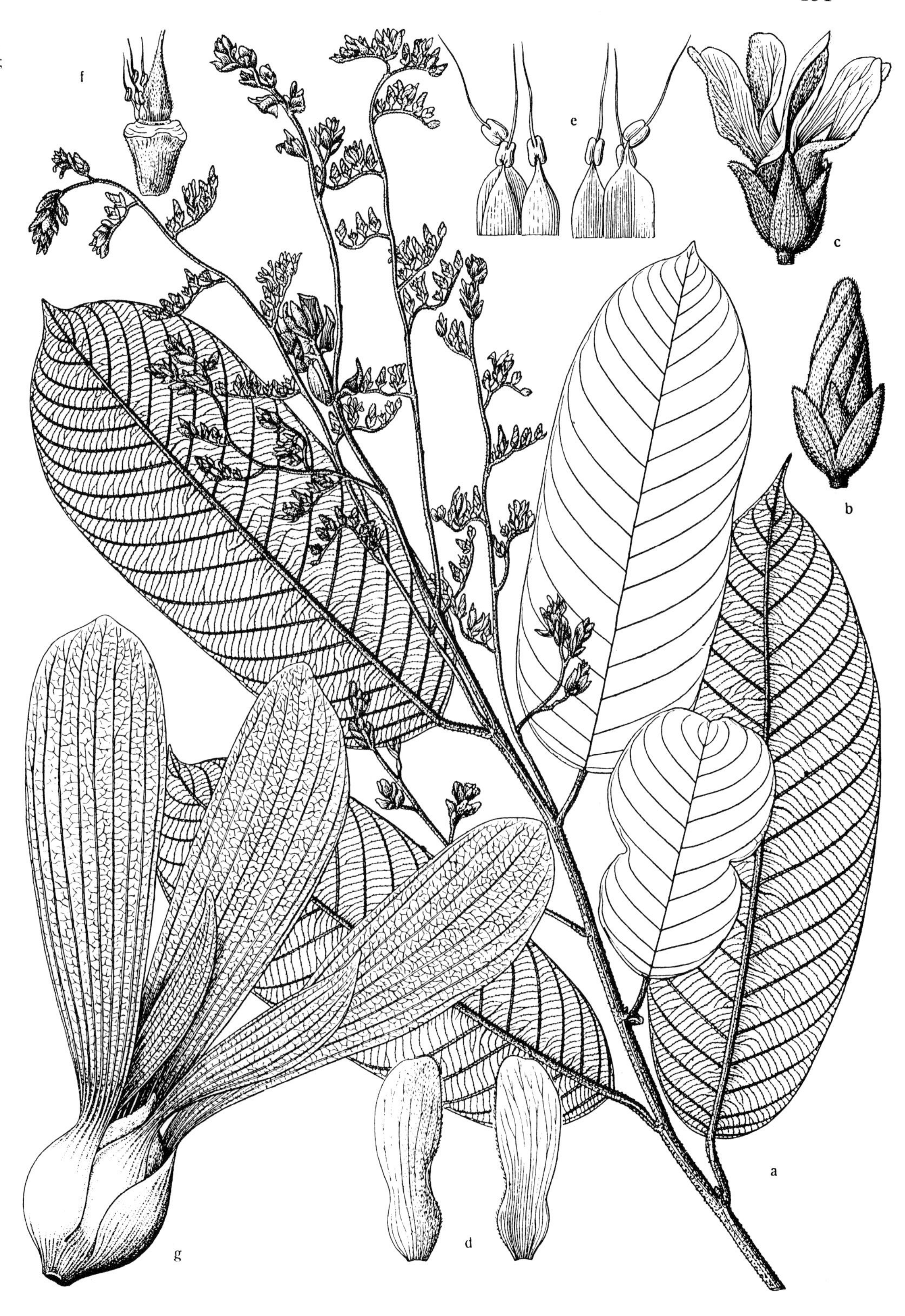

11 Tropical Africa

General

Rain forest covers at the most 9% of tropical Africa (Bourlière 1973, p. 280). As a consequence, comparatively little attention is paid to the rain forest in general studies of the vegetation of tropical Africa. Two important books, however, focussed attention on the African rain forest (Fig. 11.1), one by Aubréville (*Climats, Forêts et Désertification de l'Afrique tropicale*, 1949), and the other by Hedberg and Hedberg (*Conservation of Vegetation in Africa, South of the Sahara*, 1968). In both, the data are arranged by countries. In addition, a small book by Hopkins called *Forest and Savanna* (1965) provides an outstanding introduction to west Africa, in particular and Volumes 3 and 4 of *Introduction à la Phytogéographie des Pays Tropicaux* by Schnell (1976–1977) are an indispensable source of compiled data. The Netherlands were the last among the West European countries to start research on the tropical African rain forest – in 1955, at Wageningen, though a number of fine Dutch publications have appeared since then. I should add that the series of papers edited by Meggers in 1973 (*Tropical Forest Ecosystems in Africa and South America*), are excellent though on the whole are more of a biogeographic than ecological nature. This series includes the fauna which, unlike the flora, is as diversified as anywhere in the tropics.

Climate

Africa is a thirsty continent. As Richards (1973b, p. 24) noted, a distinct dry season occurs everywhere, even within the African rain forest limits of 9° on either side of the equator. For example, at Debundja in Cameroun with a mean annual rainfall of 10.170 mm, there is a 2-month period when less than 200 mm falls, while at Eala (Zaïre) no less than 17 rainless periods, each lasting 20 days or longer, were recorded between 1930 and 1952.

The total annual rainfall is also comparatively low. Aubréville (1949, p. 209) collected annual averages for marginal forested areas and found them to vary between 1250 mm and 1450 mm and described 1600–3000 mm as a "pluviosité très forte". He claimed, for good reasons, that 1400–1500 mm of rain may sustain a magnificent forest in Africa, but only if it is more or less evenly distributed throughout the year. It is the length of the dry season, which must not exceed 3.5 months at its very utmost, that will preclude the development of rain forest; annual rainfall, though a mean of at least 1350 mm is required is not the decisive factor. Past human interference also plays a part, of course. In Guinea, there is no rain forest at either Bayle or Kankan though both enjoy rainfall of over 1770 mm: it is in this region that the ancient empire of Ghana (no connection with the present state of Ghana) was situated, around 1100 AD. An area previously inhabited and deforested for a long period will remain savanna, and annual fires will ensure that forest trees have no chance to grow. Only large, completely abandoned savannas may develop into forest, though often of low quality (Letouzey, pers.commun.).

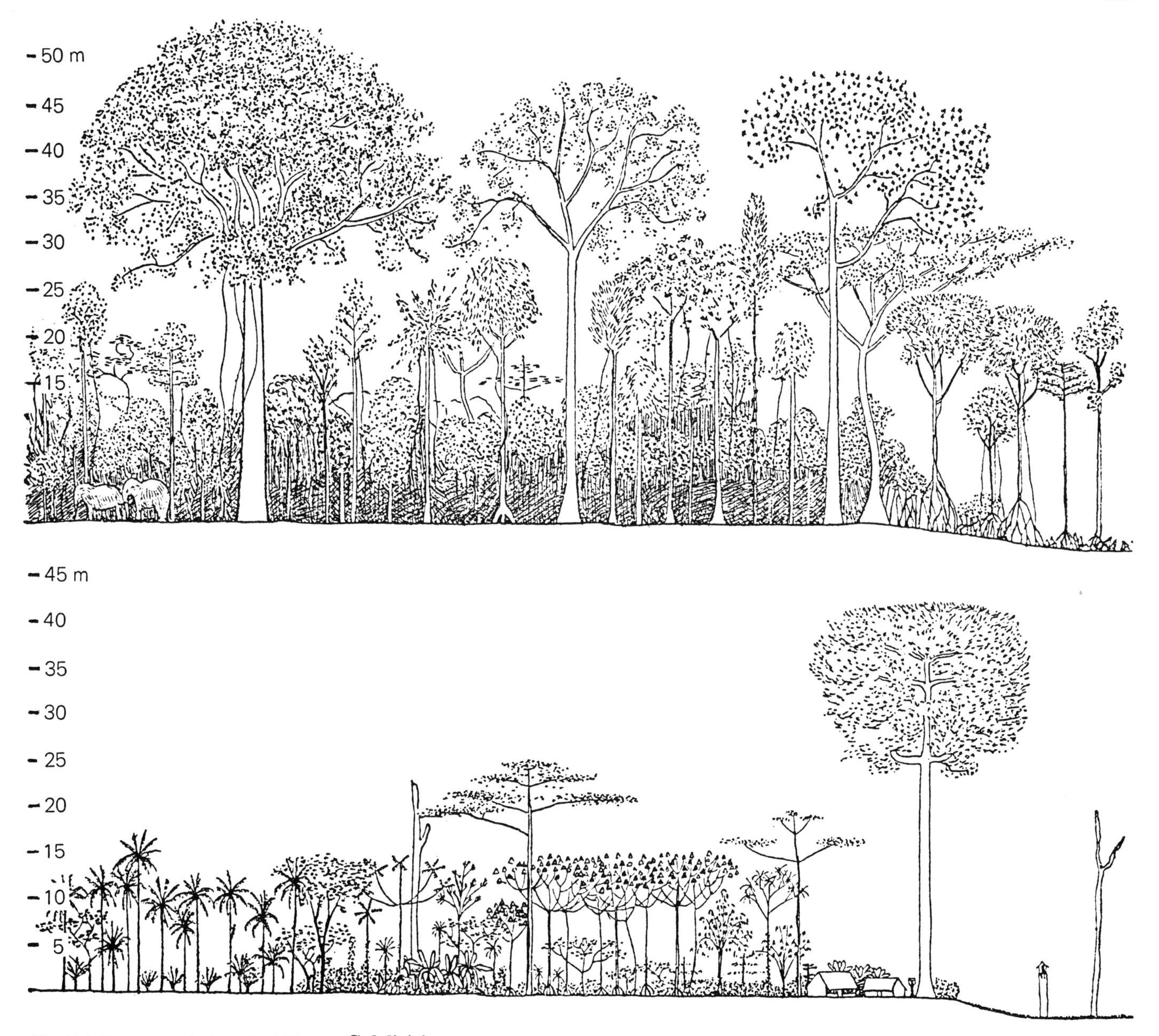

Fig. 11.1 Evergreen rain forest in Africa. *Above*: Original state, with a transition to swamp forest on the *right*. *Below*: Same place, after the forest was converted into cultivated land (Aubréville 1949). See also next figure (Aubréville 1949)

Subdivision

African rain forests can be divided into four distinct vegetation units. Even though it is impossible to find any two ecologists who agree on either the principles or methods to be employed in defining vegetation units, in the case of Africa, however, most authors seem to recognize Aubréville's delimitations although sometimes under different names. I will therefore follow Aubréville here. Aubréville defined:

1. Guinea Forest. An irregular zone bordering the Atlantic between Freetown (Sierra Leone) and Lomé (Togo), 1800 km long and 120–360 km wide. The eastern boundary is the Dahomey Gap. This is a zone of a drastically changed vegetation of apparently recent origin though recent surveys have shown that the number of species restricted to the area west of the Gap is greater than was supposed; Hall and Swaine (1981) are therefore of the opinion that the Gap is older than most previous authors thought.
2. Nigeria Forest. Like the former, a zone bordering the sea, between the western and eastern borders of Nigeria, 800 km long and less than 200 km wide, curving into the interior behind the Niger Delta.
3. Cameroun-Gabon Forest. In the west a narrow junction with the Nigeria Forest, in the east (where the Sangha River, at 16° E is the limit) a wider junction with the following fourth area. This forest is coastal and follows the equator, a somewhat irregular unit about 700 km in diameter.
4. Congo Forest. This unit extends from the Sangha River eastwards approximately between 4° N and 5° S of the equator. It is almost 1500 km long and 800 km wide, a rather irregularly shaped unit interspersed with extensive marshlands near the Congo River. The great lakes of Tumba and Mai-Ndombe form the eastern border.

Within the Congo Forest unit, there is the 400 km wide (14°–18° E) Sangha River Interval, where the mean annual rainfall remains below 1600–1800 mm, lower than that to the east or west. The boundaries of this Interval are hardly discernible by forest types, however, and the forest is not interrupted by it.

The origin of the Sangha River Interval is uncertain; possibly deviations of rain-bearing ocean winds may be involved.

Although the floras of the four areas just described are markedly different, they do have a large number of species in common, and altitudinal differences are far less frequently met with and far less important than in Malesia. A considerable part of all four is below 500 m and another large part does not rise above 1000 m.

The large tree species of the African rain forest were studied by Voorhoeve (1965), who closely studied their morphology and contributed many new biological data.

The Poverty of the Flora

Richards (1973b) pointed out in some detail why the African forests (Fig. 11.2) must be described as 'poor' when compared to the rain forests of Amazonia and Malesia. The *Flora of West Tropical Africa* (1954–1972) numbers about 7000 species, whereas on the Malay Peninsula, so much smaller in size, there are about 8500 species; both areas have been so well explored that further research is not expected to reveal significantly different figures.

Though many West African forest species occur throughout the whole of the tropical rain forest region, many do not, and species whose distribution is limited are less numerous than in Amazonia or Malesia; they are mostly found in the more humid habitats. As Richards points out, the number of species on 1 ha may be comparatively high, for example up to 109 species of trees with a trunk diameter exceeding 10 cm on 1.56 ha in Cameroun, but it does not follow that the African rain forest as a whole is rich in species. The rain forest of the Ivory Coast consists of about 600 tree species. De Koning's (1983) survey, *La Forêt du Banco* (a Forest Reserve near Abidjan), describes a great number of the trees found there.

Why some large pantropical plant families, e.g. Lauraceae, Myrtaceae and Palmae, are so poorly represented in tropical Africa remains an unanswered question.

Palm species are (or were) more numerous in Madagascar than on the whole of the African continent. If Europe and Arabia are included there are only 117 palm species (16

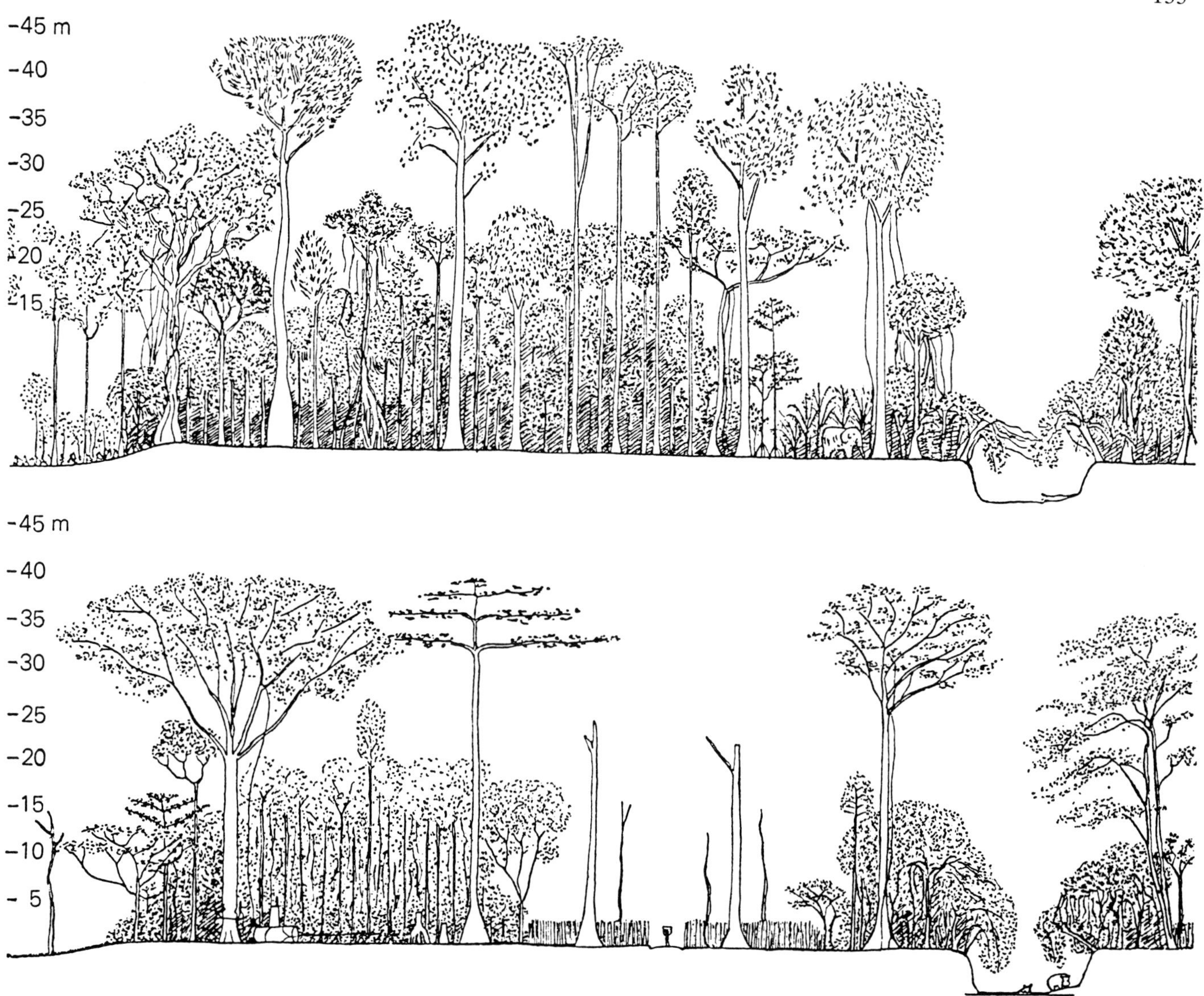

Fig. 11.2 Deciduous rain forest in Africa. *Above:* The original state. *Below:* The same place, after the forest was logged and used for primitive agriculture by triumphant man

genera) indigenous to this part of the world. Within the palm family this amounts to 4.2% of its species and 7.5% of its genera (Moore 1973, p. 69). The tropical rain forests of America contain 81 palm genera (ca. 1150 species), and of Australasia 97 genera (ca. 1400 species). There are no trunk-forming palms in African rain forests; the well-known oil palm is a denizen of the coastal swamps of West Africa and is not found in, and cannot survive in, closed rain forests.

As for orchids, most species of which are confined to the rain forest, the 403 species (58 genera) of west tropical Africa compares poorly with the Malay Peninsula's 927 species (107 genera), Holttum's (1953) estimate in *Orchids of Malaya*. While in Zaïre almost 400 orchid species have been discovered, far more than 5000 have been found in botanical Malesia. Obviously, the Orchidaceae are far less diversified in tropical African rain forests than in those of the Asian and American tropics. The flowers of African orchids are, moreover, generally very much smaller and less spectacular.

In addition, African forest is also comparatively poor in lianas and epiphytes (Richards 1952, pp. 110, 113). Bromeliaceae, a characteristic of tropical American epiphytic growth, are absent in Africa. In comparison to South America, species poverty is also true of the African savanna.

Apart from a considerable number of excellent timber-producing trees, very few useful plants of economic importance originate from tropical African rain forest. As already noted, the oil palm (*Elaeis guineensis*) is a marshland species, but the genus *Cola*, comprising ca. 60 species, is indigenous in the rain forests of West Africa; two species, *Cola nitida* and *C. acuminata* produce the Kola 'nuts' known worldwide for their stimulating properties.

Coffee, a species of *Coffea* (Rubiaceae), is also a rain forest plant but the best-known species, *C. arabica*, is found in Ethiopia, growing only in forested areas with a high rainfall above 1500 m in altitude.

Two causes for this natural poverty were suggested by Richards, who nicknamed Africa "the odd man out". The first cause was mentioned above: the climate. This already being relatively dry, only small changes in the past, e.g. during the Pleistocene glacial periods, could have had a strong effect on the flora. Alternating drier and wetter periods in the Quaternary are the possible causes for the disjunction of many forest species (West Ivory Coast – Cameroun – Zaïre) as was shown by Guillaumet's excellent research (1967–1971). Bourlière believes that, whatever came to pass, the north of Gabon and adjacent Cameroun always remained a humid refugium.

It is, however, not at all certain that the environmental changes in the past are entirely responsible for the differences which exist now, either locally in Africa or in the rest of the tropics. For example, *Nypa fruticans*, the common and typical palm of the Malesian mangrove swamps and in general of coastal Asiatic rain forest, is entirely absent from African coasts.

Its absence could be explained because *Nypa* was an element of the Laurasian coastal flora north of the Thetys sea. Fossil fruits are found in Western Europe together with other elements of that flora that do not occur in Africa like *Mastixia* and *Symplocos*. *Nypa* never has been able to pass this sea southwards and settle in Africa.

Human Influence

The second reason for species poverty in Africa, according to Richards and confirmed by all later authors, is man.

Africa is the continent where mankind began. New fossil discoveries point to an evermore remote African past for the origin of our species. It is an established fact that man has lived at least half a million years in equatorial Africa harvesting fruits and greens, hunting and burning the vegetation. "Even in the depth of the so-called primary forest there is often evidence of former human occupation in the form of pottery and charcoal fragments in the soil", Richards wrote (1973b, p. 24), and he concluded that human influence on the African forest has been so strong and so continuous that no African rain forest can be accepted as untouched or 'primary'. Evergreen high forest in Africa, Richards alleged, is as a rule old secondary forest. Several tree species, in particular the highest trees, remain without reproducing themselves, presumably because the closed canopy prevents the germination of their seeds which require light; because of this, all African forests are in some way transitional, not only with rêgard to their structure – rain forests are always changing their structure, everywhere – but in composition.

In the past, when the population remained sparse and poorly equipped, the forest at least had a chance to recover from the damage done to it. But man became the deadly enemy of the rain forest when the whites introduced the medical means to save human lives and organized large-scale settled agriculture, plantations and timber exploitation. When roads and bridges were built, this made the forest even more accessible and vulnerable.

In this century, man has so violently attacked the natural vegetation of Africa that it has justly been named 'the devastated continent'. Furthermore, according to Miracle (1973, p. 340), 73 species of exotic animals and crop plants were introduced into Zaïre (Congo) between 1830 and 1960. Nevertheless, considerable stretches of rain forest in Zaïre are as yet intact, and accordingly in need of protection.

In Nigeria, the most densely populated African country, 90% of the whole rain forest area (Aubréville's second 'domain', see above) had by 1968 (l.c. p. 92) been converted into farming grounds or was being destroyed by lumber companies; of the original rain forest nothing was left outside the boundaries of the 11 small reserves, each between 130 and 259 ha, sad remnants of what once existed.

Where destruction of the forests proceeds rapidly, sometimes even the forest reserves do not escape. New reserves have, occasionally, been established, but too few and far between to improve the overall picture significantly.

In 1959, sponsored by UNESCO and AETFAT (the organization of all botanists studying the African flora and vegetation), Keay published a vegetation map of Africa south of the Sahara. A new map for the whole of Africa, sponsored by the same organizations, appeared in 1983, edited by White.

In the preceding pages the main unsolved or partly answered questions concerning the origin, nature and future of the African rain forests have been roughly outlined. Much research is still going on, but it cannot yield results quickly enough to keep ahead of new and pressing problems.

A few considerations of a general, philosophical nature come to mind when one ponders over the rain forests and the part they play in the relations between man, plants and animals.

People usually try to understand the life and habits of the rain forest by studying environmental factors and their effects; four factors, to be exact. These four are the cosmic elements of classical Greek science: air (here: the properties of the winds reaching the forests), water (the amount and distribution of falling rains), fire (sunshine and temperatures, daily and annually) and earth (the mineral composition and structure of the soil). This classical cosmic approach seems indeed worthy of the majesty of the rain forest.

Another analogy could be made. A living organism is dependent on its environment. The increasing complexity and differentiation of the animal body, from lower to higher animals, results in a proportionally reduced direct influence of environmental factors. Mammals have a balanced internal physiology that maintains itself in dynamic equilibrium which admits environmental supplies when required and counteracts environmental threats or damage. Such exactly is the way of life of a tropical rain forest, that astonishingly vigorous community of living organisms. Tropical rain forests and mammals can, in their own spheres, each be seen as a peak of evolution.

If we accept this metaphor for the sake of argument, it means that the rain forests' internal ecology displays equilibrium and is a dynamic and balanced life community with a measure of independence from outside influences, able to bring about a selective interchange with its surrounding environment. Is it desirable, unavoidable perhaps, to approach some questions and lines of research from this point of view, if one tries to obtain a deeper insight?

12 Relationships of Plants and Animals

The Numbers of Animals

Under normal circumstances, a botanist does not encounter many animals. His interest is focussed on other things and his activities are generally diurnal. Yet he naturally meets the many signs of animal activity: termite mounds, nests, tracks, dung, the opened shells of hard fruits and other traces of animal feeding. If he moves about cautiously, he need have no fear of snakes. Mosquitoes and leeches are merely familiar inconveniences, though approaching too close to a wasp nest is an ever-present occupational hazard.

In fact, the biomass of animals in the forest is not substantial. Medway (1978) gives the following estimates for mammal densities in a Malayan forest: diurnal squirrels, 0.75 ha^{-1}, and the same for nocturnal flying squirrels; bats, 5–7.5 ha^{-1}; the six species of higher primates together, 0.5 ha^{-1}. His cautious and conservative estimate for the entire biomass of mammals therefore amounts to a mere 10 to 15 kg ha^{-1}, or the equivalent of about three European house cats. Bourlière gives a somewhat higher figure for Panama: 8.25 kg ha^{-1} for the ground-dwelling mammals, and 9.5 kg ha^{-1} for arboreal mammals. It must be realized, however, that these figures represent the estimated biomass of the scientists' study areas, a small island in the ocean of the rain forest where no hectare is the equivalent of another. A rough comparison of biomass estimates of only higher primates in different study areas of the rain forest results in estimates ranging from 0.2 to more than 7 kg ha^{-1}; while the density of diurnal squirrels may be as high as 10 to 12 kg ha^{-1}.

The following Table gives some data on the numbers of large mammal species commonly found:

Area	Total No.	Bats	Rodents
Malaya (Medway 1978)	206	86	55
Panama (Hershkovitz cited by Bourlière 1973)	196	100	48

The numbers of bird species recorded in two extensive rain forest areas give a rather similar picture (Amadon 1973).

Area	No. of Genera	No. of Families	No. of Species
Congo basin	15	37	212
Amazon basin	16	49	529

Amadon also refers to figures from data obtained in western Brazil over a long period of time: in 250 ha rain forest at 350 m altitude, he observed 320 species of birds, of which 210 were true rain forest species (or possibly 225).

For New Guinea forests, Schodde (1973) arrives at a total of 620 species of land and water birds (freshwater only), of which 45% or 200 species were confined to rain forest habitats, while an additional 25% or 155 species were frequent rain forest visitors. Schodde emphasizes New Guinea's great variety of avifauna and relates it to the great variety in forest types. The diversity of bird species, which can be monitored more easily and quickly than floral diversity, he considers to be indicative of the richness of species in other taxa. If this should be proved correct, it would give biologists a convenient criterion for the evaluation of habitats or areas for conservation purposes.

The Distribution of Animals

In the rain forest a high percentage of the animals other than birds are arboreal. According to Davis (1962), 45% of the non-flying, non-gliding mammals of northern Borneo are arboreal, as compared to only 15% in the temperate forests of Virginia (USA). That such a large number of mammals, roughly half the total number, lives in the canopy is a typical characteristic of the rain forest.

Several authors present evidence that certain animal species are restricted to particular altitudinal zones. Thus, Schodde (1973, p. 130) mentions for New Guinea several bird species that are characteristic of a particular zone, as well as for particular height zones within the forest structure (see the following Table):

Number of bird species in primary forest in New Guinea

Altitude above sea level	in the canopy		on the forest floor
	high	low	
0–200 m	7	8	7
600–1200 m	8	8	5
1400–2100 m	8	8	7
2300–3200 m	8	8	5

But comparative data on species density at different heights from the ground are still insufficient, even though the way many mammals occupy the three-dimensional space of the forest and the temporal distribution of their activities are both well documented. Medway (1978) notes that the majority of forest animals are active only at night, notably the bats, tree shrews, lorises, the flying (gliding) squirrels, rats, porcupines, civets, small cats and mouse deer. The ground shrews, the other squirrels and all the monkeys and apes are active during the daytime. Insectivores, wild dogs, bears, martens, the larger cats and the ungulates cannot really be classed as diurnal or nocturnal since their greatest periods of activity are typically dusk and dawn (Fig. 12.1).

The figures would seem to indicate that species diversity, at least insofar as avifauna is concerned, does not vary much with the altitude zone; indeed, species diversity and the correlated relative scarcity of any one species of flora or fauna in a particular area is a rain forest characteristic. This creates problems in the exchange of information (communication) between animals of the same species just as it does for the exchange of genetic material between plants. Not only must one animal cover large distances to meet another of the same species; they are both inhabiting a place where visibility is severely limited, usually the leafy crowns of trees. For this reason, sound replaces sight for the most part, and the rain forest cacophony is a startling contrast to the 'whispering pines and hemlocks' of the temperate zone. One could set one's watch by the sound of the six o'clock cicada; other cicadas imitate strangely the sound of a chain saw (that unfortunate harbinger of some ill-advised form

Fig. 12.1 Most conspicuous animals in the rain forest in West Malesia, located in time and place (After Whitmore 1975)

of 'progress') or of a pizzicato on a heavily-rusted piano string, and frogs too make an extraordinary variety of sounds. The birds on the whole, despite their diversity, contribute little to the concert of meowing, but not by cats, or bellowing, but not of hounds, or the sound of a siren here or there a rasping, drumming and hooting. And the same sound may stop suddenly in one location only to begin equally suddenly in another. In one's tent at night one is surrounded by sound, a chorus only subdued, with obvious reluctance, by the advent of heavy rainfall. The analogy of a rain forest to a Gothic cathedral must have been partly inspired by this massive, seemingly inexhaustible production of sound by creatures proclaiming honour to the Great Conductor from their every niche and territory.

An Overview

The rain forest owes its complexity to the opulent diversity of the biological relationships between its plants and the resident animals, notably those significant to the fulfillment of the life cycles of either party. The roles played by mammals, notably primates and bats, and by ants in these relationships will serve as an example.

A crucially important pattern of relationships which supports the animal world is called a food chain, or, rather, a food web. At the beginning of any such chain or web one invariably finds plants, since only plants via chlorophyll can utilize light (energy) to transform inorganic matter into organic tissue. The organic compounds thus formed are subsequently transferred from small plant-eating animals to large meat-eating predators who in their turn, by dying, return the matter with the help of decomposer organisms to the soil. But plants are not only used by animals as food; animals, particularly in the rain forest environment, are the pollinators of plants and the dispersers of their seeds. Furthermore, the interactions between plants and animals in the rain forest environment is continuous, i.e. there is no seasonal inactivity, so that plants, though needing and using animals, also need to protect themselves

Fig. 12.2 Nangka, or jackfruit, *Artocarpus heterophyllus* (Moraceae), on a small farmyard in Jambi, Sumatra. See text (Photograph MJ)

and their genetic material, the fruit, from wastage. The constant and intricate play of biological attraction and repulsion is vitally important to the life of the rain forest plants. Let us look at a few examples.

Big, Free-Hanging, Odorous Fruits

Cauliflory, or the bearing of flowers and fruits on the trunk of a tree rather than at the ends of the branches, is a rain forest phenomenon we have discussed before (Chap. 5). It can occur on both tree and liana life forms, and the advantage seems to be that the plant can produce larger, heavier fruits on such stout supports and that these are also being produced where the larger, heavier animals can more easily reach them. The biggest of all such cauliflorous fruits is the jackfruit or nangka, *Artocarpus heterophyllus* (Moraceae), a member of the big *Artocarpus* genus whose 47 species are indigenous to Asia from the Indus River valley eastwards to the Solomon Islands, though some species have been found to be so useful to mankind that they are now found pantropically. The jackfruit probably originated in the rain forests which once covered southern India; currently, it is cultivated throughout southeastern Asia (Fig. 12.2). The tree rarely grows above 15 m in height; the yellowish-brown fruits borne on the trunk and heavier branches may attain a length of 100 cm and a weight of 50 kg when ripe, a process that takes about 8 months. The entire fruit is covered with short spines, but as it approaches maturity and begins to give off its telltale and distinctive odour (similar to overripe bananas) the spines are insufficient armour against animals. At this point the human owners of the trees protect the fruit by wrapping it in jute or plastic bags to ward off the attentions of the kalong, *Pteropus vampyrus,* the largest fruitbat, with a wing span of about 1.50 m. These giant bats roost during the daytime in high trees, usually many thousands together. They hang, typically bat fashion, upside down in the blazing heat, every now and then fanning themselves with their wings to cool down. At dusk they set off in great flocks, flying with their heavy, slow wing beat to a distant location many kilometres away, in search of particular flowering trees from which they eat the pollen, and others of which they eat the fruits.

Another spectacular and highly prized fruit is the durian, *Durio zibethinus* (Bombaceae). Durians grow to 40 m in height, and the fruit may reach a maximum length of 30 cm; a veritable box fruit with its five sturdy sections covered with heavy spines, so sharp that one can hardly carry the fruit in unprotected hands. The durian fruit also develops on the trunk and heavy branches of the mother tree; it smells quite strongly, but the taste is unforgettable. Many Westerners loathe it, but Alfred Russel Wallace thought it worthy of a special voyage to the Asian tropics. The seeds (up to 4 cm in length) are embedded in a pale-coloured, unguent-like layer of some 1 cm thickness, which is the delicacy. The genus *Durio* includes 27 species of which 19 are found in the wild in Borneo, 11 in Malaya and 7 in Sumatra, but of these, none are cultivated, while *D. zibethinus,* the single cultivated durian, is not found in the wild. Nevertheless, several of the wild durians are attractive to man as well as other animals; Corner (1964) has described the competition which occurs during the fruiting season between the elephants, tigers, tapirs, deer, rhinos, monkeys, bears and squirrels crowded around the trees, unless some forest-dwelling humans have camped there first. *D. dulcis, D. kutejensis* and *D. oxleyanus* are all delicious, and it is surprising that they are not cultivated. For such large, heavily armoured fruit the assistance of animals for dispersal is vital; even the slow and lowly tortoise is, by report, of assistance to *D. testudinarium* of Borneo, which bears its fruits at the very base of its trunk (see the photograph in Corner 1964, Pl. 41 c and d)

As in the jackfruit, the creamy or slightly resilient, edible layer which surrounds the seeds is in fact the aril, and the aril as well as a similarly structured layer called the sarcotesta, is usually the edible part of a tropical fruit: the mangosteen *Garcinia mangostana* (Guttiferae), the duku, *Lansium domesticum* (Meliaceae) and the rambutan, *Nephelium lappaceum* (Sapindaceae) are all examples, and there is also the aril of the nutmeg which is the spice called mace, bright red in its fresh state and more fleshy than the dried product would indicate.

Fig. 12.3 Petai, *Parkia speciosa* (Leguminosae-Mimosaceae), free-hanging capitula on which the pods develop. Malaysia, where the tree is native (Photograph MJ)

The forest animals are not only useful dispersers. The pollination of the flowers, equally vital, is also performed by animals since wind pollination plays little or no role. The relationship of the durians to a single species of bat, *Eonycteris spelaea,* responsible for pollinating an estimated 45% of durian flowers was only discovered in 1976 by the patient research of the Malaysian scientists, Soepadmo and Eow. In the same year, Start and Marshall reported that the only roosting site for this tiny bat in Selangor Province appeared to be the Batu Caves near Kuala Lumpur, which would mean that the bats may have to extend their feeding trips as far as 38 km from their daytime roosts.

Many other pollinators were at first suspected since durian flowers open late in the afternoon and by their strong smell attract all sorts of insects to them. It was discovered, however, that the stamens are only receptive around 8 PM when it is dark and diurnal species such as bees are asleep or inactive. Bats other than *Eonycteris* were suspected, but these only ate or damaged the flowers without pollinating them. Then, after *Eonycteris spelaea* had been identified as the sole pollinator, the other 55% of durians being self-pollinated, the puzzle was to find out on what the bat was feeding during those parts of the year in which there were no flowering durians. There appear to be two fruiting seasons for durians in Malaya, but it is doubtful if individual trees fruit biannually: in northern Sumatra it took individual wild durian trees between 4 and 6 years to produce successive crops. The bat is, however, versatile, as a thorough examination of its faeces has revealed, and feeds on the pollen and nectar of a wide variety of forest species: *Artocarpus* species, *Bombax valetonii* and *Ceiba pentandra* (both Bombacaceae like the durian), *Duabanga grandiflora* (Sonneratiaceae), *Oroxylum indicum* (Bignoniaceae) and *Parkia* (Mimosaceae). With the exception of *Ceiba,* the kapok tree introduced from South America and now thriving beside its Asian cousin the *Bombax,* and of *Oroxylon,* which is a pioneer species, the other food sources of *Eonycteris* are truly rain forest trees, confined to that environment. The inescapable conclusion is that in order to continue to taste the durian, that most famous delicacy of the east, we must preserve the natural forest species which feed the bat when durians are not in flower. It is, of course, equally important to preserve the Batu Caves which, if some authorities had had their way, would now be mined for limestone for cement production. That part was preserved for the bats is due to the strong stand on the subject taken by the Malayan Nature Society.

Another spectacular, highly prized fruit, though borne on thinner branches, is the *petai, Parkia speciosa* (Leguminosae-Mimosaceae) (Fig. 12.3). The tree can attain a height of 45 m and has a flat, umbrella-shaped crown and can be found growing in any lowland forest throughout Malesia; it is also commonly cultivated. The inflorescence is robust, about 5–8 cm in diameter and suspended on a long, heavy stalk about 50 cm long. The yellow flowers smell like sour milk and are pollinated by bats, thereafter forming bunches of broad, hard, green pods up to 50 cm in length. These are sold on the market as a spice or condiment. The seeds have a distinctively strong odour, so much so that like garlic it eventually permeates the body of the eater. It is held in much esteem in the local cuisine.

Another method of attracting animal attention in the rain forest is by a display of eye-catching colour. To a certain extent, some temperate species do so as well: an example is the spindle tree, *Euonymus europaeus* (Celastraceae), a shrub several metres high commonly found along the dunes of European sandy coasts. Its fruits are quadrangular and blunt-angled, about 1 cm in diameter and brick red. In October, they burst open to reveal whitish seeds covered with a brilliant orange aril that hang suspended from the capsule by short 'threads', the funiculi or umbilical cords. The bright colour contrast of the orange against the dull red and their free-hanging suspension are undoubtedly adaptive traits, and one would expect that this shrub surely originated in the tropical rain forest.

Fig. 12.4 Fruit of *Sterculia* (Sterculiaceae), free-hanging, with deep-red valves and shining black seeds of 1 cm

A bigger and better example from the actual rain forest, where everything is bigger and more impressive, are the fruits of the genus *Sterculia,* represented by 50 species in Malesia and about 25 species in other regions (Fig. 12.4). Sterculias are robust trees with firm, rather elegantly formed flowers; the fruits, however, consist of one somewhat fleshy carpel that opens along a single suture, gaping widely to display the shining black-clad drupes against the carpel's brilliant red (Corner 1964, Pl. 24). Birds apparently like these fruits: next in importance to orange, red and black, yellow also attracts birds who feed more by eye than by taste. In *The Life of Plants* (1964), Corner has included photographs of the colourful yellow and black fruit of the Connaraceae, a family of lianas closely related to the Leguminosae in which there are so many fruits with amazing colour contrasts. *Abrus precatorius* 'beads', one-third black and two-thirds bright red, are one example. The Meliaceae family, however, more often sports red-white contrasts.

These examples are not isolated oddities, indeed, there would be little point bringing them up if they were. Along with the already mentioned different types of box fruits and berries, their freely suspended, colourful or odorous seeds are typical of many rain forest genera.

The Man of the Forest

The Malay word for man is *orang,* for forest *hutan.* The ape of that name, endemic to Sumatra and Borneo (Fig. 12.7), does spend most his life in the trees of the rain forest. Those who have ever had the privilege of watching orang-utans in the wild, are likely to disbelieve their eyes. The Dutch biologist Rijksen studied the role of this large, red-haired ape in the forest, and devoted some 450 pages to a description of his way of life, entitled *A field study on Sumatran orang-utans* (1978).

An infant orang-utan has a long, well-cared-for youth; his mother carries him for at least 2 years, during his first months of life, almost continually supporting him, though later the infant ape can cling by himself to her long reddish coat. Only after the age of five does the youngster begin to explore self-reliance, while retaining more or less close contact with his mother until his seventh or eighth year; a remarkably long period of 'education' for a wild animal.

Orang-utans find some 51% of their food plants in the forest canopy, usually between 10 and 25 m above the ground, and consequently spend most of their time in that zone. They also rest and sleep at that height, in nests constructed each night of branches and twigs. Rijksen observed that the orang-utans of the Gunung Leuser National Park near the Ketambe River utilized 92 different species of fruits, 13 different kinds of leaves and vegetative plant parts such as sprouts, bark, bulbs of orchids, of 22 other species, altogether comprising 114 species of plants. These belonged to 85 genera in 46 families, of which 21 species were Moraceae, 10 Euphorbiaceae, 6 Annonaceae, 5 Meliaceae, 4 Asclepiadaceae, 4 Leguminosae and 4 Sapindaceae. Of these plants 52% were tree-like in life form, 28% lianas, 8% epiphytes, 7% strangling figs and 4% herbs. Rijksen estimated that about 90% of the food plants are to be found predominantly in the primary, i.e. climax, rain forest growing on alluvial plains; for this reason, it seems unlikely that orang-utans would survive in completely secondary forest, a hypothesis which appears to be supported by observations. About 52 species, almost half the total amount, are important food plants; 12 of these being rain forest giants in which several apes may find enough food to sustain them for periods up to 3 weeks. 42% of the tree species, however, only provide bulk food for a short period of time and the remaining species are only supplementary sources. In the equable climate prevailing in their entire distribution range (i.e. North Sumatra and large tracts of Borneo between 6° N latitude and 3° S latitude),

Fig. 12.5 Orang-utan opening a durian with a stick (After Rijksen 1978)

Fig. 12.6 Orang-utan regaling himself on wild figs, his main food, the fruits of which develop directly on the trunk (cauliflory) (After Rijksen 1978)

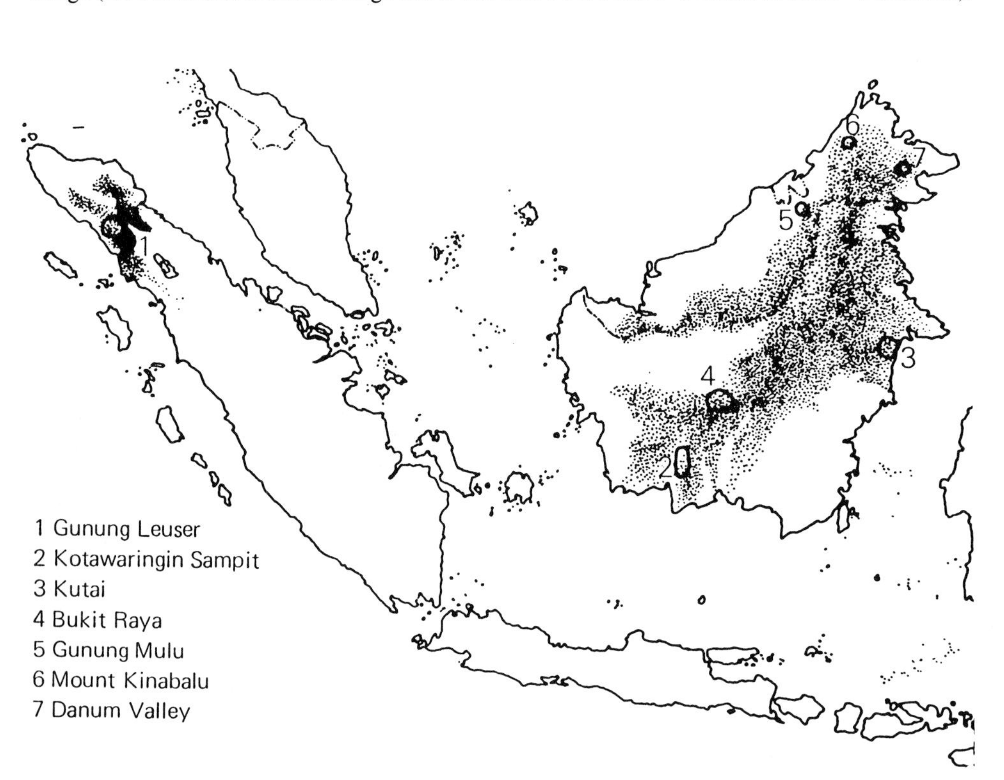

Fig. 12.7 Distribution area of the orang-utan (Rijksen 1978)

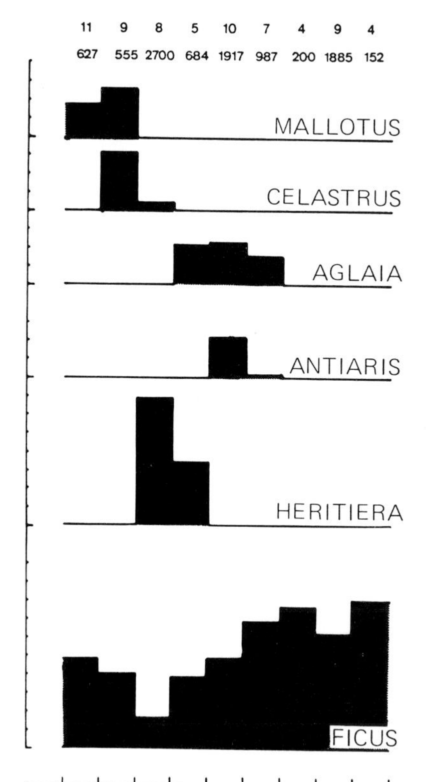

Fig. 12.8 Main sources of food for orang-utans in North Sumatra, over several months; note the large part of *Ficus*, and of *Heritiera* in the 1 month in which *Ficus* is scarce (Rijksen 1978). *Upper row of numbers:* utilized foodplant spp.; *lower row of numbers:* orang-utan × feeding duration

the apes can find some food all year round (Fig. 12.8). In the Ketambe area during the leanest month of the year, a giant *Heritiera elata* (Sterculiaceae) tree provided a steady, nutritious meal for the orang-utans; its nut-like fruits had such a hard skin that only orang-utans could crack them. The apes broke off whole branches laden with the fruits, taking them away from the tree in order to eat them at leisure, undisturbed by competition, sometimes as far away as 150 m from the source. Yet the bulk of the orang-utan's diet consists of highly nutritious fig fruits, which happen to be available almost the year round, as more than eight different species of, mainly strangling, figs provide the fruits favoured by the apes.

Orang-utans are, of course, thoroughly familiar with the forest in which they have been born and raised; Rijksen saw them make special detours to inspect a fruit tree on its state of ripeness. They were also seen watching the flight patterns of the larger hornbill birds who favour the same fruits. If they noticed many hornbills consistently flying in one particular direction, they set out to follow their course in the hope, or knowledge, of finding a plentiful source of food, usually ripe figs. (Fig. 12.6) The durian, described in Chapter 13, is a fruit highly favoured by both men and orang-utans, yet it poses a special challenge to the intelligence because of the tough spiny shell. Only fully adult orang-utans were capable of handling the fruits without preliminaries; apparently the hardened skin of their hands had become impervious to the sharp spines. Young orang-utans had to recruit all their ingenuity in order to open the fruit. Holding it firmly in both hands the ape grasps one or two spines between his spade-like incisor teeth and pulls the spine out of its tough fruit wall, repeating the procedure until he can bite through the wall without injury to his lips and gums. Young apes sometimes used sticks and dead leaves to fix or hold the fruit (Fig. 12.5), while domesticated orang-utans, and those being rehabilitated to fend for themselves under wild conditions, readily applied acquired skills by using sticks or stones to blunt the spines and penetrate the fruit's armour.

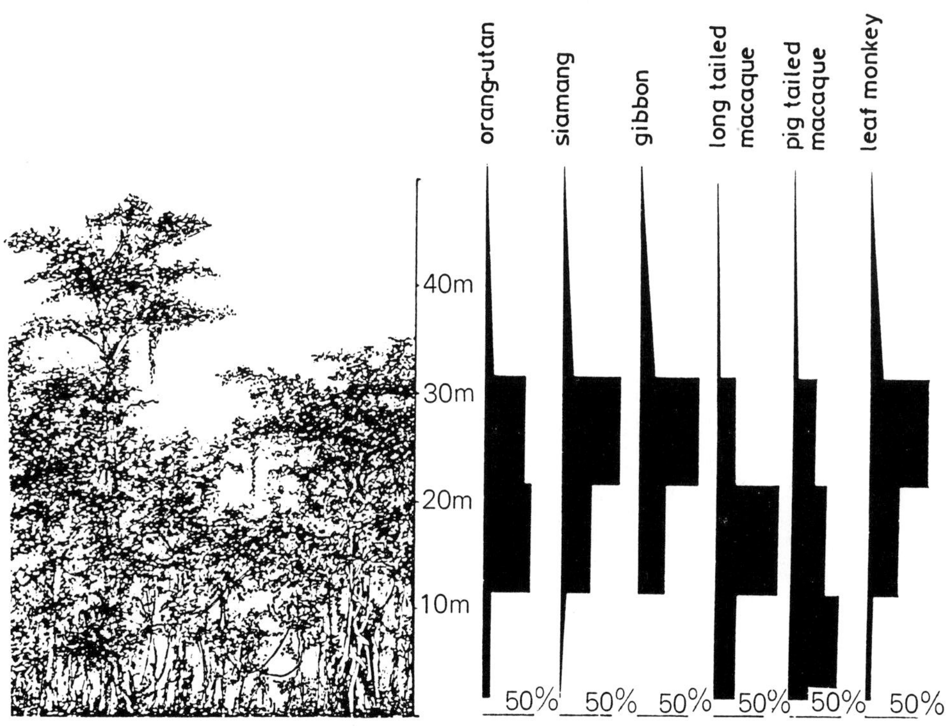

Fig. 12.9 Distribution of large primates over canopy and understorey, all in the Ketambe study area (Rijksen 1978)

As it is for all animals, the orang-utan's way of life is largely dictated by the need to range over larger distances in search of high quality food. The distribution pattern of the food sources in time, space, quality and quantity has resulted in this ape's adoption of a mostly 'solitary' life style, which appears to be more marked in Borneo than in Sumatra (Fig. 12.8). Rijksen favoured the term "limited gregariousness" for the apes he studied in North Sumatra, in order to emphasize the wide range of social options employed by this magnificent relative of man (Fig. 12.9).

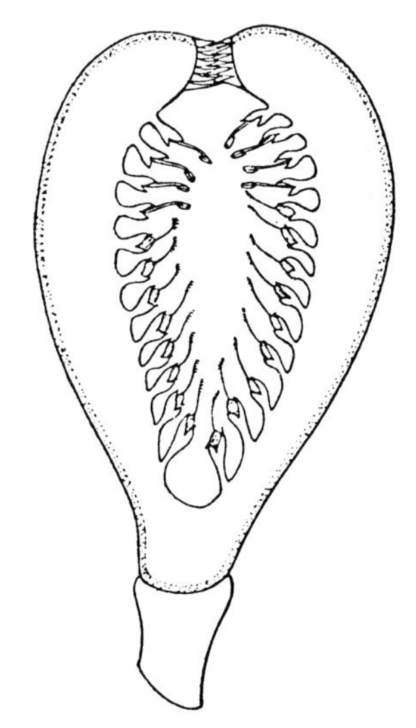

Fig. 12.10 Cross-section of a monoecious fig, schematic. At the *base*, female flowers (see next figure); at the *top* male flowers. The opening at the upper end is closed with scales (De Wit 1963)

Reproduction of Figs: A Case of Co-Evolution

Ficus (Moraceae) is a pantropical genus comprising many hundreds of species; the edible fig of the Mediterranean region, *Ficus carica,* is the best known. Some 450 species of figs are found in Malesia; and according to Corner (1970), who may be considered the world expert on this family, an average of 2% of all the plant material ever collected by botanists consists of fig species. The genus is taxonomically highly varied with four subgenera and about 13 sections; its range of life forms comprises trees varying in size from giants to dwarfs, stranglers, shrubs and (about 20%) lianas. As one can see by cutting an edible fig in half, the flowers occur along the interior lining of a jar-like, hollow pseudo-fruit, of which the aperture is closed by intersected fleshy scales (Fig. 12.10). The flowers are tiny and almost naked. One can recognize (1) male flowers producing only pollen, (2) female flowers with a long style on a small ovule, and (3) gall flowers with a short style on a large, yet infertile ovule-like organ. The gall flowers provide habitation for small, highly specialized wasps which pollinate the female flowers with pollen from the male flowers.

In general, the strangling life forms of the fig family have figs with the three flower types on the same tree; others usually have figs with either male flowers and gall flowers, or female flowers; a difference one can only detect after cutting a pseudo-fruit in half. The size of figs varies according to species, ranging from approximately 7 mm to 10 cm in size. A fig wasp is about 1 mm in length and every species of *Ficus* is pollinated by a specific species of wasp. The fig species imported to add to the collection of the botanical garden of Singapore never produced ripe fruits as the species-specific wasps were absent, as Corner remarked (1940, pp. 660–664).

If one notices swifts and martins swarming above and over a fig tree, it is the sure sign that the wasps are breaking loose from their fruit prisons, finally emerging on the outside of the fig as fertilized adult females, laden with pollen, to swarm into the wide, hostile world in search of another flowering fig tree of the same species. Once found they colonize the young pseudo-fruits through the orifice, struggling and squeezing themselves between the tightly set scales, a journey which usually costs them their wings and often a few legs as well.

Once inside a gall fig the female deposits her eggs by inserting her ovipositor into the short style of the flower, repeating the procedure until she is exhausted. Some females may thus produce up to 300 eggs. Within each gall flower, the cycle starts again. If, however, the female has entered a fig with only female flowers, she goes through her routine of egg laying, but cannot deposit her eggs because the styles of female flowers are too long. Moving from flower to flower, she probes vainly time and again, until she dies of exhaustion. Nonetheless, in her fruitless wanderings along the tops of the female flowers she deposits the fig pollen stuck to her body onto the stigma, thus fertilizing the seeds.

In the gall fig after some time the wasp's offspring develop and emerge: the males, small and wingless, scramble to copulate with the females, and after having accomplished their 'duty', they die in the fruit that nurtured them. Also at this time the fig's male flowers ripen, and thus the female wasps, before they set out to search for a place to lay their eggs, are first able to load themselves with fig pollen, storing it in special receptacles on their bodies. They

Fig. 12.11 Scheme of the pollination of a fig (in this case the dioecious, edible fig, (*Ficus carica*). The old fig **a** has been galled by the wasp *Blastophaga*, the eggs now have hatched. After fertilizing the females the males die inside the fig. The females fly out either to young figs with gall flowers **b** where they deposit eggs, or to young figs with female flowers **c** which they pollinate; the latter grow to one-seeded fruits in **d** (Wiebes 1976). *Below*: The fig wasps: **e** a male, with rudimentary wings; **f** a female; **g** parasite with long ovipositor with which the wall of the fig is pierced (Corner 1940)

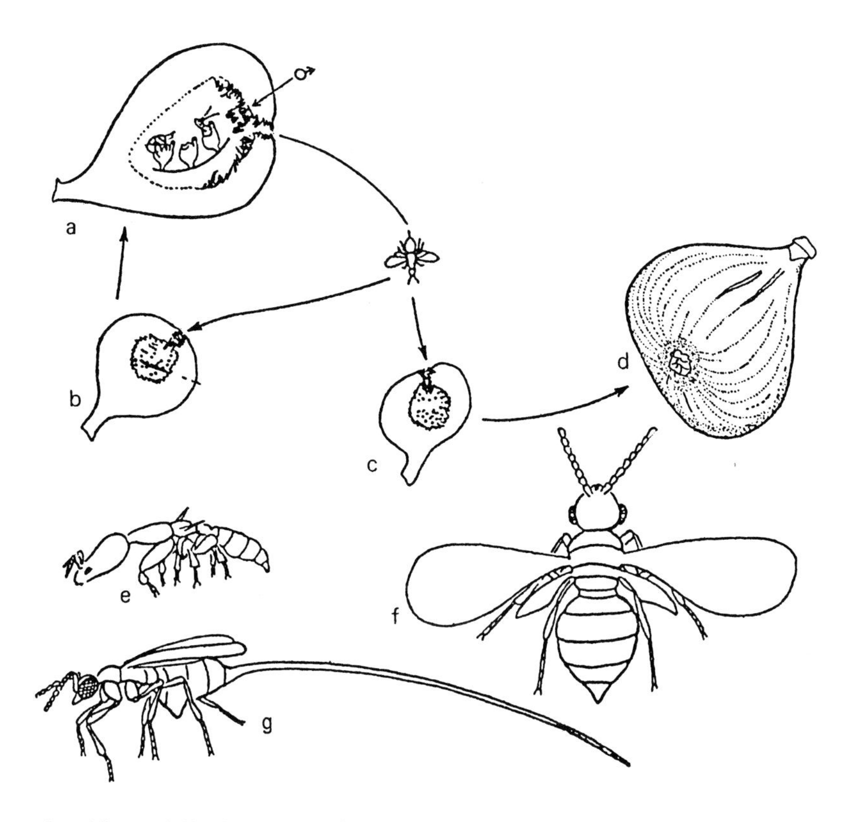

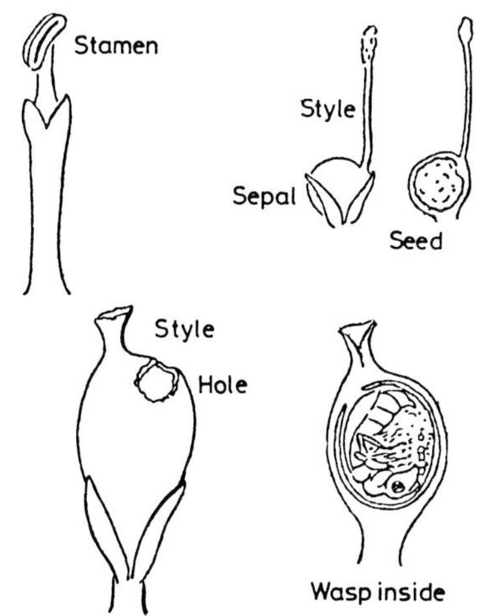

Fig. 12.12 Flowers of a fig (Moraceae), all with two sepals. *Upper left*: Male flower; *right* female flowers, with long style and later on one seed (shown in cross-section, *extreme right*). *Below*: Gall flowers, with short style; at the *left* one with a hole, left by the wasp that developed inside (shown in cross-section on the *right*) (Corner 1940)

then bite and dig themselves through the fruit wall of their fig nursery, to fly to another fig tree of the same species where they continue the cycle of fig-wasp life.

In the monoecious strangling figs, where every fig contains the three types of flowers, the female wasp is able to deposit her eggs in gall flowers while at the same time pollinating the female flowers. By the time the young wasps emerge, the anthers of the male flowers are ripe and dust the scrambling wasps with pollen. Consequently, the number of wasps per strangling fig may be rather small without impairing their fertilizing function.

There are 17 genera of fig wasps among which *Blastophaga* in the Agaonidae family ranks as the most important genus, and each kind of fig has its own species of pollinating wasp (Fig. 12.11). There are also interesting parallels in the systematics of both the *Ficus* genus and the Agaonidae in the sense that the number of taxonomic sections of *Ficus* are associated with particular, more or less related groups of wasps (Fig. 12.12). Wiebes (1976) has demonstrated that as more specimens of both figs and wasps accumulate, new relationships are revealed. Sometimes a new specimen of wasp induces a revision of the taxonomy of figs, at other times the relationships of the host plants elucidate the taxonomic status of their wasps.

There is also the rather peculiar fact that the wasp-fig relationship is parasitized by other wasps. The so-called inquilines deposit their eggs in the gall flowers from outside the fig, using their very long ovipositors to penetrate through the wall of the fruit, without rendering any pollination service to the plant. One species of these wasps does have hollow pollen receptacles in its body, but never gathers pollen in them, unlike its fig-wasp counterparts. Nevertheless, the taxonomy of these fig-wasp parasites also shows striking parallels with that of the figs.

These parallels in the taxonomy of figs and their pollinators and their pollinator's parasites indicate that they have all evolved in conjunction or have co-evolved. This conjunction has eventually led to almost complete interdependence, while at the same time being remarkably successful: fig distribution is pantropical, the number of species is vast and the life forms greatly diversified. How such co-evolution took place, and where it started must remain a secret of nature.

Herbivory

Janzen has intensively studied the consumption of seeds by animals, which he considers to be a kind of 'herbivory'. It occurs in many different ways: birds equipped with strong bills cut straight through the stone-hard shell of *Canarium* (Burseraceae) fruits as though employing steel pincers; rodents easily gnaw holes in the centimetre-thick walls of the stone fruits of Elaeocarpaceae and Palmae; weevil-like beetles deposit their eggs in immature carpels, and their larvae feast on the developing fruit from within, leaving behind little more than an empty shell and some pulp. The fig wasp also fits into this scheme, albeit as a specialist. According to Janzen such common plant-animal relationships have engendered a greater diversity in species. It is a well-known principle that a greatly abundant host permits an increase in numbers of the parasite; scarcity of a particular species of plant decreases the chances that the parasite will proliferate. Because of this fact, sparseness of distribution may be so advantageous for a species that it outweighs the possible disadvantage, with respect to cross-pollination, of a greater distance between individuals. A small population density also implies more space and opportunities for other species.

The even climate under the rain forest canopy offers animals and plants many opportunities to form permanent associations. The development of biological relationships is never inhibited by winter or other seasonal phenomena, as it is in temperate zones. Consequently, a greater array of specializations can evolve, specializations that imply greater prodigality but which will make the organism less robust with regard to environmental changes.

Richards (1973a) supposed that seeds have a better chance of survival the further away they have been deposited from their parent source of dispersal. Rain forest tree seeds are usually threatened by a great variety of animals, ranging from the larger herbivores to the minutest insects. Both classes have been studied by Janzen, who noted that a tree with ripe fruits attracts a host of consumers, and that a seed's chances of survival close to its source does indeed appear to be smaller than at some distance away from it, where fewer seeds are being deposited. The optimal survival chances will therefore be determined by whether or not a seed has some capacity to get away, or to get carried away out of the danger zone. Alternatively, the mother tree may swamp consumers with so many seeds at one time that inevitably some are left over to germinate.

Ants in Plants

The most numerous and common animals to be found in a rain forest are ants and termites. They occupy all strata from under the soil surface to the highest canopies. On dead logs and living branches one can see endless streams of these small insects marching along in a particular direction, and back, as though on a highway. The ants search for food and shelter in every crevice and cavity, and some epiphytes, such as *Lecanopteris* ferns and the Rubiaceae *Hydnophytum* and *Myrmecodia*, have partially hollow fleshly rhizomes or bulbs in which ants can make themselves a suitable home and to which they carry their prey. The wasted parts of their food combine with their excrement to form a mineral-rich debris at the bottoms of the hollows, into which the plant gratefully sends its roots. The hollow centres of some twigs seem to have evolved for the purpose. Usually the end of such a twig is somewhat thicker, having a small entrance hole somewhere in its side. Among some 830 specimens from New Guinea collected between 500 and 2000 m altitude, I found 376 trees of more than 5 m height, of which 22 had some biophysical association with ants (i.e. 6%). The tree species belonged to 16 genera in 11 families; the highest was 25 m.

Janzen has noted that ants also inhabit the large hollow thorns of *Acacia* species in Mexico, a shrub that occurs outside the rain forest; and the same is known of some *Acacia* species of the African savanna. The ants appear to protect the plants from being overgrown by lianas, as they crop all tendrils that come near their host (cf. the photographs in Richards 1970, p. 154, 155). It has also been noted that the ants attack and repel small insect herbivores on their host: in the African savanna the ant-*Acacia* symbiosis may even serve to diminish, together with the host's sharp spines, massive damage from larger ungulate herbivores.

The mineral matter accumulated in the bodies of forest-dwelling animals originates, or has been transferred from, the mineral cycles of the forest floral system. Return of that mass is, of course, localized to the place where the animals die, and hence irregular. Since the ever-wet climate constantly impoverishes the soil through leaching, the mineral cycle is consequently a delicately balanced process, and each forest ecosystem must determine precisely the permissible amount to be stored as animal biomass at any one time, i.e. the combined weight of the animals in kilograms per hectare or square kilometre.

Scarcity of Minerals

The animal biomass of a rain forest, no matter how varied, is small in comparison to the biomass of an African savanna. Curry-Lindahl (1972, p. 158) has given some data showing that the average biomass of wild savanna ungulates is 50–150 kg ha^{-1} (5000–15,000 kg km^{-2}), some 7 to 7.5 times greater than that in a rain forest. The mineral cycle is also different in the savanna, as the periodic droughts cause an upwelling of minerals from the deeper soil layers which more or less compensates for the leaching process of the wet season. The savanna with its unique fauna is no doubt also a product of long evolution, and consequently one should not expect that a quick conversion of rain forest to a savanna-type habitat (e.g. the extensive *Imperata* grass steppes of Sumatra and Borneo) will result in higher bio-productivity and hence a greater biomass – on the contrary!

For one thing, the animals adapted to the 'new' habitat type are absent. This said, however, most rain forest soils apparently have an excess of minerals large enough to sustain an impressive fauna; yet, some areas with seemingly lush forest are so poor in fauna that it strikes the explorer. The famous traveller Spruce (1908, p. 268) noted long ago that the Rio Negro (the 'black river', a tributary of the Amazon) might as well have been called the 'dead river' for its apparent lack of life-sustaining minerals. Even mosquitos were absent and he was unable to catch a single fish in the water the colour of too strong tea, dark from the dissolved humus acids that give it its deadly low pH. Similar rivers, though of smaller size, are to be found in Borneo; if they do not originate from peat swamps, which are the usual sources, they originate in areas locally known as 'kerangas', a name denoting 'unfit for agriculture' in a Dayak language. The soil is too poor to support rice plants and indeed it is almost pure sand. The kerangas is an interesting marginal habitat within the rain forest complex which will be described in greater detail later; its equivalent in South America is called the *wallaba* habitat. What is most striking in these poor, open forest types is the great richness of debris; a strange phenomenon in an ever-wet climate where one would expect quick decomposition. The reason is that the plant debris itself is too difficult for organisms to metabolize because of its high concentration of poisonous, phenol-like compounds. There is some decomposition, but the process is very slow which delays the recycling of the already scarcely available minerals. These phenol compounds also cause the synthesis of humus acids. Janzen, in his article entitled *Tropical Blackwater Rivers* (1974) has speculated that on such extremely poor soils the abundance of phenol compounds inhibits herbivory, thus banning animals from competing for the already scarce minerals. This also applies to the peat swamp forests of the very wet lowland rain forest regions, where animal life is so scanty that one is struck by the great silence.

Janzen also draws our attention to a problem that has intrigued many biologists for decades, namely how and why do dipterocarp trees flower so irregularly, but at the same time so well synchronized? I have already remarked on this phenomenon in the regional section on Malesia (Chap. 10). Here is a summary of Janzen's richly documented possible solution to this problem, drawn from the Blackwater River article referred to above.

One of a plant's possible strategies for avoiding predation is increased toxicity, as we have seen already. But if its seeds are to be distributed to animals, this is a great disadvantage. On the other hand, if seeds lack toxins, the chances are great that a large part of the total seed production will be eaten and destroyed. The seeds of dipterocarps are indeed vulnerable in this respect: insects, rodents and ungulates such as wild pigs, who roam over large distances in Borneo and Sumatra, are known to prey upon them, especially in lean periods, just as orang-utans exploit the fruit yield of *Heritiera* when other fruits are rare. A small, regular production of fruits is in fact a great risk for a tree, unless its seeds are small and protected against destruction in the digestive tracts of animals, its entire crop may be harvested and destroyed. The dipterocarps have circumvented this risk by their irregular pattern of flowering and fruiting: they save their energy over the years until a drought triggers massive flowering which commonly leads to a massive ('mast') fruit production. The vast quantity of fruits then produced outweighs the feeding capacity of any assembly of exploiters, other than humans, so that enough seeds remain to form a carpet of seedlings and to assure reproductive success. Janzen has wondered how the dipterocarps evolved this strategy (1974, p. 89). I believe such mass flowering is rather common in the family to which they belong. Temperate oaks and beeches also alternate abundant seed production with lean years, and it is not strange to find such a habit to be a family characteristic. It is probably the ecological adaptation that has allowed them not only to survive on very poor soils, but has also enabled them to proliferate and diversify so successfully. By means of their highly efficient mineral economy they could invest their surplus in sturdy trunk building as well as ensuring their reproduction by swamping possible predators with seed at regular intervals.

The Usefulness of Hollow Trees

Foresters commonly assert that a rotten core in an adult tree is fatal or at least a great disadvantage to the tree. Janzen (1976) contends, however, that a hollow core may in fact represent an adaptation, notably on poor soils. A hollow core appears to attract animals, inviting them to make nests and deposit mineral-rich excrement and organic debris inside which enriches the soil immediately under the trunk. It is likely that a tree beyond a particular diameter has enough support from its outer trunk tube to dispense with its core, at least in those regions where wind squalls are rare. Thus, the tree may utilize an otherwise useless part of its anatomy. This is not to say that a hollow trunk is without risks for the tree: Corner (1978, p. 53) has noted how in Malaysia the elephants may scrape out the hollow cores of trees to get at the wet, rotting wood, and even attempt to push the tree down in order to finish the whole resource. Be that as it may, a hollow tree reflects one intricate evolutionary relationship between plants and animals; a relationship which, for such animals as the hornbills, is crucially important. These spectacular birds would not be able to survive without the occurrence of hollow trees (Fig. 12.13). The female breeds in the cavity, the entrance of which is almost closed with a cement-like substance composed of debris, rotten wood, soil and perhaps excrement, applied by both the male from the outside and the female from within. The male subsequently feeds the female through the remaining slit throughout the incubation period. She is apparently safe from predation in her 'prison' and when her offspring

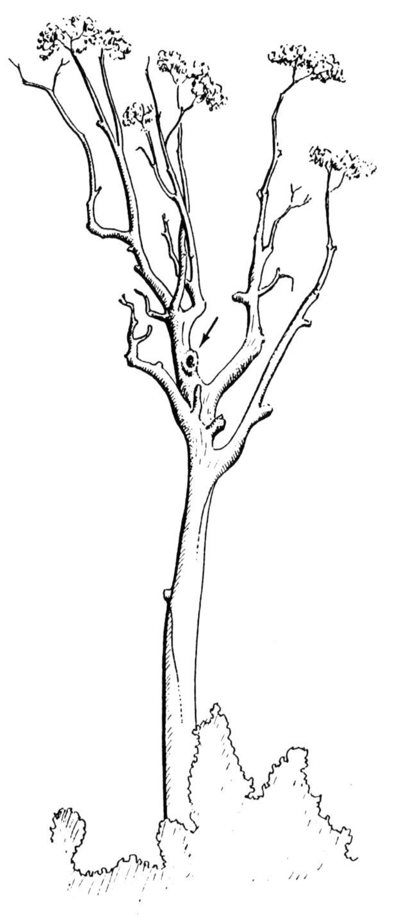

Fig. 12.13 The value of hollow trees. This declining 'tree of the past' harbours a hornbill's nest (*arrow*) (After Bartels and Bartels 1937)

becomes too large for the cavity to hold both of them, she breaks out with the help of her mate and subsequently closes the entrance again (Fig. 12.14). Then they both feed the young bird until it has developed into a fledgling (Bartels and Bartels 1937).

It is worthy of notice that the hollow trees, as we shall see shortly, are being destroyed by the commercial forester in order to ‘improve’ the quality of ‘his’ resource. Yet because hornbills cover large distances, flying low over the forest canopy, they are responsible for the dispersal of a great variety of tree seeds.

Fig. 12.14 A male ‘rangkong’ hornbill, near the hole in which the young are left after the female has flown out. It has been feeding its young and is about to fly away (After Bartels and Bartels 1937)

Defence Mechanisms

Although it would be impossible to enumerate all the relationships between plants and animals in a rain forest, some publications by Van der Pijl (1960, 1966, 1972) are remarkable attempts to do just that. Here, I can only mention the role of birds in pollination, for instance the specialism of the Neotropical hummingbirds and the sunbirds of the Old World. The role of large ungulates in the distribution of *Rafflesia* (Rafflesiaceae) has already been discussed in Chapter 6. All such relationships are aspects of the whole immensely complex system called the rain forest: all forest-dwelling animals play some role or another in the intricate web of life whether it is in the pollination of a plant or in the distribution of seeds or some other aspect. Of course, such relationships are not confined to the tropical forests and savannas, but as in no other system the rain forest offers cases of incredible variety and complexity. In Chapter 13 (Evolution) we shall elaborate some of these further but I cannot end this chapter without noting that certainly not all animal-plant relationships are symbiotic. One must keep in mind that almost every relationship contains ambiguity: the case of the orang-utan and its favoured fruit, the durian, illustrates this point.

The matter is quite simple: the larger the plant part, the greater the nutritional reward to the consumer. Large seeds and large edible leaves of flowers are therefore at great risk. Thick-trunked plants with a large growth centre full of stored energy reserves are especially vulnerable to herbivory, human predation included. People consider the hearts of palms as a great delicacy, even though the palm must be killed to get it. Hence, it should not surprise us to see the vegetative shoot of many palms heavily covered with a barrier of woody sheaths which are often armed with sharp spines and thorns. This applies to pandans and the ancient *Cycas* (Cycadaceae); bananas (*Musa* spp., Musaceae), being giant herbs unable to produce wood, protect their growth shoot with a sticky, bitter slime. Figs, until the moment they are ripe, are inedible due to the profusion of latex in their rind.

In a similar way, the protective function of hairs becomes clear; one needs only to look at the mechanical obstruction experienced by an aphid struggling along a hairy leaf. A leathery skin is another mechanical protection, although it takes time for a young leaf to develop such a protective layer. The nutmeg, *Myristica fragrans*, is one of many plants to have a dense cover of protective hairs on its young parts, adult parts have shed their hairy cover and find protection in their tough skins. A collector would do well to observe and record such phenomena.

While the success of every form of defence is only relative, it is perhaps no coincidence that the mangrove forests of some tropical coasts, being easily accessible and hence vulnerable, have leaves that are extremely rich in tannic acids. It seems that such mangrove forests are not under heavy pressure from herbivory; even though the Bornean proboscis monkey is able to exploit this resource. Every successful defence diminishes a predator’s chances of survival, and although the latter has enough time to find an answer to the problem, he will inevitably stay behind in the great evolutionary race.

Flora Brasiliensis, tabula XXXI. '*Artocarpus integrifolia*, in its own shadow.' Cauliflory. The branches bear epiphytes. This species, with edible fruits, has its origin in Indo-Malesia. Drawn by B. Mary in 1856 (Martius 1840–1869)

13 Evolution

A Car Drive into the Past

From the western part of The Netherlands a continuous auto route leads to Paris, by way of Antwerp and Lille. The road is quiet at night, enabling the driver to keep the car in fourth gear. We are looking for a comparison to illustrate the passage of a million years. Running in fourth gear, when a large Volvo achieves about 110 km h^{-1}, the crankshaft rotates 4000 times a minute, 240,000 times an hour or a million times in 4 h and 10 min. If we use one rotation of the crankshaft to symbolize a year, less than 30 s, or about 0.9 km of road, will have returned us to the beginning of the Christian era. After 4 h and 10 min, covering approximately the distance between Paris and The Hague, we will have gone back a million years.

In 1 million years a species from the rain forest could, at a supposed speed of 1 km a century, travel back and forth three times over the whole length of Sumatra. The oldest dipterocarp pollen grains found in Borneo date from 30 million years ago, and it is thought that the development of the modern tropical rain forest in Borneo started then. To bridge 30 million years with our car, we would have to make at least 15 round trips between The Hague and Paris.

To grasp the temporal aspects of evolution, one must become accustomed to such long-term thinking, and the general inability to do so forms, indirectly, the greatest barrier to the understanding of the biological basis of our existence. The time factor puts everything into new perspective; over several millennia changes are realized that would seem only infinitesimally probable in some lesser period. Our task in this chapter will be to look back to that improbable beginning from our present reality.

The Approach

During the Carboniferous period, 350–270 million years ago, forests already existed whose wood became fossilized into coal. Most of the vegetable material consisted of cryptogams, but a few seeds from that period have been discovered. During the Permian period, 270–225 million years ago, the gymnosperms came into being, and during the subsequent Triassic and Jurassic periods, 225–180 and 180–135 million years ago, they developed fully. Subsequently, during the Cretaceous period, the larger part of the cryptogam groups appeared between 135 and 70 million years ago. The Tertiary, the formative period of today's plant world, came to an end 1.5 million years ago. The equivalence in distance covered by car (based on an average distance of 450 km for each million rotations of the crankshaft) is 31,500 km to the beginning of the Tertiary, and 60,750 km to the beginning of the Cretaceous period (Fig. 13.1).

Where rain forest is concerned, we cannot count on finding macrofossils; wood, leaves and fruits in the rain forest decay far too quickly. Sometimes, however, microfossils consisting of pollen grains were preserved in swamps; Muller (1970) wrote an article about this subject which elucidated a certain sequence in the appearance of plant families (Fig. 13.2). This, however, does not tell us much about plant forms or their interaction with animals. Both are

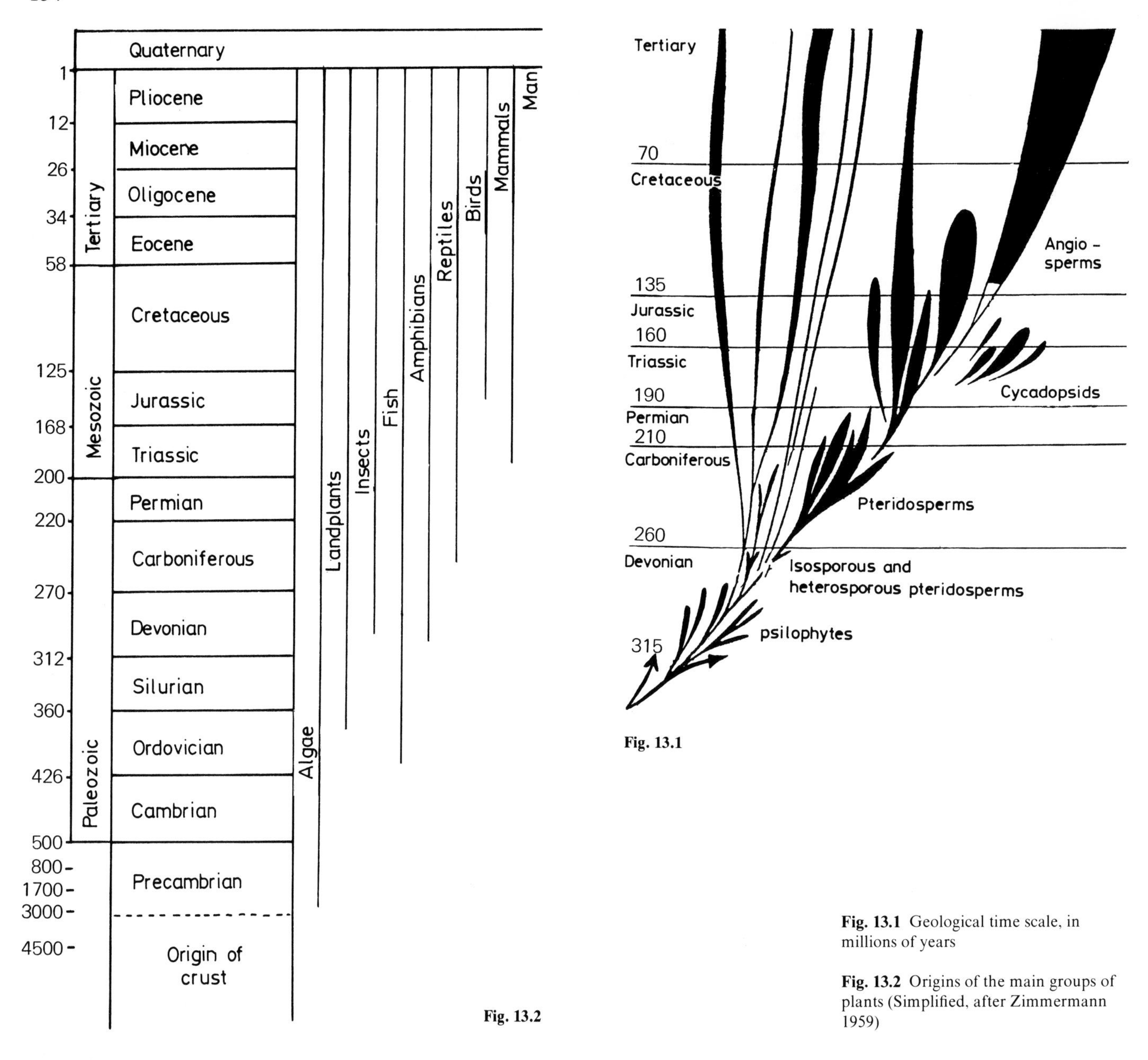

Fig. 13.1 Geological time scale, in millions of years

Fig. 13.2 Origins of the main groups of plants (Simplified, after Zimmermann 1959)

vital aspects of the picture we are trying to reconstruct; evolution does not involve a single taxon or an isolated characteristic, but rather the interaction of plant and animal life, which, as we have seen, is symbiotic as well as anti-biotic: plants attract animals only under certain circumstances, such as flowering and fruiting, while repelling them at all other times. Evolution – especially in the tropics – means co-evolution.

In such co-evolution a structure is biologically tested; a favourable result means a greater survival value, which can be statistically quantified. If the seeds of a plant with slightly juicier fruits have a 2% better chance of being dispersed over a distance 100 m longer than average, this will have an effect in the long run. Our crankshaft rotates 4000 times a minute in fourth gear while the car covers a distance of about 1.8 km, a distance which corresponds to a period of 4000 years or 100 generations of durian trees, and more than 400 generations of orangutans. To arrive at a million years – not a very extensive period, from the geological point of view – every figure has to be multiplied by 250: that is: 25,000 durian generations and 100,000 generations of orangutans. Small changes accumulating over such a period can have considerable effect.

After having studied so many structures which are only found in the tropics we should not expect much of the discoveries made outside the tropics and theories based on those discoveries only. The tropical plant world cannot be understood by extrapolating from non-tropical data. The only road to the past open to us is the road which starts in the tropical rain forest of today. Basing ourselves on structures and relations which still exist in today's forests we can try to reconstruct the main lines of development.

In which direction, however, should our thoughts go? How can we discover what is called by Corner "the arrow of time", pointing from original, primitive structures and relationships to modern, derived ones? It would not be the first time that we discovered that a beautifully postulated series of development stages could just as easily be read the other way round.

Much work, of course, has already been done in this field: for more than a century morphologists and taxonomists have done their best to arrange all sorts of characteristics in series. Thus, it was agreed that the occurrence of a certain organ in high, irregular numbers is more primitive than the occurrence in low, but constant numbers: the flower of the water lily with its many petals (arranged, moreover, spirally, and not divided into a calyx and a corolla), is considered more primitive than that of the buttercup with its five petals (arranged in a whorl and clearly discernible from the calyx).

The use of such terms as 'primitive' nevertheless risks confusion. A structure may be evidently primitive and be recognized as such without this being the case for the whole taxon within which the structure occurs. Even if we could always with certainty discern primitive from derived, it would become clear that no single family or genus possesses only primitive characteristics. Concentrations may sometimes be pointed out: complexes of primitive characteristics which invite attention.

If one is on the track of such a complex, the question of its possible function with respect to evolution does arise. This question concerns such matters as investment and return (from the point of view of the plant), the effects of restricting factors such as drought and cold, and how to deal with those factors with the means that are available.

The Thick-Stemmed or Pachycaul Form

Large seeds can only be carried by stout twigs. A twig is a young, non-lignified branch though in discussions of a general nature such as we are engaged in, the general term axis is more appropriate. An axis is every organ of a plant which carries other parts: a stalk with leaves attached to it, or the axis of an inflorescence.

A certain ratio exists between the width of an axis and the size of the organs it carries. The thicker a branch is, the broader the growing point at the top which, dividing itself, produces the leaves. The leaves produce 'assimilates': the organic substance needed to build the plant body. A thick axis with big fruits needs a large quantity of 'assimilates' and only big leaves are able to produce them. The coconut palm, *Cocos nucifera* (Palmae), is a well-known example of such a type of plant.

The concurrence of big fruits, heavy axes and large leaves has serious consequences. (1) To produce a large leaf the plant has to make a big investment; therefore, the leaf must go on functioning for a considerable period of time, much longer than the 6-month temperate climate during which deciduous trees carry their leaves. The lifetime of a coconut palm leaf is estimated by Corner (1966, p. 45) at 5 years. (2) Branching becomes problematic. A thick axis with large leaves is proportionally heavier than a thin axis with small leaves, and weaker, too: when a body grows larger, the volume, and accordingly the weight, increases more than the length, but the strength does not increase accordingly. Thus, a thick axis will break off much more easily from its point of attachment than a thin one. The thicker the axis, the greater the risk. (3) A large growing point is more vulnerable to crippling circumstances. It is possible for plants to put a small growing point out of action when unfavourable circumstances such as drought and/or cold make this necessary, but this is obviously not possible when the volume of the active and very sensitive meristem is bigger. The consequence of all this is that pachycaul plants cannot grow far beyond the warmer regions of the earth. For this reason, palms in the botanical gardens of northern Europe are brought inside in October, and the giant rhubarb, *Gunnera chilensis* (Haloragaceae), is packed in thick layers of dry leaves. The thick rhizomes of the European water lilies are protected by the mud and the relatively deep water in which they grow. Incidentally, one should not infer from this that all leptocaul plants are hardy; cold is just one of the restrictive factors for pachycaul plants. (4) Large growing points are very much sought after by large plant-eating animals and, therefore, need special protection, which means that the plant must also produce thorns and/or heavy, strongly lignified leaf sheaths closely surrounding the growing point (Fig. 13.3). (5) A large leaf is vulnerable to wind and rain. The leaf of the banana tree, *Musa* (Musaceae) generally tears along the lateral nerve to within a few centimetres of the midrib shortly after it unfolds. A compound leaf, such as that of the coconut palm, is better equipped since the leaf blade is divided into separate leaflets, each with its own midrib, all of which can bend separately.

In the rain forest there is no lack of plants with relatively short, thick, simple or only slightly ramified trunks and large (often divided) leaves whose 'heart' is protected by an armour of lignified leaf sheaths and/or thorns. In Chapter 6 we became acquainted with them as *'Schopfbaum'*: many palms and pandans belong to this group.

How did such plants come into being? Are we not used to accepting as self-evident the idea that all big structures must have started from small beginnings? In such cases, however, there are temporal difficulties. The cryptogams started to develop during the Cretaceous or perhaps even in the Jurassic period; long before that, however, during the Carboniferous era, trees of considerable size already existed. The ultimate 'small beginning' of these trees must be sought before the Devon period, which started 400 million years ago. During the approximately 300 million years which preceded the Cretaceous period – that is to say, a 135,000 km drive at 4000 rotations a minute, which covers twice our Christian era – the plant world produced the forms which may have been the starting material for our seed plants. During that period, the rise and fall of the vegetation out of which the coal layers were formed took place. As for the matter of 'large' originating from 'small'; the palaeontologist Zimmermann (1930), whose arguments were adopted by Corner, theorized that the oldest land plants consisted only of small stems which divided themselves constantly in two, thus creating groups of little V's. In those groups several processes occurred which led to differentiation. One-half of the V may have become more developed, pushing aside the other half, a phenomenon called 'overtopping'; or, the small stems put themselves on the same plane: 'planation'; or, between the stems in the plane, tissue developed: 'webbing' (Fig. 13.4). The essential point of the process described here is to sketch the possibility that basically small organs begin to form combinations, and together start to fulfill one and the same function.

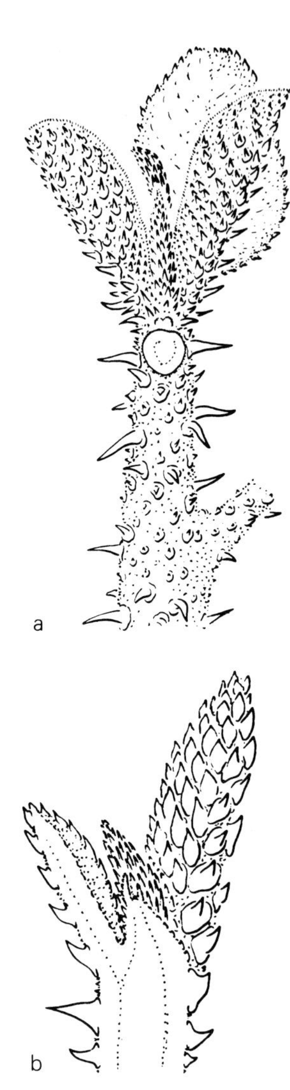

Fig. 13.3 a,b. Growing point of thick-stemmed forest treelet, protected by scales: *Saurauia* (Saurauiaceae), New Guinea. The older leaves have been removed. **a** Surface view, × 2; **b** cross-section, × 4 (Corner 1964)

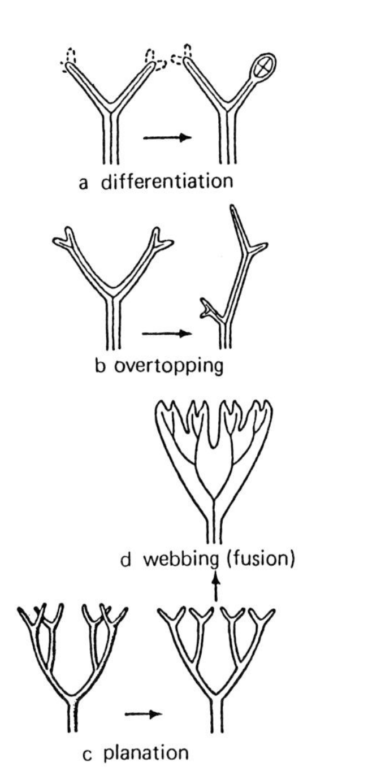

Fig. 13.4 a-d. Some of the processes having occurred during the early evolution of the land plants. **a** Differentiation: one axis bears a reproductive cell, the other one remains sterile. **b** Overtopping: from an equal-sided, forked branching one branch develops more strongly than the other one, accordingly a system develops with main stems and side branches. **c** Planation: all axes develop in one plane. **d** Fusion: between the axes tissue is formed (Lam 1962)

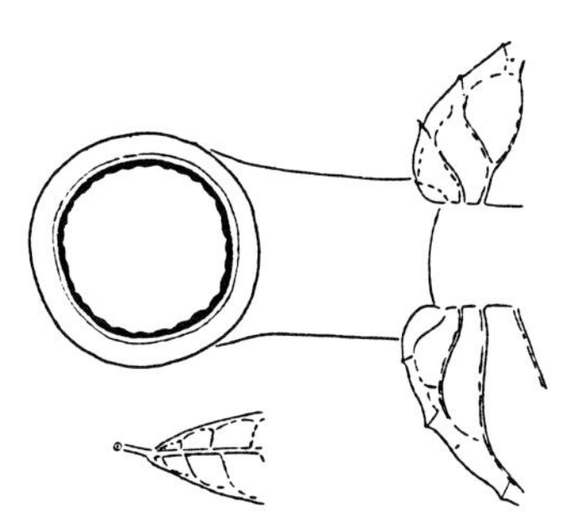

Fig. 13.5 Pachycauly and leptocauly. Leaf stems of two different species of *Ficus* (Moraceae), the largest from a thick-stemmed tree with leaves of 1.80 m length, the smallest from a thin-stemmed tree with leaves of ca. 10 cm, 85% of natural size (Corner 1964)

The Thin-Stemmed or Leptocaul Form

The pachycaul form grows slowly, and hence continuously, for it cannot afford to let its growing point hibernate or rest during a period of drought. It is also unable to become widely ramified. Its material investments are considerable, and each is bound to certain periods of time. To armour its growing point takes up extra material, which stays unproductive. In short, such a plant lacks flexibility. It can neither take advantage of an opening in the canopy which is situated sideways (for it can only grow straight upwards) nor can it try to survive unfavourable conditions by throwing off its foliage; the investment was too large for that. The thick trunk becomes top heavy after a certain period of time, which puts a limit to the height it may reach. For a trunk with a diameter of 40 cm, 20 m is already a considerable height.

The plants that chose another strategy are well known to us: the birch, *Betula* (Betulaceae) is an example par excellence of a tree that is the opposite of a pachycaul. The seeds are small, grow on thin twigs which are abundantly ramified, with small leaves, i.e. a small investment, that fall off in winter. New side branches are quickly formed if others are damaged. From the same quantity of material used by a pachycaul, a birch succeeds in forming long internodes, and is thus able to extend its useful surface in all directions. Thanks to all those provisions the birch is highly flexible in its struggle for life.

To be able to do this, however, the leptocaul tree has four abilities which a pachycaul tree lacks: (1) Secondary growth occurs in girth, i.e. in the layer of cambium tissue right under its bark, which makes the axes gradually thicker with the result that their strength becomes equal to the weight they will have to carry. (2) High quality material is used to build a slender construction which reaches as high as possible in order to catch a maximum of light, for which long and tough ligneous fibres are needed. (3) It has a superior system of transport for water and assimilates, through thinner vessels to greater heights, and in larger quantities. The tree must invest in a mass of foliage within a short period of time, and as it is short-lived, it has to give a quick return. In the spring, the birches actually bleed when a branch is cut off: the stream of sugar-containing sap is that strong. Fortunately, the leaves themselves do not need a thick outer skin as a protection against an unfavourable season and can evaporate plenty of water and produce assimilates. (4) The leptocaul tree has a root system which is able to pump the necessary water to the height that is needed.

Trees thus equipped have multiple advantages over the pachycauls. They are able to grow higher, to spread wider in order to catch more light, to withstand unfavourable times and to invest in small packages, including the production of small (= easily transportable) seeds in large quantities, seeds which can also survive under unfavourable circumstances.

Thus, the picture of the evolution of the tropical rain forest becomes clear. The oldest forests must have consisted of pachycaul forms with large leaves; Corner believes that they grew mostly in swamps because of their large evaporation rate and not very well-developed internal system of water transport (Fig. 13.5). Because there is an evolutionary premium on the flexibility inherent in being a leptocaul, natural selection worked in that direction. Trees with thinner trunks, which because of this were able to ramify, developed and took advantage of the possibilities offered; they gained in height, and acquired pioneer qualities with their small seeds. One success brought on another. Giant forest trees became the dominant life forms, together with lianas, which are the leptocaul forms par excellence, and epiphytes, whose minute seeds could be easily dispersed over long distances so that they became pioneers in the structure which had now become possible: the canopy. The pachycauls thus had to be content with more and more shade, and in many places they lost the silent struggle. They are still occasionally dominant in places where they receive enough light: thickets of banana trees, palms and pandans are normally found along coasts and river banks and in other open places; and many plants still retain typically pachycaul structures.

Evolutionary development could only have been from pachycaul to leptocaul forms, and not the other way round, which would have meant going against the direction of specialization, i.e. the development of the four facilities just mentioned. The biological advantages lie too evidently on the side of the leptocauls to doubt this.

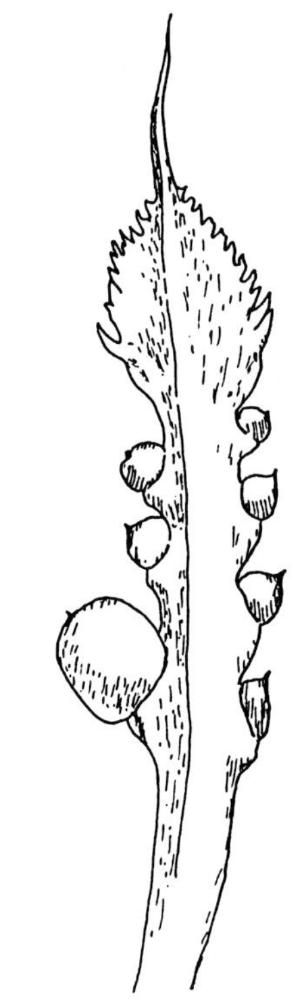

Fig. 13.6 Ovule-bearing leaf of a gymnosperm: *Cycas circinnalis* (Cycadaceae). The ovules develop on the margin of the leaf; no closed ovary occurs. This is supposed to be a primitive form (De Wit 1963)

The Development of the Fruit

After this analysis of vegetative structures we return to seeds and fruits. The flower itself gets little attention here, because its significance is small compared with that of the fruit. Flowers are never a purpose, but a means, a short-lived formation, an ephemeral pre-stage of the fruit. A plant's ultimate purpose is the production of seeds: the fruits are the packages containing them, and the seeds have to arrive safely at favourable, new locations. They will not do so of their own accord: a delivery service is needed, and for large seeds only animals can fulfill this function. Spores, which are unicellular, float on the wind; and small seeds, the end products of evolution, are also wind-dispersed. But the big, unrefined seed formed the basis for relationships between plants and animals which were of mutual advantage, and which probably came into existence during the Cretaceous period.

By means of, and in, plant-animal relationships, evolution became a process in which elements of both purpose and intentions became involved, and inventiveness became important. Only with this complex relationship in mind can the evolution of tropical fruits be understood; a resumé of some of the general notions concerning the subject in Chapter 12 might be helpful here.

The fruit is a difficult subject, and one which has not yet been thoroughly studied. Corner (1964, p. 211) supposes that there are just as many categories of fruits as there are natural groups of genera, i.e. thousands. Some fruits burst open when ripe, and it has to be assumed that this is an original characteristic of angiosperms. In most cases seed buds are formed along the edge of a specially formed leaf, called the carpel, as can be easily observed in the gymnosperm *Cycas* (Cycadaceae) (Fig. 13.6). When the edges of the carpel grow together,a little tube is formed in which the seeds lie well protected. Such fruits are found in the *Sterculia* (Sterculiaceae) already described in Chapter 12. The growing together and the bursting open of carpels later on, with the result that the seeds are displayed, are mechanisms which are too simple not to be used. Almost as simple is the growing together of three, four or five carpels at the edges, which loosen again when the fruit is ripe. The horse chestnut possesses three carpels which are easily recognizable.

At the start of plant-animal relationships, which resulted in the dispersal of seeds, the colour red must have been of importance. Red is the colour of blood, and as such well known by animals; it was the best colour of all to attract their attention to certain parts of the plant. In *Sterculia* the red colour is further accentuated by the contrast with the pitch-black seeds. But other much coveted substances had to be associated with the seeds in order to make them attractive. Such substances are, according to Corner's durian theory, mainly found in the aril. We have already mentioned arils while discussing the spindle tree, *Euonymus europaeus* (Celastraceae). Well known, too, is the aril called mace (Fig. 13.7) which partly surrounds the nutmeg, *Myristica fragrans* (Myristicaceae). The mace is ramified, slightly fleshy, red in colour and has a strong aromatic taste. It is more or less embedded in the surface of the nutmeg shell. In the durian, the seed is completely surrounded by an aril which is markedly thick, succulent and fleshy, and therefore relished. In the mangosteen, *Garcinia mangostana* (Guttiferae), too, it is the aril that is eaten; it is pure white, and has a slightly sourish taste.

Fig. 13.7 Nutmeg, *Myristica fragrans* (Myristicaceae) with the aril, better known as mace

The aril is an outgrowth of the umbilical cord or funiculus, the string by which the seed is attached to the fruit. I have not succeeded in discovering a publication which explains how this outgrowth may have started. However that may be, the fact is that the aril is found in

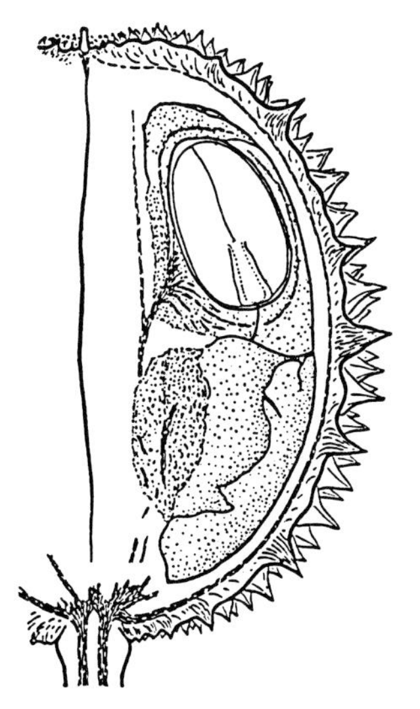

Fig. 13.8 Durian, ripe fruit, reduced ca. 3 ×, in longitudinal transection. Within the heavily thorned shell of this capsule two seeds are enclosed in the fleshy aril (*dotted*); the upper seed is cut. A strand connects the seeds to the middle part; the placenta, the aril is an outgrowth of this strand (Corner 1949)

many different families: Corner (1953, p. 474) lists more than 50, that is about a sixth of the total number of families of angiosperms. In the rain forest, arillate fruits occur among the Annonaceae, Bombacaceae, Celastraceae, Connaraceae, Dilleniaceae, Elaeocarpaceae, Flacourtiaceae, Guttiferae, Leguminosae, Lowiaceae, Marantaceae, Melastomataceae, Meliaceae, Myristicaceae, Sapindaceae, Sterculiaceae and Zingiberaceae; but overall, this represents no more than 1% of the tropical flora. It is remarkable that an aril is often found in reduced form; Corner (1964, Figs. 72, 73, 74) arranged a progression according to which the aril becomes smaller in the Leguminosae, Sapindaceae and Meliaceae respectively. In these cases it should be considered a remnant.

A fruit such as the durian (Figs. 13.8 and 13.9), which eventually bursts open to display its large, aril-covered seeds, has to be well armoured in order to keep the animals away until the seeds are fully ripe. From the durian, Corner (1964, p. 217) succeeded in deriving all the important types of fruits by way of diminution and, in connection with the development of leptocauly, desiccation, including those which, after diminution, develop the juicy wall which results in a berry. The unripe berry is armed against animals with chemicals, e.g. the tomato, with oxalic acid.

In this context, cauliflory, or the ability to flower and fruit from the mature wood of branch or trunk, which makes it possible to carry the large fruits associated with big animal dispersers should also be regarded as an ancient characteristic. Such remnants from the past are often found, however, to have co-evolved to a new position among the leptocauls where the fruits became maximally attainable for animals.

The plant-animal symbiosis became a success. Smaller animals followed the flowers and fruits high up in the canopy. And not only birds profited from their ability to fly: other vertebrates profit by a certain ability to fly or glide, such as bats, flying squirrels and flying lizards. According to Corner, fruits came to be consumed on such a large scale, that all kinds of defence mechanisms became necessary: mechanical ones such as thorns and hairs, chemical ones such as poisons, nasty tasting substances and latexes, and ecological one such as irregular flowering, and this brings us back to evolutionary selection.

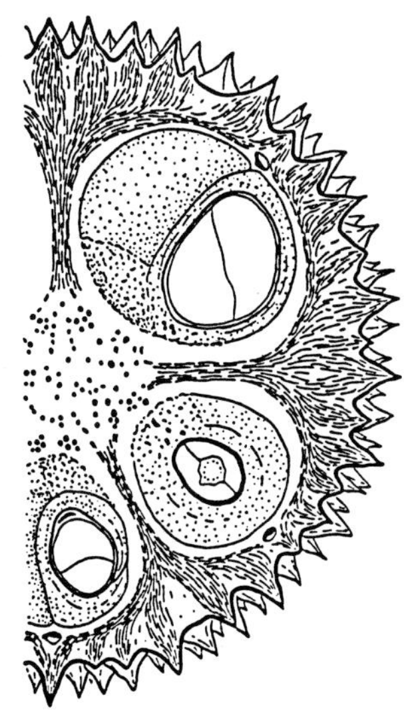

Fig. 13.9 Durian, ripe fruit in transverse section, with three seeds, all of which are cut, each enclosed in its own aril. The fruit is about to open with five valves (Corner 1949)

Old Characteristics and New Functions

Every family consists of a mixture of original (or primitive) and more advanced characteristics, as we mentioned in the beginning of this chapter. Some characteristics have more survival value than others; we can call them 'biologically superior'. What we should expect is that where such characteristics develop, the old, biologically inferior characteristics disappear. This undoubtedly usually happened, but not always.

For everywhere in the vegetable kingdom we run into structures which have no demonstrable function, and whose existence is incomprehensible, i.e. unless we do regard them as organs no longer functional, very likely maintained because the plant was not hampered by them in its new biological functions. These structures no longer possess any survival value; their existence can only be understood by supposing they had a function in the past. Possibly, the function took place in another part of the world, before a migration. One example is the spininess of the fruit of the horse chestnut, *Aesculus hippocastanum* (Hippocastanaceae), which has been described by Corner (1964, p. 220) as "a small durian that has lost its aril but retained its spines, large seed, and fleshy rind".

Another example is found in the arum family or Araceae. They form one of the most recognizable families because of their inflorescence which consists of a fleshy stem with minute flowers (called a spadix) carrying at its base a spatha. Usually the spadix is protectively enveloped by the (colourless) spatha. The European wild arum lily, *Arum maculatum*, is a good example: the spatha does not open until the time of flowering. In *Anthurium*, however, the spatha is flat and of bright scarlet colour; the spadix stands on it at a right angle. The red spatha, we assume, lures fertilizing insects to the spadix like clients to the florist's shop. Thus, an old – for widely spread in this family – organ has adopted a new function. Such things often happen, creating a new spectrum of biological relationships and opening the door to possible future co-evolution.

Chemical Defence

To round off the subject we give Corner's (1964, p. 149–150) eloquent description of the wild relations of one of the tropics best-known cultivated plants, the papaya or pawpaw, whose fruit, a general favourite, is 20–30 cm long, shaped like a rugby ball or slightly pear-shaped; its taste is a mixture of melon and carrot. The plant is "...

...... 10–30 ft. high, unbranched or with a few branches, which grow straight up and merely repeat the character of the main stem. It peters out at this low height; the leaves grow smaller and dwindle to incompetence; no crown is constructed. The trunk is soft; lignification is so slight that the trunk can be cut with a penknife; secondary thickening is weak. As trees they are so feeble that they are often regarded as overgrown herbs. They have been dubbed umbrella trees and rosette trees from the umbrella-like rosette of large leaves. Their timber is useless; their appearance is unornamental; their parts are too large for the herbarium; they are intractable and generally unwanted; they are very little known and discarded as curiosities. They occur, usually with rarity, among diverse families of plants where the student of trees can recognize in them the same primitive rarity as the monotreme in zoology or the cycad in botany. One kind nevertheless is familiar as the American pawpaw or papaya (*Carica*) now widely cultivated in the tropics for its melon fruit. There are other species of *Carica* that are better trees, up to 50 ft. high, more branched and, as one has now learnt to expect, with smaller and even simple leaves".

Outside the Rain Forest

Earlier in this chapter we suggested that the plant world of the tropics cannot be understood by studying temperate plants. The opposite, however, is possible. With the help of Corner's series, the development of the plant world outside the rain forest can, in principle, be completely derived from that within the rain forest. But in this connection, a few things have yet to be made clear.

One should not suppose that the old pachycauls specialized only in the direction of leptocauly. Specialization in exactly the opposite direction also took place: from thick to thicker. For the seed in its most primitive condition, Corner postulated a diameter of about 1 cm, maybe a little bit more, about the size of some wild species of nutmeg, *Myristica* (Myristicaceae), the seed which he in fact pointed out as the most primitive (Corner 1976, Vol. 1:55). Evolution works in two directions: towards big size, such as in the durian (Fig. 13.10), the avocado and other typical shade germinators; and to small size, such as in the pioneers, the typical light germinators, which sprout on the bare earth and swarm out to far outside the rain forest, with the minuscule seeds of orchids as a final stage; the orchids which also assumed pioneer behaviour and became epiphytes, staying in the rain forest, but high up in the canopy.

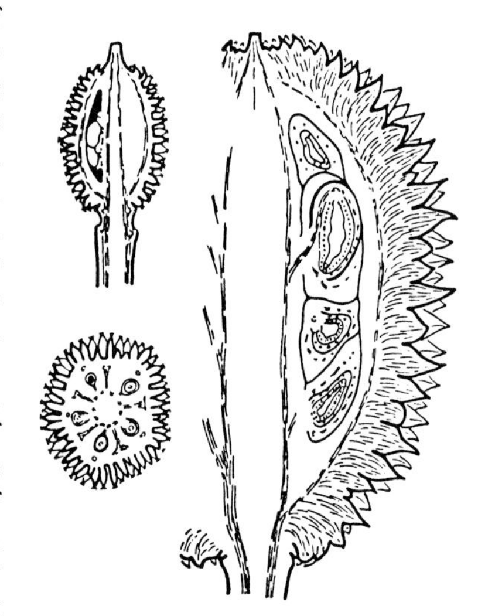

Fig. 13.10 Durian, developing fruit, in longitudinal and transverse section. The thorns are formed before the aril (Corner 1949)

Something similar happened to stems. While most plants evolved towards leptocauly, a few went in the opposite direction. A very wide trunk, which can contain a large quantity of moisture, is of no value in the rain forest, but extremely valuable in dry climates. The succulent form can thus be considered as a specialization with extra development of pachycaul characteristics. The result can be seen in the sometimes bizarre bottle trees in deserts, on small isolated islands or high up on relatively old mountains in the tropics, notably in the Andes and on mountain tops in equatorial Africa. Corner (1964, 30, 31) gives striking pictures of *Pachypodium* (Apocynaceae), *Adenium* (Apocynaceae) and *Dendrosicyos* (Cucurbitaceae) on Socotra (Fig. 13.11), and *Espeletia* (Compositae) in the Andes. Mabberley (1977), who deals with the *Lobelia* (Campanulaceae) and *Senecio* (Compositae) of the

Fig. 13.11 Hyperpachycaul curiosities, a species of *Adenium* (Apocynaceae) on the island Socotra (After Corner 1964, photograph 316)

east African mountains, speaks of 'hyperpachycauly'. Their occurrence in isolated spots indicates that they lost the competition against taller plants in the more hospitable areas, and also that they are of a very old origin, developed from life forms which formerly occurred worldwide.

The leptocauls, which in the rain forest gained in height, evolved towards more ecological versatility and adaptability, thanks to the fact that their leaves and seeds became smaller. First of all, this enabled them to settle in open places in the forest area, as 'biological nomads' with an r-strategy, discussed in Chapter 8. This strategy, moreover, proved useful for the survival of drought and cold, because small seeds can, and sometimes must, have a rest period.

Reduction of the whole plant body was the next logical step, a step which we find realized everywhere on the mountains in the tropics and the dry subtropics. White, for instance, wrote an article about the remarkable "underground forests of Africa" (White 1977) where close relatives of forest and savanna trees form woody, tuber-like bodies in the earth with sprouts standing above ground that perish in case of fire, while the tuber stays unharmed. Compared to them, several species of grass, such as the alang-alang, *Imperata cylindrica* (Gramineae), are much further developed towards leptocauly. They possess of fast-growing network of slim, subterranean rootstocks, exactly like the well-known weed *Triticum repens* (Gramineae), couch grass. Around the Mediterranean all intermediate forms from trees, via 'half-shrubs' (with stalks sprouting from a woody base just above ground), to real herbaceous plants may be found. Among the herbaceous plants perennials, biennials and annuals are discerned; in western Persia I noticed a predominance of perennials on the mountains, where the climate is relatively cool and humid, whereas the annuals were predominant in the desert. The biennials were found in between.

In the plant world of Western Europe there are pachycaul organs for which a new function has been found, namely that of storing reserve food for hibernation. The onion genus, *Allium* (Liliaceae) is an example of bulbous plants indigenous in Europe where a thick part of the stem is, in the words of Reinders (1949, p. 214) surrounded by "the swollen and fleshy lower ends of the closed leaf sheaths". The characteristic smell of onions undoubtedly deters many animals, and may be counted as a way of chemical defence, a mechanism which, again according to Corner, represents a recent direction in evolution, a direction taken independently by many different plant groups. A number of large plant families, which in the tropics are mainly woody, are found in western Europe with a modest number of leptocaul representatives. The Apocynaceae, for instance, with a world total of 300 genera and 1300 species (according to Lawrence 1951) are represented in The Netherlands by one creeping species: the periwinkle, *Vinca minor.* The Araliaceae, with 65 genera and over 800 species, have the ivy, *Hedera helix*. The Umbelliferae are so closely related to this family that botanists regard them as a herbaceous relative of the Araliaceae, which has populated especially the colder part of the northern hemisphere. The Euphorbiaceae, with 283 genera and 7300 species, many of which grow in the rain forest, have two wild genera in The Netherlands, with about a dozen species. The Guttiferae, with 35 genera and as many as 1000 species, are restricted to the tropics, but are represented in the Low Countries by nine species of the Hypericaceae, which are closely related to them. Of the Leguminosae, with a total of 550 genera and 13,000 species, the Caesalpiniaceae are completely woody and mainly restricted to the tropics, the Mimosaceae mainly woody and found in the tropics as well as in the subtropics and the Papilionaceae mostly herbaceous and found in all parts of the earth, in The Netherlands with 14 genera and about 54 species, of which only the broom (Genista) and its close relatives are more or less woody. The Rubiaceae, with 400 genera and 5000 species, for the main part woody, have 3 genera and 13 species among the Dutch wild flora.

Of course, the specific relationships are not so simple that we should imagine the plant world of the temperate zones to be derived directly from that of the tropics. All kinds of elements from the subtropics, for instance, have found their way into the tropics. One example is the genus *Rhododendron* (Ericaceae), on which detailed information can be found in Sleumer (1966). Of the 850 species, 500 are found on the Asian mainland and 280 in Malesia, for the most part on the mountains, with an occasional representative in the lowland forest and even among the mangroves. But it is generally true to say that the plant world originated in the tropics of the earth. Exactly where in the tropics, however, is still an unanswered question.

The tropical rain forest is a matrix of life forms; from there, lines can be drawn, probably via the secondary destruction and repair stages of vegetation, to the herbaceous plants and the few small leptocaul trees that we still find in our latitudes. If somebody tried to project the lines in the opposite direction, he would have to go against the direction of the specializations which have been discussed. Should the plant world outside the rain forest be destroyed, biological nomads would ultimately succeed in recolonizing drier and colder regions of the earth, since they already possess the adaptations needed to do so. But if we destroy the tropical rain forests, they could never again come into being.

What Does a Geologist Say?

The book by Flenley, *The Equatorial Rain Forest: a Geological History* (1979), is based on fossil finds, particularly those of pollen grains. These 'microfossils' are obtained from samples drilled out of the earth in places where for long stretches of time pollen grains rained down and quickly sank into a layer of mud.

Palynologists, as such specialized paleontologists are called, look for such formations mainly on the banks of lakes, and in peat bogs. The outer layer of pollen grains is extraordinarily durable, and rich in microscopic characteristics. Nearly every family has its own type of pollen; often, even genera can be discerned. The herbarium is a source of

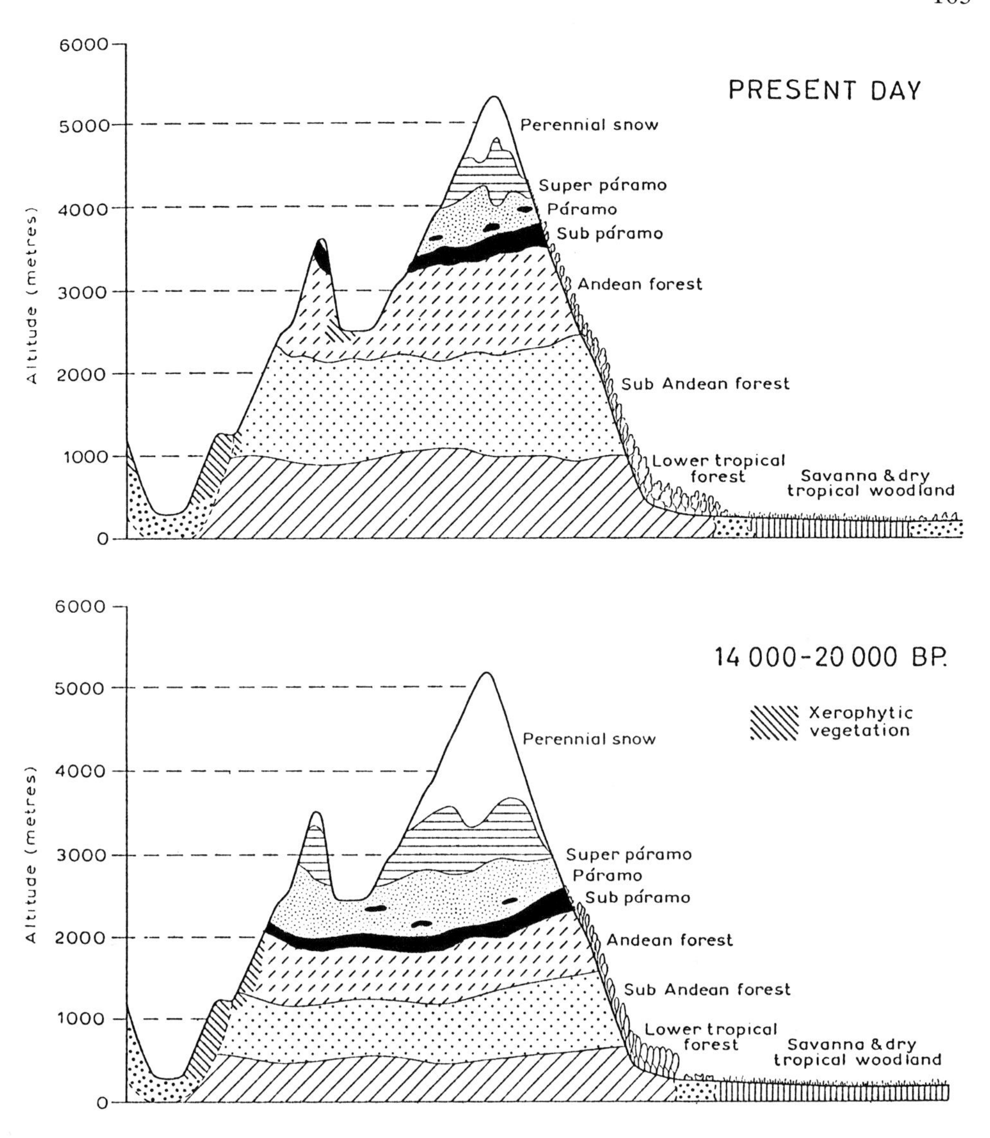

Fig. 13.12 Vegetation zones on the Andes, South America, now and during the Glacial period: the zones are shifted telescopically, but the lowland forest has remained stationary, leaving the richest, lowest regions intact (Flenley 1979)

information regarding the pollen of recent plants which is correlated to that of the fossil finds in order to determine them. With the help of radioactivity the finds can even be dated; a palynologist may discover in this way which plants subsequently have grown in a certain location. Then he has a look at the geomorphological information, and can thus reconstruct the picture of the changes in vegetation over as long a stretch of years as possible. The last 5 million years yield the best information; accordingly, Flenley pays mostly attention to the Quaternary period even though he devotes one chapter to earlier history.

Flenley is opposed to the picture of the evolution of the rain forest taking place only under stable circumstances. Already in his preface, he states: "The equatorial environment is now believed to have changed markedly in the past, and ice ages in temperate areas were, on the whole, times of aridity in the tropics, not pluvials. The rain forest, and other vegetation of equatorial regions, is now shown to be in a state of considerable instability. The effects of the last major climatic change are still reverberating there, particularly in montane areas". And on p. 28: "Hence the suggestion which is sometimes made that our present tropical rain forests are of great geological antiquity can be only partially true. The vegetation of equatorial regions has probably been in a dynamic state for a very long time indeed".

Whoever reads this interesting book will soon notice, however, that it hardly touches upon *our* subject: the tropical rain forest below 1000 m. It does indeed deal with vegetation along the equator, but almost exclusively (high) above the lowland. There on the mountains, considerable shifts have indeed taken place. During the Ice Ages, all vegetation belts shifted to lower altitudes, which made more space available, for example, for the migration of cold-loving plants from temperate areas to the mountains of Malesia. That the shifts should also have influenced lowland forest is not self-evident. In Flenley's Fig. 4.27 (Fig. 13.12), which shows sections of the Andes near Bogotá at the time of the Glacial maximum 14–20,000 years ago, and today, the small zone of the 'sub-páramo' has only moved from 2000 to 3300 m, and the upper boundary of the 'lower tropical forest' from 500 to 1000 m.

The 'seral changes', i.e. the changes in pioneer vegetations and successions, to which Flenley devotes his Chapter 6, are indeed considerable, but they do not concern the rain forest. As far as the lowland rain forest is concerned, Flenley is not very detailed throughout; the subject is rarely mentioned and he offers no explanation for the scarcity of information.

The probability that the rain forest area of South America was much smaller during the Ice Age has already been mentioned in Chapter 9. That the present climatic humidity makes today's rain forest area considerably larger than it was in former days seems plausible. But it is not plausible that such an abundance of species and variety of life forms as is characteristic of the rain forests of the Sunda Shelf could have evolved other than under the very circumstances which exist today, over a very long period of time, and more or less in this very area (Fig. 13.13). I do not maintain an unchangeable configuration of land, sea and climate, but only this particular unbroken evolution. That does not exclude the possibility of fluctuation of the rain forest area. Even if my haphazard estimate of the migration speed of the rain forest is too high by a factor of five, the rain forest could still bridge a distance of 2000 km in a million years. Moreover, the obvious differences within Malesia between the richness of the Sunda Shelf, which has stayed relatively stable, and the poverty of the geologically unstable Celebes, The Philippines and the Moluccas, does seem sufficiently convincing.

Fig. 13.13 Durian, *Durio zibethinus* (Bombacaceae) on a roadside stall. Port Dickson, Malaya. (Photograph MJ 1980)

14 How Species Are Formed

Problems

The preceding chapter was devoted to macro-evolution, the kind of evolution which over long stretches of time produces obvious differences: a historical and unrepeatable process. In this chapter we turn to evolution as it is revealed by a magnifying glass: evolution in full action, happening at the taxonomic level of species or even lower, where processes, in principle, do allow experiments. On this level, as far as herbaceous plants in temperate zones are concerned, Stebbins (1950) has assembled a large amount of knowledge in his classical work. For forestry purposes, several species of trees with commercial value have been studied, although their long seed-to-seed cycle makes this considerably more difficult than in the case of herbs. In a rain forest the problems are still greater. Trees in botanical gardens, which are always single examples, are of no use whatsoever if one wants to know what happens in the forest itself, where trees of one species are often few and far between and do not always flower at the same time.

Wind pollination is extremely rare in the rain forest; this means that every transmission of pollen which takes place involves an animal, a zoological link, with all the complications this adds. Without sound taxonomic knowledge of all the species that may be encountered in the course of the experiments, useful research is impossible. There are also technical difficulties. Even if contraptions are built to enable the researcher to work at the surface of big treetops, this does not imply automatic access to all the factors involved; although in the canopy the treetops are vast, and flowering (when it occurs) is abundant, in the lower levels the conditions are just the opposite; and this suggests that processes of a different character are in action.

For this reason, Ashton, who leads the research in this field, focussed his attention on both the huge dipterocarps and the low *Xerospermum* trees (Sapindaceae). This work still continues in Malaya. The articles of Ashton and his collaborators (Ashton 1969, 1973, 1976a; Gan 1977; Kaur et al. 1978) are the only ones I know of that do not approach the problem of speciation in the rain forest in an exclusively theoretical way, even though their research focusses on only one category of species, viz. those forming solid blocks. There are another three categories which must be considered. I must also dissent from the majority view that the processes of evolution in the rain forest do not differ essentially from those that take place outside the rain forest, and I offer the evidence below to justify my opinion.

Plants and Animals, Their Different Evolutionary Pace

The period of time which elapses from the moment a rain forest plant is fertilized to the moment when, after the intermediate stages of growing up and reaching maturity, a new generation comes into being – its life cycle – is usually far longer than in the case of rain forest animals. A female orang-utan produces her first young when she is about 10 years old (Rijksen 1978, p. 381); but a robust dipterocarp flowers for the first time when it is 50 or 60 years old.

A difference of a factor five or six which seems to be valid for smaller plants vis à vis smaller animals as well. Rain forest animals, therefore, produce far more generations in a century than plants do; they are also mobile, which gives them the opportunity to investigate the suitability of a certain biotope in the environment, as well as enabling them to use several biotopes more or less simultaneously.

All these factors make it easy to understand that the process of speciation works much faster in animals than it does in plants, and indeed the number of animal species is four times as large as that of plants. It is probably in the tropical rain forests that the greatest differences in evolutionary rates are found: the reproduction rhythm of the animals is never interrupted by unfavourable seasons, and the largely woody character of the vegetation prolongs the life cycle of the plants. There is also, in an even climate, a tendency for plants to flower irregularly, followed by resting periods that may extend for years. Ashton (1969, p. 187) quotes Stebbins' estimate that under maximum selection pressure a new species can develop in a minimum of 50 generations, but that 250,000 generations seems more likely in short-lived plants. It should be remembered in this respect that most rain forest trees are extremely long-lived, at least an estimated 100–500 years, and certain trees, notably *Balanocarpus* [at present *Neobalanocarpus* (Dipterocarpaceae)] and the ironwood, *Eusideroxylon zwageri* (Lauraceae) may possibly live much longer.

Genetic Exchange

Pollination is the process by which pollen grains (produced by the stamens) are transferred to the stigma of a plant belonging to the same species. In the rain forest this is done by birds, bats and, of course, insects. It could be said that they transport little parcels of genetic material from one place to the other.

Pollination may be followed by fertilization: gametes merging into a zygote. Surrounded by protecting organs the zygote can grow into a seed in which the genetic properties of both parents are stored. In the process of dispersal the seed is also dispersed. There is thus continuous year-round transportation of genetic material in the rain forest, a bit reminiscent of the traffic that goes on, once again, in a big town (the activities of the ants being approximately those of the sanitary department). If the fertilizing pollen originates from the same plant, we speak of self-pollination; if not, of cross-pollination. Thus, only cross-pollination makes the exchange of chromosomes between different individuals possible. This exchange is necessary to ensure that the genetic composition of the members of the fertilizing complex remain heterogeneous, and axiomatically, that in a heterogeneous community no two individuals will be exactly identical. Even along a lane planted with domesticated horse chestnuts, *Aesculus hippocastanum* (Hippocastanaceae), one can observe certain trees coming into foliage earlier than others, year after year. Consequently, some chestnut trees will bear fruit earlier than others. And even the fruits themselves are highly characteristic of each individual tree, or there would be no favourite 'conker' trees.

What the preservation of a variety of individuals can mean to a species becomes clear when a severe night frost afflicts the early-flowering individuals and leaves the late-flowering ones unharmed. In such extreme circumstances, a few individuals will have passed the selection 'test', if there is a heterogeneous population. To give another example, every excessively dry season will take its toll of young trees. The individuals which best resist drought will have the best chance to grow, and therefore to reproduce themselves. The wider the range of possibilities to which a species as a whole must adapt, the better its chance will be to survive and to spread itself again after a change of circumstances.

Where genetic exchange does not take place, the population becomes more and more homogeneous and the genetic potential less flexible with respect to changes that may occur.

We know that some plant populations greatly suffered during the Ice Ages. Now, although they still succeed in maintaining themselves, they no longer spread and another severely adverse change of circumstances will probably finish them off. This leads us to the subject of conservation, which will be discussed more later on.

If both cross-fertilization and self-fertilization can occur, it must be possible to express their relative frequency of occurrence in percentages. The distance between two individuals of the species will be one of the influencing factors, although in the rain forest distances will never be simply a matter of metres because of the biological associations involved (the availability of birds, bats or insects), and there are a number of other influences: rain, for instance, and predators and the simultaneous flowering of other plants. Everything that happens in the field of biological relations, however, can be expressed in percentages or curves. Mutation, hybridization, polyploidy, selection, isolation, migration; each one of the general, comprehensive processes involved in speciation with all its fluctuations, affects the percentages, and in the course of time these effects will be cumulative if they do not start interfering with each other.

Selection: Edaphic and Biotic

The content of the preceding section is part of the basic curriculum in biology. The well-known factors of evolution are always at work in the rain forest as elsewhere, but their relative influence differs considerably from that in the temperate zones. For instance, drought and cold are usually the first selective factors in the world outside the rain forest. In Europe we know this only too well; we hardly know of any others! But in the constantly humid, hot climate of the rain forest, drought and cold are of no importance except insofar as they define rain forest boundaries. Instead, other factors take the lead. The nature of the soil becomes much more important in the process of selection, and – in a different manner – so do the factors which concern the relations between plants and animals.

The soil-defined or edaphic complex of factors, Ashton (1964) gives much information on this, largely decides the character and composition of the rain forest. Since the composition, i.e. the number of different species and their relative frequence of occurrence, is determined by edaphic factors, and since each species has a range of preference for and tolerance of different soil characteristics still largely unknown to us, it can happen that poor soil can very well carry tall forests, rich in species, even though at first this appears to be an anomaly.

As the complex of biotic factors is even more varied, the possibilities for selection are proportionally more numerous. We refer to the story of the durian and the bat as one example which fortuitously has been investigated. Such a story could possibly be written about every plant or animal species in the rain forest; only a hundred of them at most are presently known.

Biotic selection never operates in a simple way. The interaction between organisms is too complicated and too varied, and it is possible that every difference between plants of the same species may have adaptive value. Generally speaking, dipterocarps can be considered poor dispersers; nevertheless, dipterocarp species which do not possess winged seeds are never the tallest, and conversely, the tallest dipterocarps always possess winged seeds. Even Ashton (1969, p. 172) has difficulty in deciding whether or not a certain characteristic has adaptive value. In this connection, he mentions the variable number of stamens in the flowers of several related species of *Shorea: S. sumatrana* in Sumatra and Malaya, has 25; *S. seminis* of Borneo and The Philippines has 30–40; *Shorea foxworthyi* has 33–41 in Malaya, but only ca. 32 in Borneo; *S. scrobiculata* has 28–31 in Malaya, but 20–30 in Borneo. Such a variable characteristic may possess adaptive value, but it is impossible to determine this in a herbarium.

Dispersal, Growth, Flowering

Wind pollination, on which 21% of the flora in central Germany depends, is of little significance in the rain forest: among 760 species in 40 ha rain forest in Brunei (Borneo) Ashton (1969, p. 178) found only one species that was wind-pollinated, and that one only on the mountain ridges where the rising air currents brush along the slopes. But the dispersal of seeds by wind is far more common. A good example is *Alstonia* (Apocynaceae), a small genus of robust trees with fruits 1 cm in diameter and 50 cm long filled with plumed seeds resembling thistle fruits. Other examples are *Cratoxylum* (Hypericaceae), *Engelhardia* (Juglandaceae), *Ventilago* (Rhamnaceae) and many other comparatively small, widely distributed genera, all with winged seeds and fruits. Wind dispersal results in settlements consisting of isolated individuals (Fig. 6.5), and wind-dispersed species generally belong to the fourth category of rain forest species. These categories will be discussed later (see p. 170).

Ashton (1969, p. 182) quotes Verne Grant as saying that the dispersal of one seed is equal in value, genetically speaking, to the dispersal of thousands of grains of pollen over the same distance. A minimal dispersal of seeds, therefore, cannot be genetically compensated for by a copious dispersal of pollen.

In the case of poor dispersers (the occasional seed of which may get a little further away, since minimal dispersal seems more likely than no dispersal at all), a colony will be formed around the parent tree. For example, after a fruiting season dipterocarp seedlings cover the ground like a furry blanket. This creates the conditions for strong competition among the individuals of a species, and places a heavy selective advantage on such characteristics as fast upward growth and large leaves.

One should not imagine such a colony to be more than an accidental concentration of a small number of trees, but nevertheless, genetically, they will behave differently from wind-dispersed trees. Within colonies, a new gene will spread slowly and systematically, allowing for time to adapt to local circumstances. As a result, there will be more differentiation than in the case of wind-dispersed species where a new gene will only spread in a criss-cross way from one isolated individual to another. Consequently, a wind-dispersed population as a whole will stay uniform.

As for the species, the seeds of which are dispersed by animals, they occupy an intermediate position between those wind-dispersed and those minimally dispersed. Their exact position will depend on the numbers of the animals concerned and on their eating, moving and resting habits.

Whatever the dispersal method, all rain forest seedlings have to overcome a multiplicity of factors which will never be completely known to us. Whether or not a sapling of a few metres high will succeed in becoming a huge tree is, to a large degree, a matter of chance: will there be sufficient space available when it is needed? And every big tree must also in the course of its growth adapt to a whole spectrum of microclimates between the ground and the highest point of the canopy.

The microclimates in the lowest and the highest layers of the forest are the least favourable for the growth of plants, the lowest ones because of lack of light and the upper ones because of extremes in temperature and moisture, either very dry or exposed to heavy downpours. For this reason Ashton expects biotic selection to take place mainly in the intermediate layers of the canopy; and certainly this is where the largest number of species can be found.

In an even climate, as we pointed out, not all the individuals of one species flower simultaneously. This creates a genetic barrier, though not a total barrier: in some years some trees of the same species do flower simultaneously, or some of their branches do. The flowering of dipterocarps is characteristically irregular in this manner. Mast flowering takes place only once in 5–13 years, depending on the species, but at the same time there are individual trees which flower separately. There is no specialized pollination mechanism in the wide open, star-shaped flowers, however, and although there is enough cross-pollination to keep a population homogeneous over considerable distances, dipterocarp range as a whole is confined both by inadequate seed dispersal and by limiting edaphic factors.

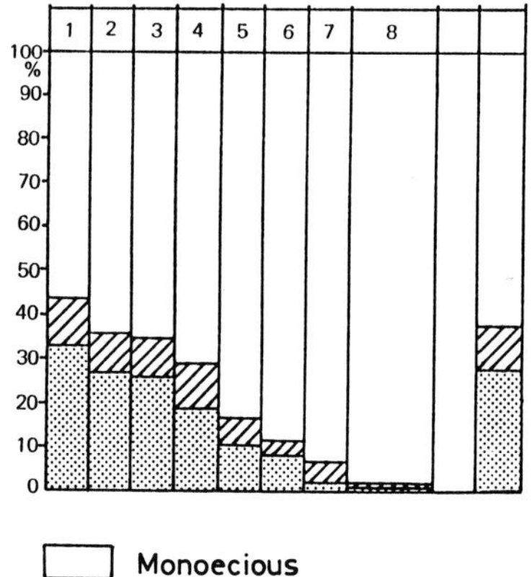

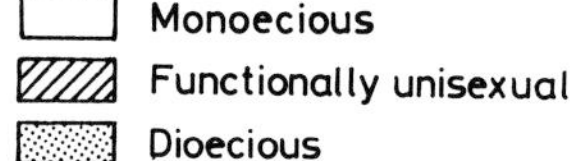

Fig. 14.1 Percentages of monoecious, functionally unisexual and dioecious trees, in rich rain forest in Sarawak, Borneo. Especially the thinnest trees are dioecious or behave as such (Ashton 1969)

Distance; Dioecism

The large number of species per hectare implies, as seen in Chapter 7, a long distance between the individuals of one species; an average of 100 m has been mentioned. This greatly diminishes the chances of cross-pollination, with all the risks involved, but without cross-pollination there is, in the long run, no chance of survival; so, cross-pollination has to be guaranteed in some way or other.

Plants have at their disposal a well-known device for this purpose: the flowers are either male or female (unisexual) and are also produced by different plants (dioecism). A temperate climate example is the willow, *Salix* (Salicaceae), with its male and female catkins. Dioeciousness diminishes the chance of pollination by 50%, since only half of the plants will carry fruit; but we must conclude that it is nevertheless genetically profitable. The risk for the species concerned that pollination will only rarely take place is apparently less that the risk that the genetic base may become too narrow.

In the British Isles 2% of the flora is dioecious. In the rain forest of Sarawak (Borneo) Ashton (1969, p. 177) found that among trees with a diameter of 10 cm or more 26% were dioecious, and that in another 14% there was either a divergency in the moment at which stigma and stamen matured, or unisexuality, or polygamy – the last term meaning that male and female flowers occurred besides the bisexual ones. Dioeciousness was most frequent among the thinner trees, i.e. those occupying the lower layers of the canopy (Fig. 14.1). This is reasonable, because this is the level at which the highest concentration of flowering species occurs.

Certain families are almost completely dioecious: Burseraceae (the family of which by the way, includes the biggest trees of the Malesian forest; nothing can be expected to be simple in this field), Ebenaceae, many Euphorbiaceae, Flacourtiaceae, Gnetaceae, Guttiferae, Menispermaceae (small lianas, mostly), Myristicaceae, Pandanaceae, Piperaceae and Rafflesiaceae. In many other families, for instance, among the Lauraceae and especially the Sapindaceae, reduction of the male or female organs occurs.

The minimum area needed for the conservation of dioecious plants is much larger than that necessary for monoecious species: consequently, they form a category that needs special attention.

Four Categories

So far our meagre survey has covered some of the traffic that goes on in our biological town, with its many methods of transport and many varieties of encounters, a town where the work never stops because of unfavourable weather. But what exactly are the results of edaphic and biotic selection in the rain forest, as far as speciation is concerned? Basing ourselves on Van Steenis (1969) we will provisionally distinguish four categories of species:

1. Coenospecies, widely distributed and heterogeneous;
2. 'Blocks' of closely related species;
3. Species of genera that are taxonomically isolated, often characterized by peculiar structures: excrescences, concrescences, discs or pitchers. Under less favourable cir-

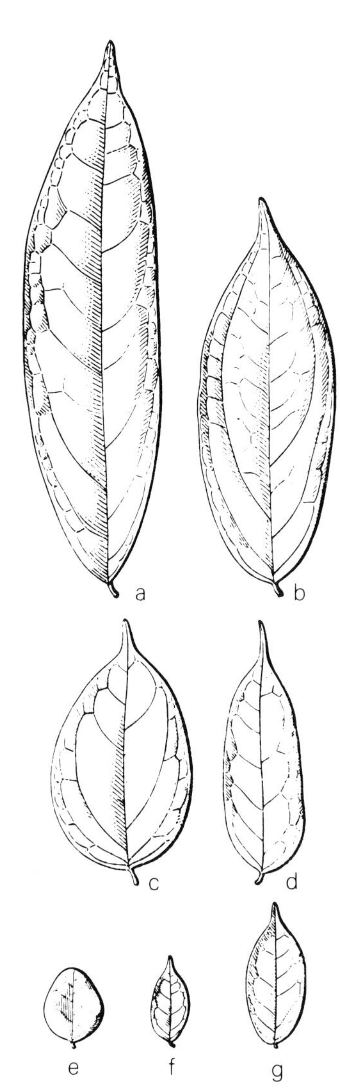

Fig. 14.2 A coenospecies, *Rourea minor* (Connaraceae): variation in leaf size and shape within a species, within which a systematist nevertheless cannot distinguish distinct taxa. (*Flora Malesiana* i 5: 516, 1958)

cumstances they would have been eliminated by selection. In the rain forest, they must be considered as 'hopeful monsters';

4. The many remaining species of which nothing is known.

Let us have a closer look at these categories.

Coenospecies

To this category belong species which are, as such, easily recognizable, but within their own taxonomic circumscription are so pluriform that the taxonomist who wants to systemize their pluriformity is completely baffled. *Pometia pinnata* (Sapindaceae) is such a species; it took me (see Jacobs 1962), a year to disentangle its variability. The range stretches from Ceylon, Indo-China and Taiwan to Samoa. The eastward frontier of the range in the Pacific is the same as that of the genus *Pometia*, and dispersal is bat-assisted throughout the range. The pluriformity of this species is greatest in the former Sundaland, i.e. in Sumatra, Malaya and Borneo. Towards the marginal boundaries of the range the differences between populations become sufficiently clear and constant to make them the base of *formae* of which an identification key can be made. In the centre of the range, this can only be done for part of the available material. Another part shows nuclei of variability from which lines to the *formae* can be projected, but the nuclei themselves can rarely be separated, although they can be discerned. Accordingly, it is not possible to give them a taxonomic rank.

Ecologically speaking, *Pometia* is a flexible and enterprising genus: originating in lowland rain forest it has spread to an altitude of 1700 m in northern Sumatra, penetrated the swamp forests in Borneo and colonized the secondary forest in New Guinea. Differentiation within this species is obviously in full swing. At the outer limits of its range crystallization of the variability has progressed further, but in the centre the pot with the 'mother liquor' is still cheerfully bubbling on the fire of evolution.

Analogous examples are provided by some lianas. It turned out to be impossible to classify *Connarus semidecandrus* and *Rourea minor* (Connaraceae) in a way that was taxonomically acceptable, even though taxa could be discerned in *Connarus*. Leenhouts (1958) indicated the pluriformity of the species with the letters a–d only, giving a short summary of the characteristic differences, and pointing out to which taxon the specimens belonged which formerly had mistakenly been described as separate species (Fig. 14.2).

Coenospecies, of course, can also be found outside the rain forest. They have been studied far more extensively in the temperate zones, and their minute differences have been treated statistically and correlated in a variety of experiments. Accordingly, much smaller taxa have generally been recognized as separate species. The genus *Euphrasia* (Scrophulariaceae), for instance, our eyebright, is considered in Europe and North America to be a complex of closely related species, as is the hawthorn, *Crataegus* (Rosaceae). In the tropics, at the present stage of scientific research, which generally has to make do with a limited quantity of herbarium material, each genus would be considered a single, but variable species; on this subject, see Van Steenis (1957).

'Blocks' of Closely Related Species

This second category of species is found in a later stage of evolution. Many examples can be found in Malesia: *Antidesma* (Euphorbiaceae), *Aglaia* (Meliaceae), *Ardisia* (Myrsinaceae), *Chisocheton* (Meliaceae), *Cryptocarya* (Lauraceae), *Dipterocarpus* (Dipterocarpaceae), *Dysoxylum* (Meliaceae), *Eugenia* (Myrtaceae), *Garcinia* (Guttiferae), *Hopea* (Dipterocarpaceae), *Litsea* (Lauraceae), *Memecylon* (Melastomataceae), *Myristica* (Myristicaceae), *Polyalthia* (Annonaceae), *Psychotria* (Rubiaceae), *Shorea* (Dipterocarpaceae),

Vatica (Dipterocarpaceae) and many more. Their contribution to the variety in species of the lowland rain forest is considerable: most of the 'blocks' just mentioned contain 30–100 species.

Ashton devoted an important article (1969) to this type of speciation among the dipterocarps; the species in this family are divided by small, exactly defined characteristics. One would not expect such characteristics to have any biological significance, but we could easily be mistaken in this respect, according to Ashton. Small-leaved species are more frequently found on dry and poor soils. "Similarly, within the *Shorea parvifolia* series, where members mainly differ in small details of bark, indumentum and leaf texture, the most fissured-bark forms (*S. revoluta* Ashton, *S. foraminifera* Ashton) occur on the most acid or driest soils, whereas the Selangor form of *S. parvifolia*, whose range occupies the most fertile edaphic habitat, is that with the thinnest, smoothest bark. Likewise *S. revoluta, S. foraminifera, S. scabrida* Sym., *S. dasyphylla* Foxworthyi and *S. ovata* Dyer possess either more coriaceous leaves or denser longer tomentum or both than the intermediate *S. rubra* Ashton, whereas *S. parvifolia* is sparsely tomentose to subglabrous in the Selangor form and has the least coriaceous leaf. These relationships again suggest that apparently trivial characters may possess adaptive significance. These species furthermore possess different growth rates and timber densities which may be expected to reflect differences in metabolic rates, light tolerances and compensation point" (Ashton 1969, pp. 171–172).

Flora Brasiliensis, tabula LV. A plantation of bananas, *Musa paradisiaca* (Musaceae), introduced from Malesia, near Rio de Janeiro. Note the false trunk, the frayed leaves and the terminal inflorescences, on which the fruits are curved upwards. Drawn by B. Mary in 1837 (Martius 1840–1869)

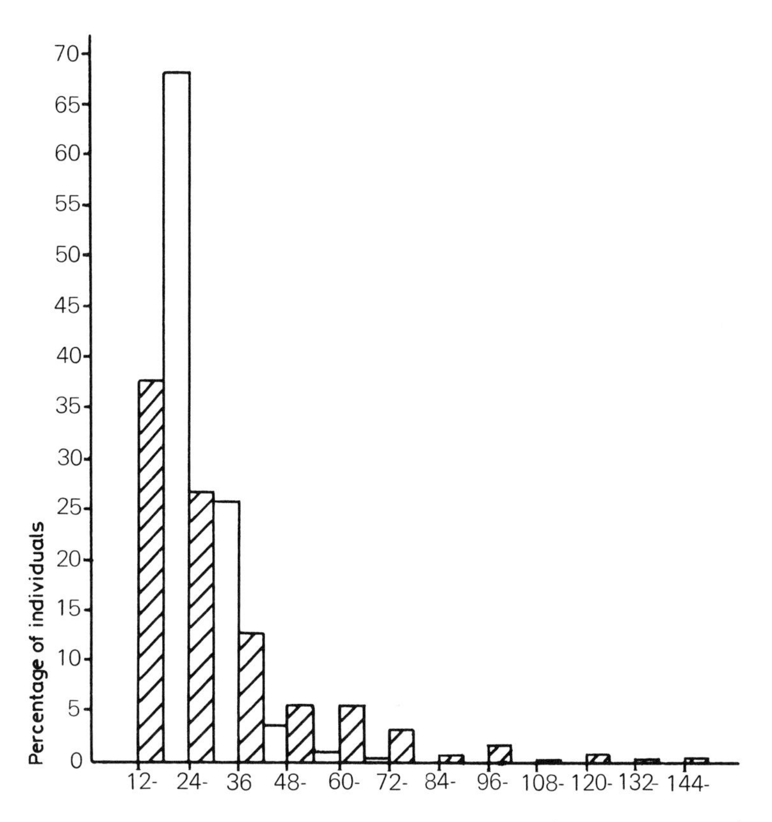

Fig. 14.3 Percentages in size classes (measured by girth of trunk in inches, each size class accordingly corresponds with 1, 2, 3, etc. × 10 cm diameter) for two subspecies of *Shorea macroptera* (Dipterocarpaceae) in North Borneo

It may be right to suppose that every difference in structure, our criterion of classification in plants, is, in principle, significant in the selection process, unless the opposite has become evident. But with regard to the rain forest, we have to reckon with such long periods of time that even a biologist is baffled. If we put a generation of dipterocarps at half a century, then 20,000 generations take us only a million years further. What do we know about the accumulation of small effects in percentages that are very small, but nevertheless greater than zero? As we have observed earlier, the evolution of plants endowed with poor means of dispersal takes place colony-wise; then new genes will linger a long time in such colonies, which results in adaptations and in marked isolation of populations. Moreover, Ashton found that in a number of cases *apomixis* probably plays a part, i.e. fertilization without pollination, a kind of asexual reproduction which is, in fact, a tendency opposite to the way in which plants try to neutralize the effects of isolation by becoming dioecious (see Kaur et al. 1978) (Fig. 14.3).

The existing isolation is promoted by the process of mutual avoidance that can be observed in closely related, only slightly different taxa which are found in one and the same area. We will come back to this subject in the section on Ecological Niches.

Hopeful Monsters

The third category consists of the ‘peculiar’, taxonomically isolated genera, sections or species. *Pterisanthes* (Vitaceae), a small genus in western Malesia, is an example. The inflorescence does not consist of a system of branches, such as found in the related vine and ampelopsis, but of a fleshy, elongated body which carries the flower pedicels (Fig. 14.4). A second example is *Monophyllaea* (Gesneriaceae), where the whole plant consists of a single

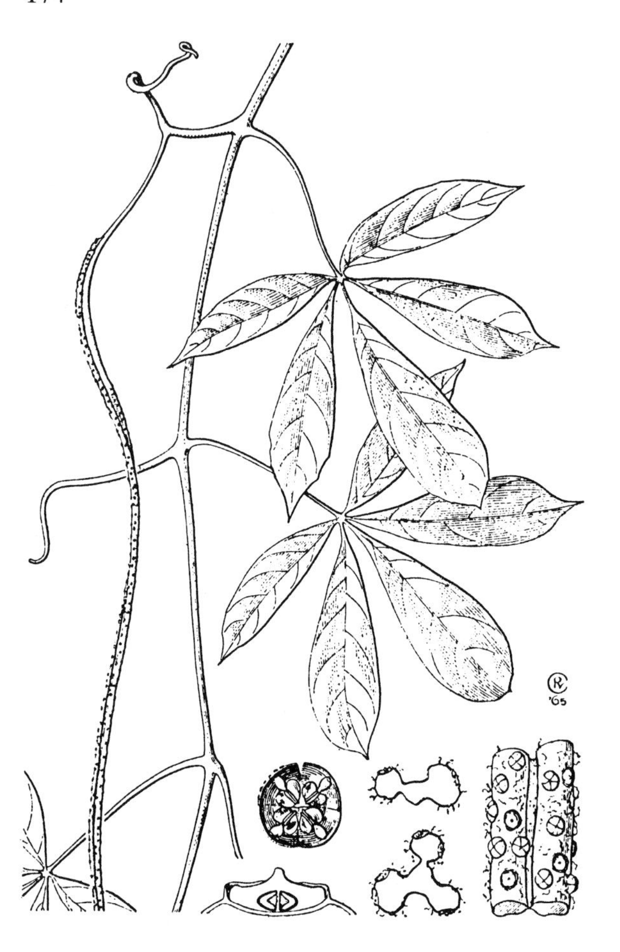

Fig. 14.4 An unusual inflorescence in the Vitaceae: *Pterisanthes linguispicatus*, a small liana from the rain forest of Borneo. The inflorescence is fleshy and strap-shaped instead of highly branched (Bot. Jahrb. 86: 387, 1967)

greyish-green leaf the size of a shoe, on which the inflorescences are inserted. The leaf is one of the two cotyledons, very excessively developed; no other leaves develop at all. The plant is found on calcareous soil under overhanging rocks (Fig. 14.5).

In the opinion of Van Steenis (1969), who based this theory of 'saltatory[4] evolution' on a large number of similar examples, such structures originally came into being as deformities. In the rain forest, their chance of survival was optimal, because such selective factors as drought or cold did not eliminate them straight away, as would have happened elsewhere. 'Hopeful monsters' is thus an apt name for them. If such species succeed in surviving, this means a step up to a relatively high taxonomic level because the taxon has obtained new and obvious characteristics. Plants, according to Van Steenis (1976), possess ample 'space' (which he calls *patio ludens*) in which non-adaptive characteristics have a chance to develop, unhampered by competition or selection. (This view obviously differs considerably from Ashton's, but it was developed with regard to another category of species.) Animals, in Van Steenis' opinion, do not possess such room for genetic experimentation, or very little.

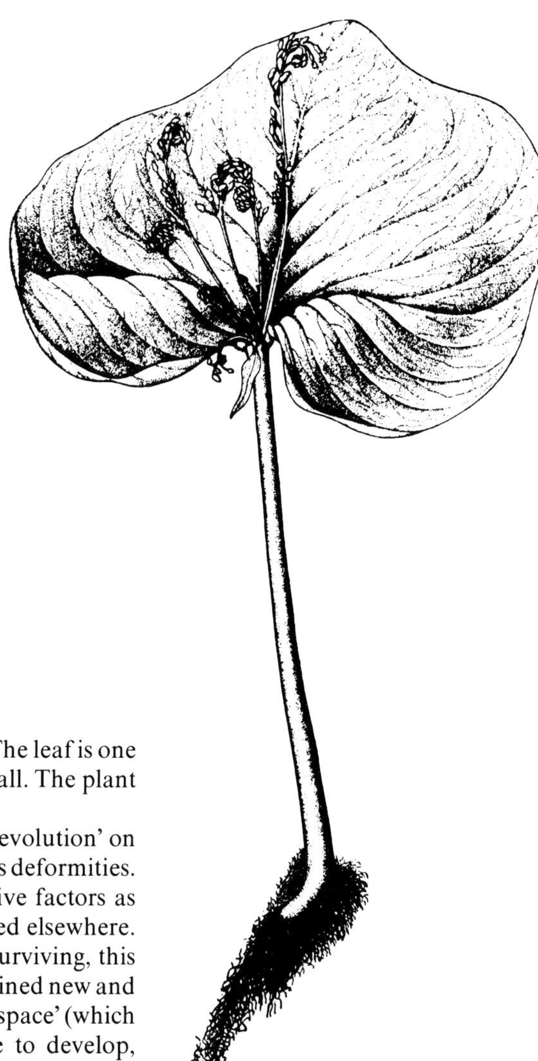

Fig. 14.5 *Monophyllaea horsfieldii* (Gesneriaceae): a plant that consists of a single leaf, on which the inflorescences are implanted. The leaf is a single, enlarged cotyledon, the other one is present opposite to it (*Flora Malesiana* i 4: xxii, 1948)

[4]Meaning jump-wise, not gradual

Remaining Species

Apart from the three categories mentioned above there are many smaller genera (up to 10, 20 or 30 species) in which the species are obviously related, but nevertheless equally clearly separated. An example for Malesia is *Rinorea* (Violaceae) with 11 species in the rain forest. Ranging from Ceylon to Micronesia and Australia, they consist of small trees, not remarkably variable and with little endemism (*Flora Malesiana* i.7:180, 1971). Other such families are the Apocynaceae, Asclepiadaceae, Celastraceae and Icacinaceae.

It is not to be assumed that in this category species will be formed according to one specific mechanism. But insufficient evidence is available to have any idea whatsoever on this subject. Each case has to be studied individually, with all its biological relations; and scientists have not even started to make an exact inventory of the genera confined to the lowland rain forest, all of which would therefore be subject to comparable conditions.

Ecological Niches

It is not important here to define exactly the concept 'ecological niche'. It could be understood to indicate the place in an ecosystem into which a taxon fits because of specific aspects of its ecology. As Ashton remarks (1969, p. 166), it is not feasible to postulate a separate ecological niche for every one of the 60 species of *Shorea* (Dipterocarpaceae) in Malaya; nor does he believe in the old adage that only one species will fit into a single ecological niche.

The curious fact has been brought to light, moreover, that the same species, when living on an island with few other species, occupies a greater number of different habitats, i.e. fits a wider spectrum of different ecological niches, than it does on an island rich in species. In this respect, Ashton (1969, p. 173) mentions a comparison between New Guinea and the Solomon Islands concerning trees and ants. In the Solomons, each one of a small number of species had a wide ecological spectrum, while in New Guinea, among many species, each one had a narrow spectrum. It is, therefore, possible to conclude that abundant speciation leads to ecological specialization, which is indeed plausible, if one keeps in mind the influence of biotic selection. Two species of trees bear fruit at the same time: monkeys and hornbills feast on them for a short time, disperse a few seeds and then move on. Serial fruiting of the two species, on the contrary, guarantees to each undivided attention and more abundant dispersal. There is no doubt that the willow profits from its early flowering, which means abundant attendance by bees and, accordingly, wide pollination. And for the animals involved in pollinating and dispersing, too, there is a premium on spaced flowering and fruiting times.

In a seasonal climate, limits are put to such spacing. In drier continental Asia dipterocarps flower punctually every year at the same time and they bear fruit a certain number of weeks later. But seasonal limits hardly exist in the constant climate of Malesia; with the result that while several species will occasionally flower simultaneously, although at random intervals, many individuals also deviate. Medway (1972), whose observations, made from a tree platform, have already been discussed in Chapter 3, discovered that individuals of *Anisoptera laevis* (Dipterocarpaceae), *Lithocarpus elegans* (Fagaceae), *Santiria laevigata* (Burseraceae) and *Shorea dasyphylla* (Dipterocarpaceae) all flowered at different times even though they stood close together.

The reason is obscure. But it is obvious that the result of such small differences will be that all kinds of biological events related to the flowering of those species, such as the movements of pollinating and dispersing animals, will now start to take place at different times or in different places, which may be situated next to, above or below each other. Mutual avoidance, resulting in the spacing of biological activity, obviously occurs.

By mutual avoidance the available space in the canopy is filled up evenly. Height counts as a premium because of the larger quantities of light available in the canopy. There is also a premium on shade tolerance in the empty space just below the canopy, where the branches of the growing tree die off. Another premium is on flowering just after the greater part of the dipterocarps, because at this time plenty of pollinators are still moving around unemployed. When two species have to share an ecological niche, they do so by mutual avoidance. If differentiation goes on, which will happen when circumstances are favourable, and four different species have to fit into a niche all four of them will start to avoid each other. And so on, and so on.

A process like this needs time, which means: millions of years. And time in this respect has to be measured under reasonably uniform conditions, i.e. in a stable climate and on soil which is reasonably stable, geologically speaking. The richer the evolutionary potential of a rain forest, which generally means: the larger its continuous range, the richer its variety in species will become in the course of time, and this almost automatically implies that its structure will become more complex if circumstances remain unchanged. The lowland rain forest of Sundaland (Sumatra, Malaya, Borneo) is the outstanding example of this process, particularly the forests of the geologically stable island of Borneo. In areas that are less rich biologically or are more restless geologically, such as the Moluccas, the Solomons, and to a certain extent New Guinea, the ecosystems will never be so richly varied.

When, at least 30 million years ago, Sundaland became populated by dipterocarps, they drove out the already present flora of gymnosperms because they were biologically better equipped and, accordingly, in a better position to compete. The area quickly became filled up with widely distributed species, a certain number of which are still there. In those widely distributed species which were initially polymorphic, just as our coenospecies still are today, local mutations occurred and spread over certain areas while adapting to the edaphic circumstances under which they grew. More species appeared in those genera which we mentioned under the 'species-block formation' category and many others. The poverty of the soil combined with the ever-wet climate (the second condition, of course, contributed to the first) encouraged the dipterocarps to evolve means of defence against seed-consuming animals, one of these being irregular flowering.

In the course of tens of millions of years the trees evolved one minimal advantage after another: and nowadays these giant trees dominate the rain forest of western Malesia (Fig. 14.6). In and below the huge forest canopy, the principle of mutual ecological avoidance also made existence possible for an ever-increasing number of species, although in ecological niches which grew progressively smaller and smaller. Gradually, species became more and more adapted to particular soil conditions and times of flowering, in a word: genetically more rigid. And while the forests of western Malesia could, because of their unique development, evolve into the richest forests on earth, they have also by the same token become the most vulnerable.

Fig. 14.6 Giant relatives of ginger (Zingiberacea) at the margin of the rain forest along a road, in Malaya (Whitmore 1975)

15 At the Fringes of the Rain Forest

Once Again: The Climax!

The natural limits of the rain forest are defined by climate and soil. The altitude above sea level influences temperature and is consequently an aspect of climate. We have already seen that 'seasons' in the tropics are primarily dry as opposed to wet seasons, which can, like altitude, impose serious biological limitations. Then there are the edaphic or soil factors. First of all is the (ground) water level. Where it is too near the surface for the greater part of the year, it is impossible for many plants to grow. Another limiting factor is the quantity of available inorganic nutrients which may be too small in some soils, even for the tolerant rain forest plants.

Any plant community optimally developed as to structure and composition within the limits set by the always present limiting factors, is called a climax vegetation, one which will not change further, if the climate remains stable, however long it is left alone. The climax community contains the maximum number of possible species, although this number may increase very slowly through evolution. Correspondingly, it can become only poorer if it is damaged. The vegetation types which will then develop are called *seres*, mentioned previously in Chapter 8. There we have seen that the accessibility of the site to seed bearers of primary forest species is a vital factor. If this factor is neglected, and the factor of human influence is accepted, while judging all sorts of vegetations, then certain types of vegetation which are, in fact, destruction stages may come to be considered climax vegetations, although this will only be so because they exist under the continuous pressure of a destructive factor (usually human) and are deprived of seed supplies from real climax vegetations. In this way standards for judging certain types of vegetation will have been abandoned. Since the areas still covered with real climax forest are dwindling so rapidly, it is not unimaginable that our knowledge of what type of vegetation could possibly grow in a certain place will become lost. On the one hand, we should learn to know or to recognize the original, and maximally different, vegetation types, assessing the possible damage. On the other hand, we should, when mapping out the vegetation, only indicate what is still left, and not what ought to grow there. Deviating from this line of conduct could lead to a too low estimation of the amount of human destructive influence. However tempting and reassuring this may be, it is the wrong basis for making adequate assessments for both exploitation and conservation because the baseline is artificial.

It was only during the 1920s that some research workers started to realize the scale on which vegetation had been destroyed by human influence. It also became clear that large-scale destruction could be brought about by relatively low numbers of people. Continuous felling and burning have caused the formation of deserts in dry climates, and have ravaged hundreds of thousands of square kilometres of moister forest. An excellent book by W.L. Thomas, *Man's Role in changing the Face of the Earth* (1956), discusses many aspects of this problem, with a historical perspective. It is difficult to believe that in some once well-forested regions only a few trees may now be left. But it is nevertheless true that where only a single tree grows at present, a forest could grow, and probably has indeed once grown. On the island of Cebu in the Philippines, 4700 km^2, there is reportedly not a single tree, yet tropical rain forest is the climax vegetation in the whole area. Nowadays Cebu is covered mainly with cogon, *Imperata cylindrica* (Gramineae), the grass known as alang-alang in Indonesia. It is important, therefore, to study the vegetation types surrounding real rain forest, keeping in mind the question of climax and *sere.*

We take the natural climax as our standard in this chapter; a climax of which the limiting factors are not human in origin. In some cases, where man has not yet, or only recently, interfered, such climax vegetations are still easily recognizable. In Africa, with its comparatively poor rain forest area, and with its long history of human interference, they are less easily recognized. In south and southeast Asia, too, one has to interpret, with the help of fragmentary observations only, a mosaic in which seasonal gradations from short and dry to long and dry occur, and which varies in altitude from sea level to nearly 2000 m, and which includes examples of all degrees of destruction and regeneration. In the scope of this chapter, however, where we restrict ourselves to the rain forest proper, we have to ascertain the limiting climaxes for that kind of forest. Many of these climaxes have their own typical destruction stage, which will also be discussed.

The differences and the relations between primary and secondary forest have already been discussed in Chapter 8. We will, however, return to this subject in Chapter 17 (Damage and Destruction).

Coastal Forests

The clayey accreting shores of the tropics are the habitat of the mangrove, a forest type poor in species, but rich mainly in Rhizophoraceae. The trees stand in the sea, on mud flats which only dry out at low tide. As the coastline moves seaward, the trees become higher and in the end a solid forest develops. Behind this forest true rain forest can and often does develop, providing the soil is suitable. In Van Steenis' fine work of 1958 the mangrove is amply dealt with. This forest type is both pioneer and climax.

On sandy accreting shores a dense forest grows without much structure and with a limited assortment of characteristic species though these are, like the various species of mangrove, very widespread. Fairly high trees can develop there, although all species in this '*Barringtonia* formation' are light-loving and rather short-lived since they are easily damaged by many causes, and age quickly. A few hundreds of metres inland the real rain forest starts, unless the terrain is exposed to floods.

Both types of coastal forests easily regenerate and can often be recognized merely by the place where they are growing. Within the scope of this book, they are of no further importance.

Riverbeds

Rivers could be described as internal borders inside the forest. They form an exclusive biotope. M.M.J. van Balgooy, on his expedition to Central Celebes in 1979 learned to recognize the river courses from a distance because their bluish-grey colour contrasted so clearly with the surrounding forest. This was due to the riverside vegetation of *Eucalyptus* (Myrtaceae) common between 1000 and 1800 m (Van Balgooy & Tantra 1986). In the Malayan lowland, Corner (1940, p.42) describes three categories of rivers: (1) Saraca streams, narrow rivers flowing tunnel-wise through the forest, and bordered by *Saraca* trees (Leguminosae-Caesalpiniaceae) together with other trees associated with them; (2) neram rivers, which are wider, and are fed by the Saraca rivers; on islets formed by boulders in these shallow riverbeds, the neram or *Dipterocarpus oblongifolius* (Dipterocarpaceae) grows abundantly; and (3) rassau rivers, which are freshwater streams of coastal areas which are also influenced by tidal fluctuations. They take their name from the screw pine, rassau (*Pandanus helicopus,* Pandanaceae), which occurs in abundance on their banks, in close association with many other species. Thus, each river type has its own more or less characteristic combination of species along its banks and, wherever possible, in the riverbeds.

The possibility for vegetation to settle and grow in riverbeds is limited by the quantity of water which passes through at irregular intervals. If heavy rains have fallen upstream, a huge volume of water plunges through the bed during the next few hours or even days. At normal low level, such riverbeds are wide and mostly dry; but if one looks for the high water line, one will be surprised since the difference in water level may be as much as 3 to 4 m. It is, therefore, at all seasons, extremely hazardous to camp in riverbeds whatever the season; and flood water arrives so suddenly that its approach makes a sound like rolling thunder.

One realizes that the plants which grow in riverbeds between the low and high water level live in a habitat with bizarre variations. In the neram riverbeds (type 2) not only does a climate prevail which is as extreme as on any large exposed area, but periodically such plants are also exposed to forces which are, according to R. Geesink's estimate, 30 times as strong as a hurricane. Nevertheless, a certain number of species survive, and there is even a category of plants, the rheophytes, which occurs exclusively in this habitat. Beccari noticed these plants for the first time in 1860 in Sarawak, and called them 'stenophyllous plants' because of their conspicuously narrow leaves. In 1981, Van Steenis published a voluminous study on this type.

Rheophytes typically possess an extensive root system and tough flexible branches, and they are often branched in such a way that they assume the form of a brush in running water. Sometimes they are even shorn on one side, like the wind-pruned trees which occur along western European coasts. In some places rheophytes are numerous, particularly, it seems, in those areas which are geomorphologically old. Yet no more than 35 species, in the strict sense, are found in Malaya, in a flora of about 8500 seed plants and ferns. They belong to various (large) families such the Euphorbiacea, Myrtaceae, Rubiaceae and also to the Podostemaceae; the latter grow on rocks in running water and only come into flower when they dry out in extremely dry years.

Besides the true rheophytes, which are not found outside riverbeds, this biotope is also inhabited by plants which do occur elsewhere. Rheophytes are also found in the subtropics; examples are the oleander, *Nerium oleander* ((Apocynaceae), and several species of *Tamarix* (Tamaricaceae). Willows also have the characteristics of rheophytes.

How rheophytes germinate, attach themselves to the riverbed and disperse their seeds is unknown. Also unknown is the function of river courses as traffic arteries in general for birds and other animals who move through the forest, fulfilling the role of pollinators and dispersers of seed.

And there are also odd cases such as that of the *Morinda* (Rubiaceae) which I discovered in eastern New Guinea growing between the boulders of a riverbed, in completely rheophytic surroundings below the high waterline. The shrub itself was full of grasses and little branches, but curiously enough, the habit of the plant in no way deviated from that of the *Morinda* shrubs growing on dry land, and its leaves, as large as a hand, were elliptic, not long and narrow at all. Nothing in tropical botany is simple (Fig. 15.1).

Mountain Forest

It is not only the decrease of the average temperature by about 0.61°C per 100 m increase of elevation which decides the aspect of the vegetation and the composition of montane flora. Mountains "suck the heavenly clouds down to earth", the regent Max Havelaar told the chiefs

Fig. 15.1 Shingled streambed of a river with occasional sudden floods (banjirs); New Guinea. Rheophytes grow between the low- and the high-water lines. The plant in the *middle*, a *Morinda* (Rubiaceae) does not have the typical narrow leaves and the wind-shorn shape; the unknown plant on the *right* does. On the slope in the back a light strip is visible: the mark of a landslide (Photograph MJ)

of Lebak in western Java (as Multatuli, the famous Dutch novelist, wrote). Mountains cause the winds to ascend to cooler layers where the vapour condenses. Most mountain summits are covered by clouds during the greater part of the day in tropical rain climates, and the effects of a dry season are milder on their slopes. The moisture favours the growth of mosses and epiphytes, which are, however, no indication of altitudes themselves. Moreover, the quality of the soil is of importance, which is why magnificent, heavy forests cover the volcanoes of Sumatra and Java, at least the inactive and uncultivated ones.

The influence of temperature, moisture and soil has a remarkable telescope effect, well known in all tropical regions (and elsewhere, too). On an isolated low mountain all zones of flora and vegetation are narrower, while on an isolated high mountain they are broader (Fig. 15.2). In his magnificent book, *The Mountain Flora of Java* (1972), Van Steenis paid great attention to all such ecological and plant-geographic factors.

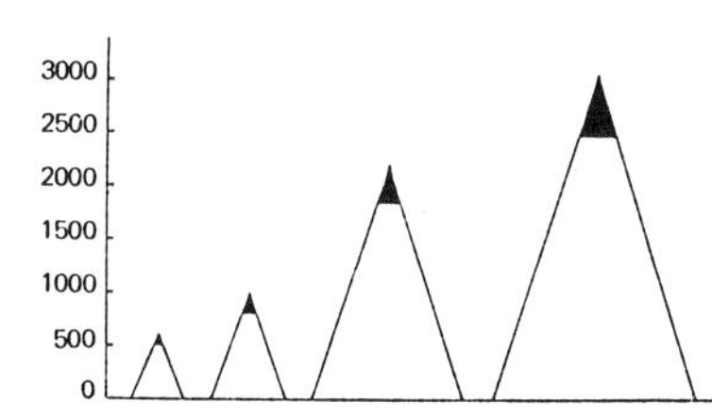

Fig. 15.2 Telescope effect of the altitudes of mountains: whether high or low, the summit of a mountain always has a vegetation of mossy shrubwood (Van Steenis 1972)

Richards (1952, Chap. 16) gives a good survey of the differences between the real tropical rain forest in the lowland and the mountain forest. The demarcation line between them, commonly located at an altitude of 1000 m, is an average which, due to the telescope effect, may actually vary between 750 m and 1650 m (on mountain ranges the patterns are more complicated). From Chapter 7 (Composition), we recall the great impoverishment in variety and numbers of species which can begin at 300 m above sea level and is certainly noticeable at 800 and 1100 m, when the character of the forest changes even more markedly. In western Java, although rasamala, *Altingia excelsa* (Hamamelidaceae) 50 m high or more still occurs at 1500 m, together with two equally tall species of *Podocarpus* (Coniferae), few specimens of those forest giants are found, and the number of species in general diminishes. The average height of the trees decreases (Fig. 7.8) and even Richards no longer distinguishes two layers in the canopy. Buttresses and cauliflory become rare phenomena, as do large leaves, a characteristic typical of the lower layers (10 to 20 m) of the lowland rain forest. The percentage of trees with compound leaves also diminishes.

In Malesia, the best-studied area, many plant families are represented by only a few species above 1000 m, or do not occur at all: Anacardiaceae, Annonaceae, Apocynaceae, Araceae, Asclepiadaceae, Burseraceae, Combretaceae, Connaraceae, Dilleniaceae, Dipterocarpaceae, Flacourtiaceae, Malpighiaceae, Meliacea, Menispermaceae, Myristicaceae, Palmae, Sapindaceae, Sapotaceae, Sterculiaceae, Verbenacea, Vitaceae, Zingiberaceae and others. Conversely, a smaller number of families is better represented in montane than in lowland forest: Aquifoliaceae, Balsaminaceae, Begoniaceae, Cyatheaceae (tree ferns), Ericaceae, Fagaceae, Symplocaceae and others. The montane forest is also indicated by the term 'fago-lauraceous', although the Lauraceae are found in the lowlands as well.

Which succession follows after destruction depends on the climate. In Malesia, with an ever-wet climate, a mixture of shrubs, later forest, develops in mountain areas, with many Euphorbiaceae and Urticaceae, or a tangled wilderness of rèsam, *Gleichenia,* a scandent (climbing) fern with very tough stems. A totally different type of vegetation, however, develops in a seasonally dry climate, such as that of northern Sumatra, northern Luzon, eastern Java and the Lesser Sunda Islands: savanna with pines or montane casuarinas, cemara. In Sumatra, the pine is *Pinus merkusii* (Pinaceae) which has its needles in pairs; in Luzon and mainland southeast Asia, it is *Pinus insularis* (= kesiya) with its needles in sets of three. Trees of 70 m high with a diameter of 2 m are known. These two pines are southern hemisphere outposts; the *Pinus* genus has its main distribution in the northern hemisphere. It is a typical light-germinator, and at 10 years of age, it develops a fire-resistant bark. The fallen needles, which decay only slowly, become dry and therefore inflammable just as the grasses under the trees. Periodical burning in the cool mountain climate therefore favours the growth of *Pinus.* It can grow down to sea level as well, but in the lowland more rapidly growing trees of other species generally smother the young *Pinus* trees. If one could prevent the burning of the *Pinus* savanna in the mountains, which destroys seedlings and young trees, broad-leaved trees would gradually settle between the grasses and the *Pinus.* Alternatively, if mountain pine savannas were burnt even more heavily, the *Pinus* trees too would vanish, leaving only grassland.

As already observed (Jacobs 1972), spontaneously growing *Pinus* is unable to retain the soil, and accordingly rainfall is followed by erosion which is vastly augmented by frequent burning (Fig. 15.3). Frequent burning also increases the area of mountain savanna at the expense of the original forest. Again and again a strip of forest is penetrated by grass, burned and subsequently eroded. This has resulted in the deforestation and erosion of Luzon's mountain areas on a massive scale and the inhabitants do not seem to understand the amount of damage they cause. Eroded material has filled the lake behind the Ambuklao dam which, as a result, will have approximately half the life span originally estimated.

The cemara, *Casuarina junghuhniana* (Casuarinaceae), from an Australian genus, is found from the Lawu volcano (111° E) on Java eastwards to Timor. Superficially it resembles a pine tree, but instead of needles the tree has thread-like twiglets with scale-like leaves; the twiglets, like pine needles, are shed as a whole. Ecologically the cemara resembles *Pinus* in every detail: the cemara, too, has with man's help been spread into tall, although rather sterile, forests, covering the slopes of the volcanoes, except where these, too, have in their turn been denuded of cemaras by wood collectors or by too frequent burning. Van Steenis (1972) gives a well-illustrated detailed description of this type of ecology.

Fig. 15.3 Secondary forest of *Pinus insularis* (Pinaceae), on the periodically dry mountains of Luzon, Philippines, at ca. 2000 m. Originally dense broad-leaf forest grew here. The pines cannot hold the soil, and erosion results (Photograph MJ)

Subtropical Rain Forest

Where a more or less continuous lowland rain forest belt extends to the north or south of one of the tropics, the tropical rain forest only very gradually loses its typical structure and floristic composition. This is the case in eastern Australia (Richards 1952, pp. 368–372), and eastern South America (Hueck 1966, Chap. 20). Isolated subtropical rain forests can be found along the foothills of the Himalaya, in Southern China, along the east coast of South Africa and scattered along the Gulf of Mexico as far north as Florida. Schimper (1898, Chap. IV, 1) gives excellent descriptions of various rain forest types outside the tropics.

Factors complicating this phenomenon are: cold seasons (which for a given latitude have a much stronger effect in the southern than in the northern hemisphere), an altitude above sea level, drought and plant-geographic relationships. For example, the forests of northern India contain far more Malesian floral elements than those of southeastern Australia.

Richards (1952) concluded that the overall picture resulting from comparing subtropical to tropical forest is remarkably similar to that obtained by comparing mountain to lowland forest within the tropics. The richness of species diminishes, and the tropical lowland taxa disappear to be only partly replaced by others, which are often related to non-tropical genera. Buttresses and cauliflory become rare phenomena, while the forest on the whole becomes less tall and tree diameters are reduced. All the same, this type of forest is still very impressive, especially to those who are not familiar with the true rain forest.

Monsoon Forest and Savanna

In Malesia, the areas covered with true deciduous forest are of modest size: one finds them only occasionally in eastern Java and on the Lesser Sunda Islands. Although they are of mixed composition, they are not particularly rich in species. What makes them remarkable is their plant-geographic affinity with the vast multiform monsoon forests of Southeast Asia, from which they are separated by the ever-wet rain forest area of the Sunda shelf.

Van Steenis (1965), in his admirable and concise introduction to the vegetation types of Java, explained that during the Pleistocene Ice Ages when large parts of the Sunda shelf were dry land, a 'drought corridor' extended from Southeast Asia to eastern Java. The flora of the monsoon forests must have migrated along that corridor, which was probably composed of dry, isolated patches situated one after the other like 'stepping-stones' and surrounded by more humid forest. The plant species which entered the archipelago in this way can be divided into several 'drought classes', related to the climate in which they grow, the climate type, in this case, being expressed by the number of days of rainfall during the four driest months in succession. The smaller the number of rainfall days, the more severe the East Monsoon, and the composition of the monsoon forest changes accordingly.

One of the characteristic species of the monsoon forest is teak or jati, *Tectona grandis* (Verbenaceae). It is indigenous in India, Burma, Thailand, and perhaps also in Java where teak is most frequently cultivated on plantations. In very dry areas the tree stands bare during a considerable period of the year. In a more humid climate, the tree grows more rapidly, and the wood is therefore of inferior quality. In general, the matter of losing foliage in the dry season is a question of the percentage of the number of leaves as well as of the number of weeks the season lasts. It will be clear that in a climate with strictly marked seasons, flowering and fruiting also have strict yearly rhythms, though it is remarkable that the young leaves (and flowers) often open before the beginning of the wet season. Not all tropical tree phenology, however, synchronizes with climatic change; the end of a dry season cannot be compared to the temperate climate of spring.

Another characteristic species of the monsoon forest is the well-known purple flowering bungur, *Lagerstroemia speciosa* (Lythraceae), often planted as a wayside tree in Malaya. The nim or neem, *Azadirachta indica* (Meliaceae), which grows from India and Burma to Java and Sumbawa, is cultivated in Malaya and throughout India. The tembusu, *Fagraea fragrans* (Loganiaceae), also a beautiful wayside tree, is less clearly bound to the monsoon forest, but has seasonal behaviour (see Corner, *Wayside Trees of Malaya*, 1940). In general, it can be said that a large proportion of the species used in Malesia for street and garden planting has its origin in the monsoon forest. Their regular flowering habits, their drought resistance and their light-loving character make them better suited for this purpose than rain forest trees, although the monsoon forest in Southeast Asia contains a number of drought-tolerant species of the typically rain forest dipterocarp genera, *Anisoptera*, *Dipterocarpus*, *Hopea* and *Shorea*.

Monsoon forest, as a habitat, is of course very different from real rain forest. During the most unfavourable time of the year the tree canopy is too thin to exercise any moderating influence on the microclimate of the lower layers. Consequently, there is neither permanent shadow nor constant humidity, and bright daylight therefore has free access to the litter below, which dries out. Also, in contrast to conditions in ever-wet climates, the flow of minerals is not constantly directed downwards, but alternates between a downward and an upward direction so that the soil is less permanently poor than the soil in the rain forest. In the monsoon forest, the tree roots reach deeper.

In such forest there are many lianas, but only a few epiphytes, although the staghorn fern, *Platycerium*, is fairly common; some of those have 'nests' of nearly a cubic metre in size, and fronds 2 m long. The (relatively poor) undergrowth possesses rootstocks and seeds, and is adapted to the dry season. Therefore, grasses may easily play a predominant role, as do other plants with good fire resistance. Indeed, the hard-scaled fruits and seeds of some Leguminosae and Myrtaceae species can only burst in the heat of a fire which thus makes their dispersal and germination possible.

In short, fire effects a shift in the spectrum of vegetation types which runs from closed monsoon forest with grasses in the undergrowth at one end to treeless grassland at the other. Fire destroys the seedlings of trees, and leaves unharmed only plants which have their seeds or permanent rootstocks sunk down in the earth, and only the trees with very thick bark, such as *Casuarina* or *Pinus*, or trees with a bark rich in moisture, such as *Eucalyptus* (Myrtaceae), survive.

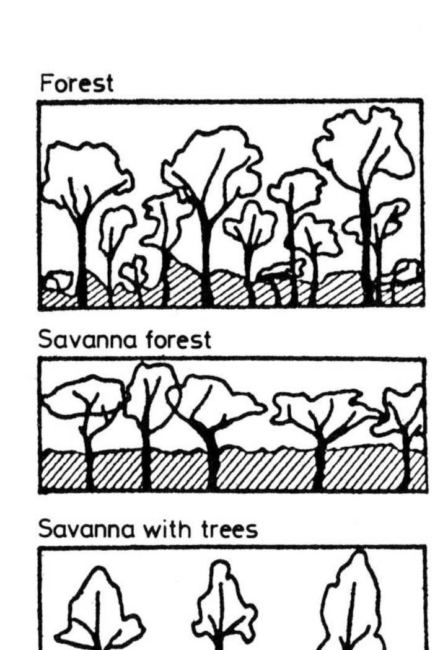

Fig. 15.4 Diagrams showing open forest, savanna forest, tree savanna and shrubby savanna (Hopkins 1965)

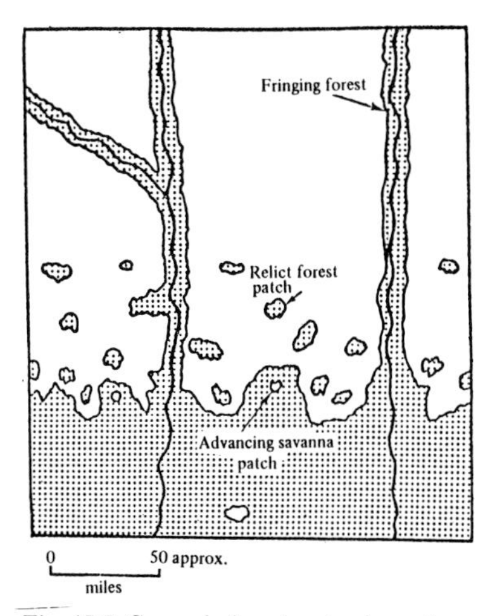

Fig. 15.5 Ground plan showing boundary between forest and savanna. The forest can maintain itself best along the rivers, as gallery forest. In the savanna, encroaching on the forest from north to south, small patches of forest can maintain themselves a while. Another view of the same process is given in Fig. 15.7 (Hopkins 1965)

What happens to vegetation along the fringes of the rain forest under the influence of burning depends more on the frequence of the fires than on the climate. Grass may catch fire after only a few dry days. Even the mossy cushions which surround the twigs and stems of a wet mountain forest burn very well when they are dry and they are deliberately set on fire (Jacobs 1972). But the rain forest itself does not catch fire or only very rarely (Editor's note: the author died before the great fires in the Borneo rain forest, early 1983 occurred) because the ever-closed canopy preserves the moist microclimate. Only if the canopy is lacking, does such forest become vulnerable to fire.

Almost all fires are caused by man, "for hunting, . . .for pestering neighbouring villages, by carelessness, for clearing land, for making land passable, for converting forest into pasture land, in short for innumerable purposes" (Van Steenis 1965, p. 33) and this has already been the case for (tens of) thousands of years, which has frequently resulted in an inextricable mosaic of destruction and recovery stages; when the latter exist close to original forest with a clear-cut boundary between them, it is always an indication of destruction. In Vietnam, rich in such mosaics, a fairly good survey with good photographs was made by Schmid (1975). Aubréville (1949) described the spectrum of stages for Africa: it runs from rain forest through savanna and steppe to true desert. The forest and savanna theme is treated in a very concise and instructive way by Hopkins (1965), particularly for West Africa (Figs. 15.4 and 15.5).

This mosaic will generally have the character of a tall grass vegetation in which small groups of trees are scattered, indicating recent destruction, or with only isolated individual trees, indicating more permanent destruction. The front of a fire moves almost just as easily down- as uphill. Yet in river valleys, particularly the deep and narrow ones, the forest is often left intact for a long time. From above one sees those long narrow green stripes twist through the landscape. This forest type is called gallery or fringing forest. The humidity caused by evaporation in these forest relics sometimes even makes the growth of epiphytes possible (Fig. 15.6).

Gallery forests are very important from the ecological point of view. Narrow rivers are completely overshadowed by them and are thus protected against desiccation. This means water for the animals in that area the whole year round, while in addition the riverbed and its forested banks form a migration route for animals (which also transport seeds). Finally, the gallery forest forms a reservoir from which the adjoining land can be recolonized with trees. For this reason, any gallery forest should be specially protected in order to avert an ecological disaster which will equally affect human agriculture and meat production.

When fire is kept out of the savanna for some decennia, and when the dangerous phase, in which the grass stands high and is very liable to fire, is safely passed, trees will gradually invade the savanna because their seeds can germinate in the shelter of the grasses. Soon young trees rise above the grass vegetation, creating shade and putting the light-loving grasses at a disadvantage. After some time a more or less closed canopy of savanna trees, and finally of monsoon forest, will develop, depending of course on undisturbed succession, the accessibility to seeds and the climate (Figs. 15.7 and 15.8).

If the area is used primarily as pasture, however, this situation creates a shortage of grass. For this reason the wildlife sanctuaries in Africa are regularly and with careful planning burned in sections. Perhaps for this reason the idea has gained a foothold that all sanctuaries should be 'managed' by the active intervention of man. In tropical rain forest, however, with its ever-closed canopy and totally different ecological system, every act of interference can only mean harm. On the contrary, the rain forest should be safeguarded against all human interference to guarantee its survival.

Fig. 15.6 Monsoon forest composed of *Dipterocarpus intricatus* (Dipterocarpaceae) in South Vietnam (Photograph M. Schmid 1975)

Deciduous Forest

So far, our attention has focussed upon the tallest, most splendidly developed types of rain forest, where the canopy is completely closed the whole year round. It is true that in such forests trees may sometimes stand completely bare, but this lasts no longer than about a week. *Pterocymbium javanicum* (Sterculiaceae), a robust tree which occurs from ever-wet West Java to East Java where there is a long dry season, only stands without leaves for a very short period in Bogor (West Java), but this leafless period is extended according to the length of the dry season in its domicile.

Fig. 15.7 Monsoon forest in Africa. Dry season. *Above*: Virgin forest. *Below*: Degradation caused by human interference. At the *left* a plantation, grading into savanna interspersed with agricultural land on which the useful trees have been left standing. The *right half* is occupied by savanna that is burned periodically (Aubréville 1949)

Fig. 15.8 Remnant of deciduous rain forest in southeast Celebes, surrounded by alang-alang, which is degraded by fire time and again (Photograph MJ)

Gradations of this sort make it difficult to indicate a limit between true rain forests and rain forests in which some trees become partly leafless during a certain period of the year under influence of a dry season. The latter forest is paradoxically called 'semi-evergreen rain forest'. Whitmore (1975, pp. 122, 126–129) devotes a few passages to this subject, and takes as a – rather arbitrary – standard to define this kind of forest that one-third of the taller trees are deciduous, and not necessarily all at the same time. He remarks that this form of rain forest is probably the most extensive, occupying much of the central part of the Amazon basin and Africa, whereas in Malesia it only occurs in pockets in those places where a distinct, but not too long, dry season prevails.

Although the species are the same as those of the evergreen rain forest – at least, that is what we suppose – the semi-evergreen forest is slightly poorer and bamboo establishes itself there more easily. In Africa, where Richards (1952, pp. 338–340) so carefully studied the relation between climate and forest, the following types can be distinguished: True Rain Forest, Wet Evergreen and Dry Evergreen, which can all be considered as facets of 'the' tropical rain forest.

In Africa, dry evergreen rain forest is so damaged by man, just as the monsoon forest is in East Malesia, that it is almost impossible to reconstruct it as a whole. Some tree species from the dry variant appear in the recovery stages of the more 'wet' forms, and only an experienced and specialized botanist is able to distinguish and diagnose these forms with their mutual relationships, derived stages and possible divergencies in species richness. This problem is particularly evident in tropical America and Africa. In trying to protect these rain forest areas one has to deal with such problems in a very clear and botanically explicit way.

Kerangas, Wallaba, Caatinga

After discussing climatological restrictions on, and spatial limitations of the rain forest, we must now define its edaphic limitations. Because rain forest can grow on limestone (Fig. 15.15), limestone cannot be considered as a limiting substrate, although in this case its composition differs considerably from that of the rain forest on more acid soils. Whitmore (1975) deals with this subject in his *Tropical Rain Forests of the Far East*, and more recently a fine article by Chin (1977) appeared on the same subject.

We will, therefore, first discuss the vegetation on very poor soils, where the subsoil water level does not reach a level harmful to the plants, and then we will discuss the vegetation affected by the subsoil water level.

The name kerangas is used in Sarawak (in northwestern Borneo) for land on which rice cannot grow, the soil being too poor. Such soil is found in ever-wet climates, predominantly in the lowlands and particularly in Borneo (which consists largely of sandstone), the isles of Banka and Biliton, occasionally in eastern Sumatra and Malacca, and, in the New World, in British Guyana and in the river basin of the Rio Negro, along the equator between 63° and 70° long W. The water draining such soil is as brown as that from the oligotrophic peat swamps in Europe; the Rio Negro owes its name to this fact. Flying low over the vegetation one occasionally sees a glimpse of bleached sand shimmering through the green canopy. In Guyana such areas are known as wallaba, which is also the local name of *Eperua falcata* (Leguminosae-Papilionaceae), a tree which occurs there in abundance. In the Rio Negro region it is called caatinga; a description is given by Hueck (1966, pp. 39–42).

By whatever name, kerangas forest is remarkable. The marvellous photographs in Whitmore (1975) give a good idea of it (see also Fig. 15.9). Trees above about 30 m never occur there, and sometimes they do not even exceed 10 m. The canopy is level in height, but so thin that it hardly deserves the name canopy. Trees of large girth are rare, buttresses or cauliflory

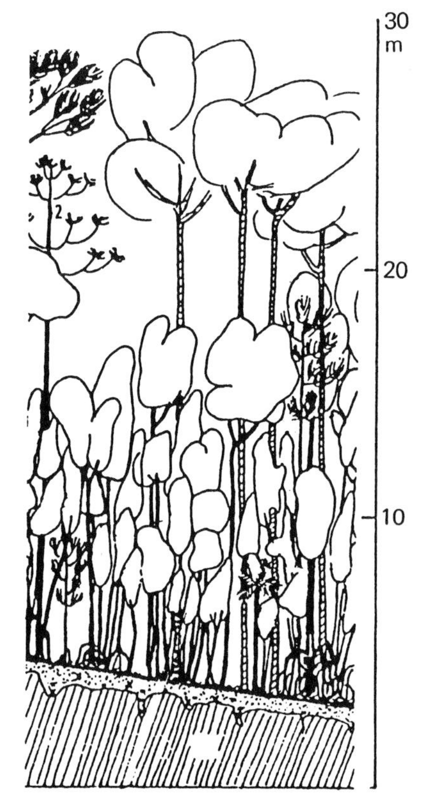

Fig. 15.9 Profile diagram of kerangas forest in Brunei, Borneo. The width is 18 m, the trees reach up to 20 m height (After Brünig in Whitmore 1975, p. 127; also includes a list of species)

are absent and the leaves are usually small and sclerophyllous. In Sarawak, the tree ru ronang, *Casuarina nobilis* (Casuarinaceae), is often found in the kerangas, as are *Dacrydium* and *Podocarpus* (both Coniferae), with scale- or needle-like leaves. The trees grow close together. Richards (1952, p. 240) counted 617 trees of 10 cm or more diameter per hectare in wallaba forest, and the shrub layer is remarkably dense. Large lianas are absent, but small lianas are abundant. The most conspicuous epiphytes are ant plants or myrmecophytes with thick tubers: *Hydnophytum* and *Myrmecodium* (both Rubiaceae). And if one searches on open moist places one will find *Drosera* (Droseraceae) and *Utricularia* (Lentibulariaceae). The occurrence of all those genera already indicates a very oligotrophic environment: the ants and other insects enrich the plants with nutrients. The small sclerophyllous leaves of the trees exactly resemble those of the trees which grow on leached ridges in the mountain forest: the minerals cannot be washed out of such leaves when it rains. The dead litter, rich in phenolic compounds, only decays slowly, which guarantees a 'leak-proof' nutrient cycle. There are few fleshy, juicy fruits in the kerangas, animal life is poor; consequently, it is silent there. In Chapter 12 the underlying ecological strategy has already been discussed.

It took some decades of investigation before the kerangas was properly understood. The flat and sandy grounds with a thin vegetation, called *padang* in Malay, had long attracted attention. These areas were so sparsely covered with vegetation that they were compared with heathland, and the bordering forest was called 'Heidewald' or 'heath forest', even though Ericaceae do not grow there. When Richards (1952) compared *padang* vegetation in Malesia to the wallaba formation in Guyana, he pointed out the extremely low supply of mineral nutrients in the soil as the cause of the poor character of both these forest types; but while stressing, once again, the remarkable similarity in their general parallel development and appearance, due to similarity of climate and soil, he noted that there is hardly any resemblance in composition.

Kerangas soil is so very poor, even by rain forest standards, that it acts as a truly limiting factor. While there are transitions to be seen between rain forest and kerangas, the extreme lack of nutrients makes it almost impossible for the vegetation to re-establish itself once the cycle has been interrupted. If a small quantity of nutrients is still available a slowly growing, low vegetation will develop, poor in species, consisting of certain shrubs and herbs.

Consequently, there can only be very limited possibilities for the exploitation of this forest type. Whitmore (1975, p. 131) states that "heath forest is easily degraded by felling and burning", but on p. 135 remarks that "heath forest, if carefully managed, can provide a continuing source of timber." This is not the only instance of an indistinct statement concerning exploitation. The fact is, that just as the poorness of the soil, the decisive factor, varies from place to place, so does the quality of the forest.

Concerning the composition of kerangas in Sarawak, E.F. Brünig, who has done much research there on this type of forest, found that in Sarawak there were a total of 849 species of trees from over 428 genera. Of this total, 220 species also occurred in the adjacent lowland dipterocarp rain forest, and another 146 species occurred in the nearby peat swamp forest, while conversely all species occurring in the dipterocarp rain forest as well as in the peat swamp forest were also to be found in the kerangas. These figures were compiled by Whitmore (1975, pp. 207–209), who also remarked that the kerangas and mountain forest on quartz and quartzite ridges have a number of highly distinctive species in common, among them the ground orchids *Arundina* and *Spathoglottis* (Orchidaceae), the fern *Dipteris conjugata* and the blue-flowering *Dianella* (Liliaceae).

Freshwater Swamp Forest

Fig. 15.10 Freshwater swamp forest along a flat riverbank at high water (only freshwater intrudes here). Seven zones can be distinguished. Details in Corner (1978)

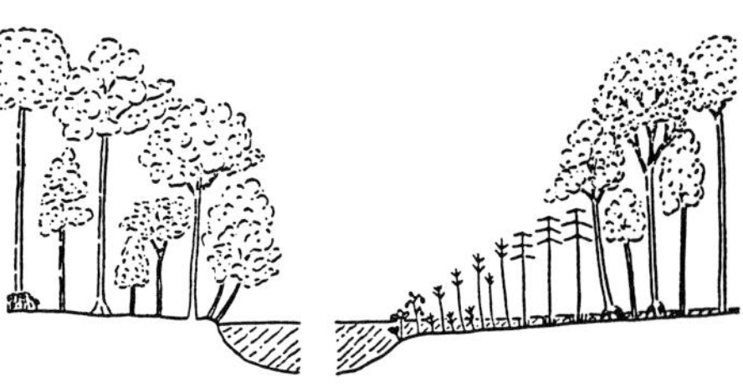

Fig. 15.11 Periodically submerged broad-leaf forest (*left*) and true, more or less permanently submerged, freshwater swamp forest (*right*). Details in Corner (1978)

Behind coasts, within the reach of rivers, along lakes in the interior and in every location where freshwater often, but not always, reaches the level of the soil surface, freshwater swamp forest (not to be confused with peat swamp forest which will be described subsequently) can be found (Fig. 15.10). Freshwater swamp forest roots stand continually or periodically in water, while in peat swamp forest this is not the case.

In an even climate the water level is usually constant. If the water level is too high for too long at a stretch, no forest develops at all, but instead a herbaceous-like vegetation of swamp plants. If there are marked differences in the water level, for example where very wet periods and very long dry periods prevail, the forest will be inundated during a part of the year, and yet will survive. In such places, seasonally, one can sail with a canoe between the trunks. There is always some sort of selection in the direction of a certain composition in freshwater swamp forest, which a single species often dominates. (Van Steenis 1954, pp. 226–229, mentions a few of the dominant species in New Guinean swamp forests.)

During very intense dry periods, however, even freshwater swamp forest can become inflammable. The tree species which benefits from such burning in Malesia is the kayuputih, *Melaleuca leucadendron* (Myrtaceae), striking because of its almost white bark. Grass or reed vegetations which become periodically dry may also burn, after which they recuperate again.

Now to discuss those swamp forests in which the water level is more or less constant, even though they do not form an invariable climax: such forests are differentiated in belts, a certain proof of succession. Corner (1978) described them briefly in his commendable book, *The Freshwater Swamp-forest of South Johore and Singapore*. Between sea and river, seven belts can be distinguished: the mangrove, then the nipa, *Nypa fruticans* (Palmae), growing in brackish water, then four belts of varying depth and composition, and finally the Saraca stream type already mentioned. Next to these belts parallel to the river, there was also fully developed freshwater swamp forest in the river mouth area (Fig. 15.11). The succession may, according to Corner, develop in two directions: the first leads to (assumed) mixed 'dry land' rain forest, the second to (proved) peat swamp forest. What actually happens depends as much on the height of the soil above sea level as it does on the height of the water table.

In the first-class freshwater swamp forest described by Corner, which has in the meantime largely been destroyed by exploitation, the average height of the trees was about 35 m, with a few specimens attaining 50 m. There were some lianas, including large ones, and a substantial quantity of epiphytes. The root systems of the trees, of which Corner has included some very fine photographs, are particularly spectacular. Water, by excluding air, obstructs root respiration. The root systems are therefore close to the surface, and there are all kinds of adaptations for bringing the roots into contact with the air. Large bunches of breathing roots sprout from the trunks of certain species almost to high water level; which means that most of the time roots hang downward above the water. Stick roots, knee roots or sinuous snake-like roots emerge from the mud. Several trees stand on a fair number of stilt roots which diverge from the base of the trunk. The variety of buttresses is legion, of which the most spectacular forms a kind of flying buttress such as found on the outside of a Gothic cathedral. Each of these root forms are hereditary characteristics of the species concerned, and even in a botanical garden situated on dry ground, such adaptations to swamp life will continue to form.

This type of freshwater swamp forest, also frequently found along the coast of East Sumatra and of West and South Borneo, is very mixed. The list of specimens composed by Corner contains 1082 names, that is one-seventh of the flora of all Malaya. Though an analysis of those figures is wanting it is nevertheless evident that the greater part of these species also occurs in the mixed 'dry land' rain forest. Yet in many groups it is even possible in a herbarium to recognize the specimens taken from freshwater swamp forest: their colour is often more deeply brown. There is, however, little doubt that the species of the freshwater swamp forest do, in general, originate in mixed dry land forest. This has consequences for swamp forest protection: although the freshwater swamp forest obviously forms a special ecosystem, protecting this forest type without protecting the bordering dry land forest as well means cutting off the freshwater forest from its most important source of seeds and therefore species variety. From the mixed dry land forest a freshwater swamp forest could be repopulated, but the reverse is impossible.

Freshwater swamp forests in Borneo and further east are the natural localities of the sago palm, *Metroxylon* (Palmae), which provides the starch that forms the main foodstuff for large numbers of the local inhabitants. The sago palm flowers only once. Just before the flowering, the stem which contains at that time a maximum of reserve food is cut and cleaved length-wise, and the starch between the fibres is then beaten out which thus provides a whole family with sufficient staple starchy food for 1 year. Another much exploited commercial species is the ramin, *Gonystylus bancanus* (Thymelaceae), a well-known timber tree.

Many freshwater swamp forests are now threatened with draining and clearing for agricultural or plantation forestry projects, with varying, often small, success. The peaty soil underneath sets strongly after deforestation; it guides the water horizontally, but not vertically, and has an isolating effect, which occasions strong fluctuations in temperature on the surface. The peat layer oxidizes soon after draining, and offers a poor subsurface for the vegetation; ill-rooted trees often fall over after some time. This has been described in the booklet, *Peat and Podzolic Soils in Indonesia* (Anon. 1977, p. 15). In Amazonia the situation is completely different. The varzéa forest (seasonal swamp forest) along the big rivers behind the shore walls is flooded yearly and the silt which the water deposits makes such soils more fertile than soils found at higher elevations; see Hueck, *Die Wälder Südamerikas* (1966, pp. 42–50).

African swamp forests have contributed the oil palm, *Elaeis guineensis* (Palmae). This palm which originated in the coastal swamp and riverine forests of Liberia southward to 5° S has today become one of the world's most important commercial plants.

Peat Swamp Forest

The name 'peat swamp forest' still reflects an initial confusion between the freshwater swamp forest discussed in the previous section and the peat swamp forest in which rain is the only source of water. For this reason true peat swamp forest is, like kerangas forest, always restricted to an ever-wet climate, to flat terrain and to substrates that are extremely poor in minerals. Extensive peat swamp forests are found in West Borneo, near the mouths of large rivers; these remarkable formations are also found in Sumatra and Malaya. The peat swamp forest of Sarawak and Brunei have been thoroughly studied floristically and ecologically by Anderson (1963). His work is summarized and illustrated by Whitmore (1975, pp. 144-156), who also mentions the rare occurrences of this type of forests in tropical America.

As has been said before, peat swamp forest is one succession stage of freshwater swamp forest. It develops on peat which in its turn often rests on stiff, impenetrable clay, originally covered with mangrove forest. Because of the low percentage of oxygen which the water contains, and the low pH which results from the lack of minerals, natural decomposition of the litter is obstructed. Gradually the peat soil formation rises above the original water level, a process which aggravates the lack of mineral nutrients. In the highly oligotrophic centre, found on the highest elevations, the small quantity of litter available is almost entirely converted into peat. In the lower areas, close to the margins, several factors (such as an occasional inundation by river water) may cause some decomposition of organic matter. The

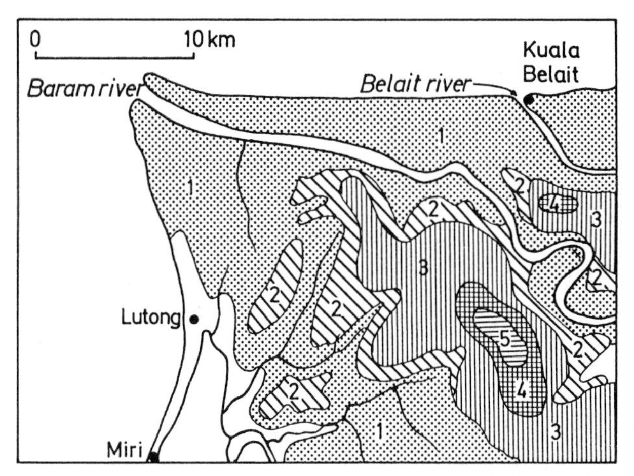

Fig. 15.12 Peat swamp forest in West Borneo, in the delta of the Baram river. Five concentric zones can be distinguished (*1* to *5* from the outside inwards), each with its own forest type, the diagrams of which are given in Fig. 15.14. Type 6, the poorest and highest situated on the high central parts of the raised bogs, occurs farther inland only (After Whitmore 1975)

Fig. 15.13 Peat swamp forest in Sarawak, Borneo, type 6: the rightmost type of the series shown in Fig. 15.14. This type is the end stage of the development, occurring on the very poorest soils. The larger trees are *Combretocarpus rotundatus* (Rhizophoraceae) (Whitmore 1975, p. 151)

result is the development of lenticular peat areas, varying from a few to many kilometres in diameter, and up to 15 and even 20 m deep (or high, which is the same). Radio-carbon dating has yielded an estimated age of 4500 years.

Each lenticular peat bog forms a separate ecological entity. The vegetation on such lenses can be divided into more or less concentric zones (Fig. 15.12). On the outer edges mixed rain

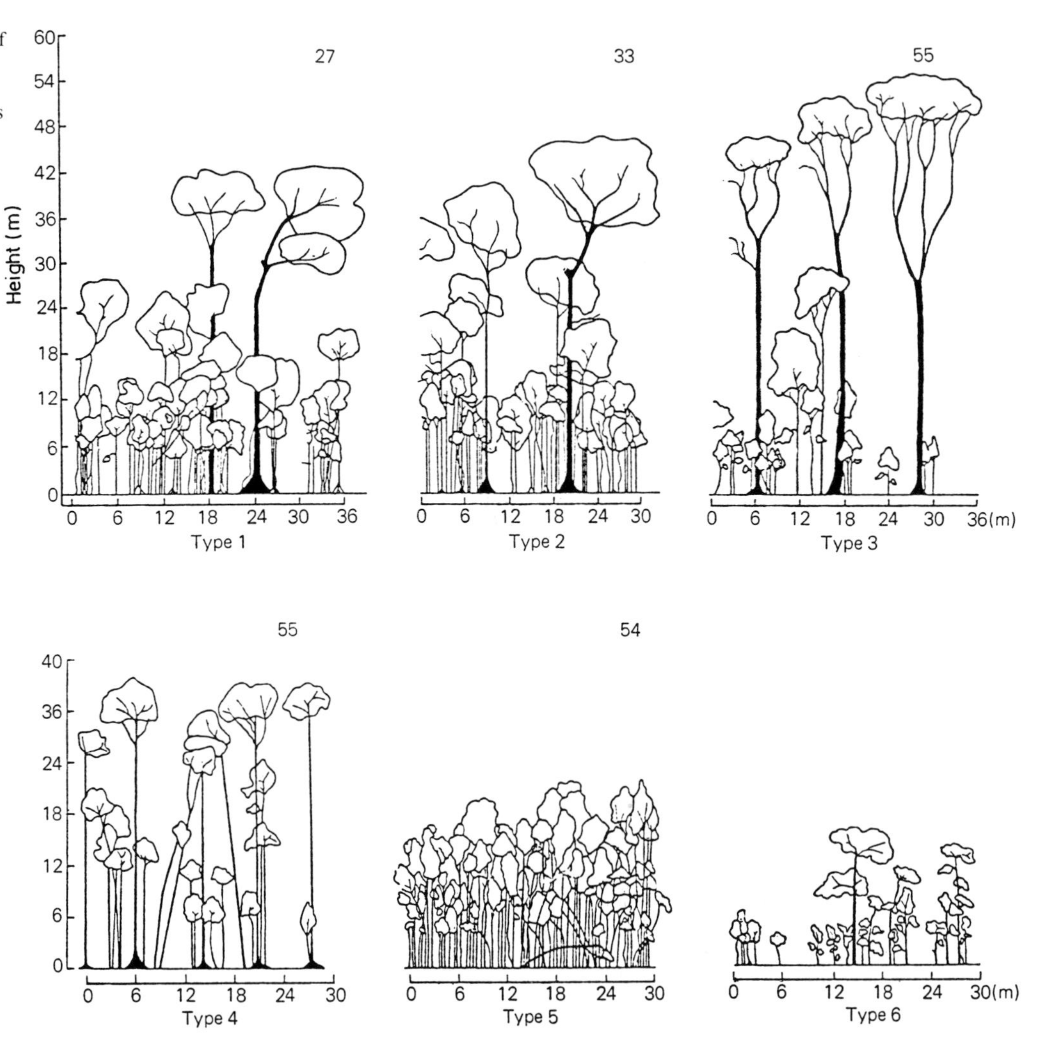

Fig. 15.14 Profile diagrams of the types of peat swamp forest as depicted in the zones of Fig. 15.12. The *figures* in the *upper right-hand corners* give the numbers of species that are also known from the kerangas (After Whitmore 1975)

forest grows, whereas in the centre, on top of the faintly convex peat formation, a stunted, extremely poor forest grows. In between a separate formation is found which (in Borneo) is dominated by *Shorea albida* (Dipterocarpaceae), of which J.A.R. Anderson showed me specimens over 70 m tall.

A further peculiarity, which I nowhere found mentioned or explained, is the fact that the tree roots form in some places a tough meshwork, covered by a thin layer of soil and vegetation, with a hollow space underneath. One can push a stick into it often as far as 1 m before feeling firm ground.

From the edges to the centre one sees the flora change in an irregular manner into a poorer type. Whitmore (1975, p. 147) gives a list of species with diagrams (Fig. 15.14) and lists 45 species for type 1; 33 for type 2; 18 for type 3; 9 for type 4; 16 for type 5; and 6 for type 6 (Fig. 15.13). Although in several cases the species are the same, we face here a whole spectrum of which the poorest, type 6, represents the climax, the natural ultimate development of the peat swamp range. It also marks the end of the usefulness of such a climax idea. Anderson (1963), who identified and enumerated all the species of peat swamp forests in Sarawak and

Brunei, listed 927 species in 224 genera and 70 families of which nearly one-third is common to the formation as a whole. Divided as to different life forms, they appear as follows:

Trees of 60 cm thickness or more	38 Species
Trees of 25 to 60 cm thickness	65
Trees of 10 to 25 cm thickness	38
Small trees	11
Shrubs of various types	21
Ground herbs	23
Epiphytes, light-loving	22
Epiphytes, shade-loving	22
Climbers, tall	20
Climbers, small	19

The similarity to kerangas is remarkable: 11 of 15 species of Dipterocarpaceae occur there as well, and we have already mentioned 146 species common to both vegetation types. Moreover, the general appearance of the vegetation in the poorest parts is strikingly similar. But as Anderson's list indicates, a great many of the species are also present in the surrounding rain forest, which has the same bearing on conservation policy as already noted in the section on freshwater swamp forest.

Whitmore (1975, p. 155) points out that peat layers over 3 m thick are of no use for agricultural purposes in any case; presumably the forest cover on those layers may hopefully remain untouched.

Long-Isolated Rain Forest

Destruction by man has whittled many once large rain forest areas into archipelagoes of biological islands. But there are also stretches of rain forest which already even longer ago became isolated from the larger areas to which they originally belonged, as the result of natural causes such as a change of climate or of geological nature. Examples are the narrow strip of rain forest along the east coast of Brazil, small sections in Ceylon, South India and possibly in South Burma, and the remnants in northeastern Australia. Their floristic affinity can be detected through plant geography.

The description of such long-isolated forests is of a type that rarely makes a concise and precise comparison with similar forests elsewhere possible, since the authors thought it generally of more importance to characterize the forest concerned in such a way that it was clearly distinguished from other types of forest in the same country. The description of the 'wet Dipterocarpaceae forest' in South Burma by Stamp (1925, pp. 23–24) is merely tantalizing because of its conciseness – I do not know of a better description.

On the whole, such rain forest-like forests are lower, simpler in structure and poorer in composition. Many species occur only very locally, with often only a few specimens surviving; they clearly have the characteristics of relics. One gets the impression that quite a number of the species in them are becoming at least locally extinct through natural causes. How this happens is still barely understood. A study of this process, which has probably been going on very slowly for a long time in such isolated forests, could yield valuable knowledge for nature reserve management and conservation policy in general.

Fig. 15.15 A forested limestone mountain, Bukit Krian, from the Bau formation in Sarawak, Borneo. The forest on the summit was burned once, probably after a stroke of lightning; it is now secondary. (Photograph by Anderson in Whitmore 1975)

Flora Brasiliensis, tabula VIII. 'Forest, shading a public road', province Sao Paulo. Natural erosion of uncovered soil (Martius 1840–1869)

16 Values of the Rain Forest

Twelve Points of Value

At least three authors have, independently, tried to outline the range of assets to the world which the tropical rain forests represent (Jacobs 1976a 1978c; Budowski 1976; Poore 1976a). Their ideas cover a wide spectrum, and it may help to summarize them here. We can reduce them to 12 points of varying complexity. The rain forests serve as:

a) Supply of wood;
b) Retention of soil;
c) Regulation of run-off;
d) Stabilization of climate;
e) Source of minor forest products;
f) Pool of new useful plant species;
g) Pool of genetic material useful in other ways;
h) Home and food source of animals, and of hunting-gathering tribes;
i) Matrix of evolution;
j) Source of knowledge;
k) Object of respect for the creation;
l) Medium for education and recreation.

We can thus see that to regard the forest as a supply of timber is to consider only one point of twelve. Many of the other points apply, of course, to all natural forests, but several are in direct proportion to a forest's richness in species and these apply to the maximum extent to the rain forests of the tropics. For the same reason, secondary forests, poor in numbers of species, have a lesser value which is discussed at the end of this chapter.

Let us now expand on these functions and values and the conclusions to be drawn from them.

a) Supply of Wood

In general, woods from conifers are called softwoods, the rest being called hardwoods. Among the latter, there may be woods that are actually very soft. Conifer wood often, but not always, comes from plantations; kauri, for instance, *Agathis* (Araucariaceae), is in Malesia mostly extracted from natural forests (see Whitmore 1977). Among the hardwood species there are some that come from plantations, e.g. teak, *Tectona grandis* (Verbenaceae) in Java, and elsewhere (see Hedegart 1976).

As for the quantities of timber extracted nowadays from rain forests, we cite a few figures from the two largest countries. In Brazil, where an estimated 45 billion m^3 wood stands on 2,600,000 km^2, in 1985 an annual harvest is expected of 165 million m^3 of logs, 13.5 million m^3 of sawn timber, 4.7 million m^3 of veneer and plywood (Brune and Melchior 1976, p. 207). In Indonesia, on an area of 1,214,962 km^2 there is an estimated commercial stand of 2.7 billion

m^3, of which 2 billion is made up of dipterocarps. The annual production went up from ca. 6 million m^3 in 1969 to ca. 26 million m^3 in 1978, but it has been going down since: ca. 19 million m^3 in 1981 and 17 m^3 in 1982.

The number of timber species exported by a rain forest country lies in the order of several dozens. For Papua New Guinea, Pape (1973) listed 30. For Malaya, Kochummen (1979, pp. 326–329) cited 43 'preferred species', of which all but 7 are Dipterocarpaceae, and 49 'acceptable species', of which 25 are Dipterocarpaceae.

An assortment of export species takes decades to build up. Both Brazil and Indonesia have some 300–350 tree species whose boles can exceed 40 cm in diameter. These are scattered in a variety of forest types, in different sizes and numbers. Those species which are fairly common and large must be identified by botanical name, and must be known by their field characteristics of bole, bark and slash. Then they must be tested for their properties; if these are superior, that species can then be introduced to the international timber market.

Commercially important properties as listed by the British *A Handbook of Hardwoods* (1956), are the following:

Weight, after drying to a moisture content of 12%.

Shrinkage, during the process of kiln drying.

Movement, after seasoning, in response to changes in atmospheric conditions. Such changes are not necessarily related to shrinkage.

Wood-bending properties, assessed by "the minimum radius of curvature at which a reasonable percentage of faultless bends can be made for a given thickness of clear material".

Strength properties, eight in number, namely: maximum bending strength, elasticity load, energy consumed to total fracture, resistance to suddenly applied loads, maximum compressive strength parallel to grain, resistance to indentation, shearing strength and resistance to splitting in various directions. Many of these strength properties vary with the moisture content of the wood.

Defects caused by wood-boring insects, such as ambrosia, longhorn, powder-post and furniture beetles, as well as termites.

Resistance to marine borers, like *Teredo navalis*, which are particularly destructive in tropical waters.

Natural durability, which means resistance to fungal decay. It makes a great difference whether the wood is applied under a roof or not, or in contact with the soil or not.

Amenity to treatment with preservatives. Sapwood is much more permeable than heartwood. The more permeable, the more durable a timber can be made by artificial means.

Working properties, in the three main directions: longitudinal, radial and tangential, on machinery with various cutter positions, speeds, etc. These properties are determined not only by homogeneity of structure and characteristics of the fibre elements, but also by the amounts contained in the wood of resin, gums and crystalline substances like calcium oxalate which has a blunting effect on cutters.

Suitability for veneer and plywood manufacturing. Veneer is produced by the rotary cutting of logs in a lathe. This enormously increases the value of a timber since it can be used to embellish cheap wood. But veneer wood must meet many requirements in these processes, and only superior logs of certain suitable species do.

Grain, flame, colour, odour are all esthetic properties which help determine market value for all timbers to be applied in the interior of buildings.

Chemical properties like the presence of waxy or fatty substances which may have a greasing effect on moving surfaces, or of poisonous, irritating saps like in some Anacardiaceae, or of disinfectants like camphor in *Dryobalanops* (Dipterocarpaceae).

Resistance to water and chemicals like oxides of unprotected iron, weak acids, lyes.

The applications of wood are too numerous to dwell upon. We only need to remember cabinet making, carpentry, cart wheels (with very different demands being made on the hub, the spokes and the rim), clothes racks (which must not stain when wet), construction purposes, crating (light and heavy), flooring, furniture, joinery, model building, musical instruments, panelling, posts and fences, rafts and life-saving gear, railway sleepers, shingles, shipbuilding, sport equipment, table and counter tops, tool handles, window fittings. Serials like *Bois et Forêts des Tropiques*, *Forest Products Journal* and *World Wood* describe many applications, as do various books.

This must suffice to explain the variety and importance of tropical timbers, but additionally something must be said on the subject of *charcoal*. Wood can be converted into charcoal by the slow process of partial combustion without fire. This can be done in closed vessels, in very simple kilns or even under a cover of earth. Charcoal has 20–30% of the dry weight of wood, but per kilogram delivers twice the amount of heat, and burns very steadily. It is much easier to transport than wood, as well as easier to trade in small quantities, and many people in poor countries are dependent upon it, mainly for cooking. Thus, charcoal alleviates the unobtrusive, but widely spread, 'silent energy crisis' in tropical countries, compellingly described by Eckholm in *Losing Ground* (1976).

But since charcoal is made best from the heavier hardwoods, tropical rain forest trees are mostly used for its production, and although this is usually done on a small scale, it is both widespread (often far from the site of consumption) and constant. Earl (1975) supplies many data and advocates the conversion of much rain forest into charcoal, of course, in his greedy eyes, all forests are 'renewable resources'.

b) Retention of Soil

Erosion caused by heavy rainfall can be observed at any open construction site in a rain forest country, where pebbles are left on dumped soil. The rain usually falls straight down and with great force which is broken by the pebbles. Grains of soil are therefore carried off between, but not below, the pebbles. This gives rise to a midget landscape of skyscrapers, some as much as a foot high, each with a pebble as its roof.

Such a simply observed phenomenon reveals the effect of soil cover – any soil cover. The cover provided by the rain forest is of a superior sort because it consists of three layers: the canopy, the undergrowth and the litter. On average, erosion under rain forest amounts to a mere 0.2–10 t $year^{-1}$ ha^{-1}. In the densest of man-made forests this becomes 20–160 t, in grassland 200 t or more, on sites under primitive agriculture 1000 t or even more. Figures from Malaya, quoted by Brünig (1977), point to an increase in the loss of soil after conversion of rain forest, which loses 24.5 m^3 km^{-2} annually, into tea plantations with a loss of 488 m^3, a factor of 20–30.

We will discuss what the effects of erosion are in Chapter 17; here, we only recall the deep weathering in tropical ever-wet climates, discussed in Chapter 4. Deep weathering leaves the lower layers as poor in nutrients as the upper, so that erosion will not uncover any more fertile soil.

Retention of soil is one of the best-known functions of the forest. It has long been the custom in land-use planning to leave the forest intact on steeper parts and summits as 'protection forest'. Where this was neglected, like in the headwaters of the Jati Luhur Dam in West Java, skeleton slopes remain, while the artificial lake fills up with shocking rapidity. And although forest can grow on amazingly steep slopes, we must presume that regeneration difficulties are proportionate to steepness.

c) Regulation of Run-Off

The 'inflow' of rain in a forest can be divided into four parts: (1) one part is intercepted by the canopy, and will evaporate again, (2) of the remainder reaching the soil, part flows along the surface into water courses, (3) of the remainder, which sinks into the soil, part will be taken up by the vegetation, and evaporate through the leaves, while (4) what is left finds its way into springs, streamlets and rivers. The percentage of the total rainfall in the four categories varies according to the quantity of rainfall: when rain is short and gentle, most will be caught in the canopy and nothing will flow along the soil surface. The degree of slope and permeability, too, are factors. So is the amount of humus, as this takes up a great deal of water.

But basically all types of forests act as sponges, quickly absorbing a large part of the precipitation, which after some time reaches the rivers.

When the forest is gone, the rainfall is distributed quite differently. Item (1), which is estimated at 20–40%, goes to items (2) and (4), and item (3) is cancelled. In brief, the twofold buffer activity of the canopy intercepting rain, and of the humus and roots absorbing and recycling it, is lost. The water that comes down runs off quickly, and only the portion that sinks into the soil is held up. The rivers emerging from a rain forest, therefore, always hold water; those coming from deforested lands flood excessively after a downpour, then quickly run dry. After the conversion of rain forest in Malaya to oil palm and rubber plantations, the peak flows were doubled, and the minimum ones were halved (figures cited by Brünig 1977). The cause of the disastrous floods in the Ganges plain and Bangladesh is the deforestation of the Himalaya slopes.

Hence, the value attached to intact forest in 'catchment areas' is to ensure the water supply of cities, and to protect man-made lakes for hydroelectric power and irrigation works.

d) Stabilization of Climate

Who can fail to be impressed by the thick banks of mist which lie in and over the canopy of a tract of rain forest in the early morning? As the sun rises, they are lifted by the quickly warming air; it does not take long before the first clouds of the day are formed from them, and the quantities of water vapour which compose them must be huge. According to Sioli's rough, but knowledgeable estimate with regard to Amazonia, about half the precipitation in that basin consists of water that has previously evaporated from it.

Evaporation greatly increases the humidity in the vicinity of a tract of forest, and the taller the forest, the more water it puts into circulation. This removes from a local climate the extremes of heat, cold and drought. A climatic map of the Netherlands makes this effect visible, although in this case it is the sea that produces the water vapour and causes the coastal regions to suffer less frost in winter and during the night in other seasons, and less heat and drought in summer. The figures shown are averages which prevail in certain zones suitable for the cultivation of certain crops.

In the tropics, too, zones which suit certain crops can be demarcated and if the supply of water vapour dwindles, the zones soon shift and/or become narrower. As a result, the areas suitable for a given crop may decrease. In Malesia the general influence of the sea is likely to mitigate the worst effects, but in the large continental masses of Africa and South America, a loss of forest cover, i.e. a loss of evaporation, will certainly affect the cost/benefit account of agriculture in the vicinity.

Apart from the effect on local climates there is persistent concern about the climatic effects on a global scale which may follow major deforestation, first with regard to rainfall patterns, second with regard to air temperature influenced by carbon dioxide content. But it is difficult to determine the share of the tropical rain forests (or of all forests) in maintaining a balance.

Air temperature is influenced by the proportion between incoming radiation from the sun, and outgoing radiation from the earth into space. Part of the sun's radiation is reflected on the earth's surface; this reflection is called *albedo*. Desertification increases albedo; melting of the world's glaciers decreases it. Thus, a variety of factors outside the rain forests have their effects on the radiation balance. Because this balance determines the level of the open sea – a rise in temperature causes the polar ice caps to melt – its stability is of utmost importance to mankind. However, as climatologists are still divided as to the actual cause of the Ice Ages, it seems impossible to define the exact role of vegetation cover or to use such an argument to save the tropical rain forests.

The carbon dioxide or CO_2 content of the atmosphere is an equally elusive factor. What makes it important is the fact that it acts as a trap for sun radiation. The higher the CO_2 percentage, the less heat will be radiated back into space, the warmer the atmosphere will become, resulting in the above mentioned rise of sea level.

Carbon dioxide represents one stage in the many chemical processes involving carbon. Much carbon is locked up in the biomass of the earth's various ecosystems. Woodwell et al. (1978) gave a few figures:

	Weight in 10^9 t	Percentage
Tropical rain forests	344	41
Tropical seasonal forests	114	14
Temperate evergreen forests	79	9
Temperate deciduous forests	95	11
Northern conifer belt	108	13
Other terrestrial vegetations	84	10
Open seas	4	0.5
Total	831	

Woodwell et al. (1978) also determined the main factors involved in the world's carbon economy. They found that burning rain forest biomass releases into the atmosphere an amount of CO_2 similar to that of burning fossil fuels. On the other hand, the CO_2 content increases far less due to these processes than might be expected. Large quantities of CO_2 must therefore be withdrawn from the atmosphere through absorption in the oceans, but the nature of this process is not clear.

The CO_2 question is also extensively examined in *The Global 2000 Report* (Barney 1980). The conclusion is that even the best experts are unable to make more than educated guesses. As matters stand, it seems equally possible that the atmosphere may become somewhat cooler, but temperature may also be stable within rather narrow ranges. No climatic argument to conserve forest can therefore really be upheld in a global context.

In addition, those who regard the tropical rain forests as the main suppliers of oxygen may not be well-informed. The atmosphere contains 0.03% CO_2, and 20% O_2. According to modern views, this has accumulated through green plant metabolism, but the rain forests have contributed only a part of this during their long existence. Besides, 20% of the atmosphere contains such a high percentage of oxygen, 600 times the amount of CO_2, that life on earth could probably do well with a couple of percent less. The 'green lung' argument has only the slightest connections with reality (Fig. 16.1).

e) Source of 'Minor Forest Products'

The products of a forest other than timber are called 'minor'. We came across a few in the quotation from Wallace describing the palm products used by the Indians in tropical America.

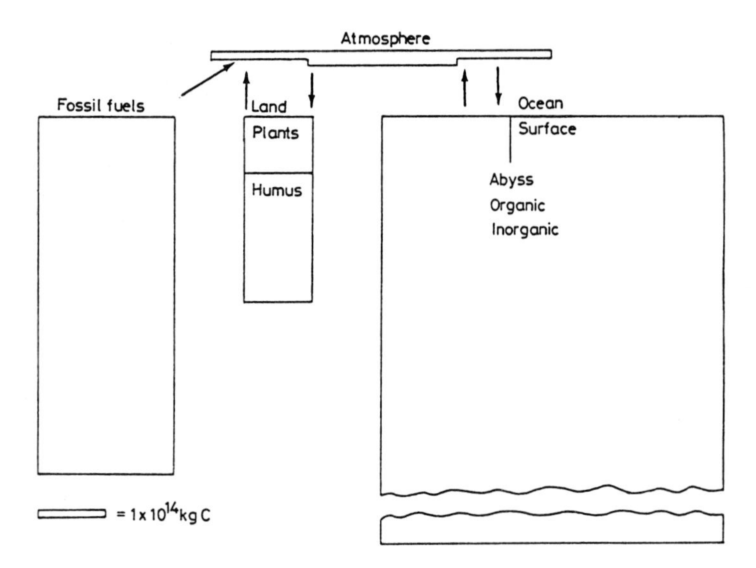

Fig. 16.1 The world carbon budget. The sizes of the boxes indicate the relative magnitudes of pools. (Woodwell 1978)

A primary tropical rain forest supplies non-timber products in small quantities, but in astounding variety. Let me list some important categories:

1. Contraceptives, like diosgenin, the active substance of 'the pill', from *Dioscorea* (Dioscoreaceae), in Central America.
2. Dyes for cloth, such as the logwood, *Haematoxylum çampechianum* (Caesalpiniaceae), from Central America; from the chipped heartwood a black-staining substance can be extracted. It has been widely grown.
3. Essential or volatile oils, distilled from various collected parts. Bark of cinnamon, *Cinnamomum* (Lauraceae), from Indo-Malesia, contains a medicinal oil.
4. Fatty oils, e.g. the seed kernels of babassú, *Orbignya martiana* (Palmae) in Amazonia; or the tengkawan nuts of *Shorea* species (Dipterocarpaceae) in Borneo.
5. Fibres, for cordage and clothing, from the bark of the climber *Gnetum* (Gymnospermae) in Malesia; of many *Ficus* species (Moraceae); and from the 'stem' of bananas, especially *Musa textilis* (Musaceae) in The Philippines.
6. Fish poisons, which, added to water, sedate or kill fish; the best known is rotenone, from the roots of a *Derris* (Papilionaceae) in Sumatra. Arrow poisons are prepared from the sap of upas, *Antiaris toxicaria* (Moraceae), throughout Malesia.
7. Fruits of *Artocarpus* (Moraceae); *Citrus* (Rutaceae); *Durio* (Bombaceae); *Garcinia* (Guttiferae); *Mangifera* (Anacardiaceae); *Nephelium* (Sapindaceae); and others, were discussed in Chapter 10 (Malesia).
8. Honey, taken from bees' nests in tall trees like kempas, *Koompassia excelsa* (Caesalpiniaceae) in Malaya.
9. Incense woods, like that of *Aquilaria malaccensis* (Thymelaeaceae), when it has been diseased by a fungus, which enters through old stumps. It is much sought after in Borneo and Malaya.
10. Latex, tapped from the bark of *Hevea brasiliensis* (Euphorbiaceae) for rubber; from *Achras sapota* (Sapotaceae) in Central America and from jelutong, *Dyera costulata* (Apocynaceae) in West Malesia for chewing gum.
11. Medicinal substances, like reserpin, a cardiac glycosid from the roots of *Rauwolfia*, (Apocynaceae) in Indo-Malesia; or kina, the classic drug against malaria, from the bark of *Cinchona* (Rubiaceae), in western Amazonia; or curare, essential in heart and lung

surgery to paralyze the muscles, from the bark of *Chondrodendron* (Menispermaceae) in South America.

12. Ornamental plants, especially epiphytes as these are often drought-resistant, for example the nest fern, *Asplenium nidus*; or early hemi-epiphytes like *Ficus elastica* (Moraceae), both from Malesia; or late epiphytes like *Monstera deliciosa* (Araceae) from Central America. Bromeliaceae, from tropical America, and most orchids are also epiphytes. But ground-dwelling herbs from the rain forests are cultivated too, usually for their variegated leaves, like many Acanthaceae and Marantaceae.
13. Rattans, which are the stems of Malesian climbing palms, used for an enormous variety of purposes, but outside the tropics best-known in furniture.
14. Resins, for illumination, paint or varnish, like damar from Dipterocarpaceae or kopal from *Agathis* (Araucariaceae), both in Malesia, where the trees are tapped.
15. Spices. The fruits of pepper, *Piper nigrum* (Piperaceae) of Indo-Malesia; the mace and nut of nutmeg, *Myristica fragrans* (Myristicaceae) of East Malesia; the flower buds of clove, *Eugenia caryophyllus* (Myrtaceae) of East Malesia; the rhizome of ginger, *Zingiber officinale* (Zingiberaceae) of Indo-Malesia.
16. Starch, from tubers of *Alocasia, Amorphophallus, Colocasia, Xanthosoma* (all Araceae); and from nearly flowering trunks of sago, *Metroxylon* (Palmae) in more swampy places, all in Malesia.
17. Stimulants like cocaine, for which the leaves of *Erythroxylum* (Erythroxylaceae) in Amazonia are masticated.
18. Strips cut from leaves of Cyclanthaceae in tropical America, Pandanaceae in the Old World and Palmae everywhere, are used for weaving mats, basketry, hats, etc.
19. Sugar and palm wine made from sap which is tapped from young inflorescences of *Arenga pinnata* (Palmae and other palm species).
20. Tannins, for making all sorts of leather and cordage durable, from a variety of barks, e.g. gambir, *Uncaria* (Rubiaceae) in West Malesia.
21. Thatch, with leaves from species of Palmae.
22. Waxy substances, e.g. from *Balanophora* (Balanophoraceae) in Malesia.
23. Wood for special purposes like axe handles from *Caryota* (Palmae) in New Guinea, or carving wood from *Manilkara kauki* (Sapotaceae), formerly in Bali. Now that this brown wood has been depleted, the black wood of *Diospyros celebica* (Ebenaceae) from Celebes is used.

This is but a modest list to give the reader an idea of the many categories; it is strongly Malesia-oriented because we have several fine standard works on parts of the region (see Jacobs 1982, for a discussion of the entire subject).

One is Heyne's, on the useful plants of Indonesia (1927, reprinted 1950); it adopts 40 main categories of products. Another is I.H. Burkill, *A Dictionary of the Economic Products of the Malay Peninsula* (1935, reprinted 1966). The rough-and-ready classification of the products which I once attempted to make, resulted in 102 categories.

A number of minor forest products are from vegetation types other than rain forest, like bamboo, which is from seasonal or disturbed forest. Of those that have their origin in the rain forest, some are commercially grown in the country of origin, like durian, or in different parts of the world, like rubber. They are traded regularly through well-administered channels. Others are gathered from the wild by collectors, like rattan and sago. Although some rattan is exported, e.g. from Indonesia, in the late 1970s, as much as 50,000 t $year^{-1}$; such products are more often traded in village and town markets. They support the domestic economy in countless, unrecorded ways, sometimes on a bartering level.

Thus, the rain forests have given support to human beings for centuries, especially in Malesia. Nevertheless, the minor forest products are a neglected research subject. This is due to the following factors:

1. Research on them is inordinately difficult. A thorough knowledge of both scientific botany and ethnobotany is required. The knowledge natives possess must be identified, understood and 'translated' into international scientific terms.
2. Research takes time. The field is extensive, and experience plays a great role.
3. Research will not result in large monetary profits. And although it may make a rural population more aware of this important contribution to its self-reliance, the rain forest must have already been saved from logging by other means.

For these reasons, good books on minor forest products are few and far between. It will be evident, however, that because of these products there are innumerable relationships between rain forests and man, and that deforestation destroys a main source of material culture in rain forest countries as well as the source of many valuable substances that are used in some form in non-tropical parts of the world. Loss of this source will inevitably impoverish the quality of human life, gradually and unobtrusively perhaps, but over a very wide range. Norman Myers, in *The Sinking Ark* (1979) tells the same story in different words. We shall come back to it.

f) Pool of New Useful Plant Species

The search for new useful species in the rain forests is far from complete. We know a small number thoroughly, and possess indications about a fair amount. But many data are old, and sophisticated chemical analysis has much to confirm or to reveal about the application of already known substances in new fields. For example, following a ban on saccharin and cyclamates, demand arose for substances both sweet-tasting and safe. Attention fell upon two species in the African rain forests, *Dioscoreophyllum cumminsii* (Menispermaceae) and *Thaumatococcus danielii* (Marantaceae), vaguely known for these properties. Studies on them were published by Holloway (1977), Most et al. (1978) and Summerfield et al. (1977). But although much useful ethnobotanical research has been published, considerable knowledge is still locked in the oral traditions of many forest tribes. The botanical identity of several long-known, almost classic products has been uncertain till recently. Only after long, tenacious efforts by Kostermans (1973) was fertile material of a local cinnamon species in Ceylon obtained. This was *Cinnamomum capparu-coronde* (Lauraceae), described formally by Blume in 1836 from a single leaf. It had been mentioned in the literature in 1919, but never properly identified. Yet only after proper identification is it possible to take measures for species protection and propagation. The splendid Brazilian timber species known as Sebastiao de Arruda was only identified as *Dalbergia decipularis* (Papilionaceae) by Rizzini in 1978 after great effort.

Balick (1979) took a different approach. He scrutinized the literature for palm species with a high content of fat in the seeds. By this method 25 species in Amazonia alone were singled out, and in his paper Balick stressed the importance of retaining the entire spectrum of variability within those species.

A large program to screen plants for carcinogenic and anti-cancer substances, conducted by Perdue and Hartwell, was reported on in 1976. Of the 25,000 species examined, a tenth of all the seed plants in the world, 2000 were positive, gymnosperms scoring twice as high as angiosperms. This work has barely touched rain forest plants, but of them four families have so far turned out to be notably promising: Burseraceae, Celastraceae, Icacinaceae and Rutaceae (see *Fl. Males. Bull.* p. 3079–3080. 1978).

The journals *Economic Botany* and *Lloydia* regularly carry notes about such discoveries.

g) Pool of Useful Genetic Material

The Russian geneticist N.I. Vavilov convinced agriculturists of the importance of wild relatives of cultivated races for use in hybridization. The early efforts concentrated mainly on such annual crops as beans, grains, potato and tobacco. Between 1920 and 1940 the Soviet Union alone organized 140 domestic and 40 foreign expeditions to collect wild strains; 200,000 were obtained. The Vavilov Institute nowadays distributes 50–60,000 samples a year for propagation among local experimental stations (Brezhnev 1970).

Through hybridization it is possible to introduce into cultivated races such properties of wild relatives as suitability for certain soils or resistance against disease, as well as desirable characters for the crop itself, for instance tall or low stature. In tropical agriculture such work has a long history in the cultivation of oil palm, rice, rubber, sugarcane, tea and other important crops. In forestry, work concentrated on widely used genera of the savanna or secondary formations, particularly *Eucalyptus* (Myrtaceae) and *Pinus* (Pinaceae). Research on trees, however, takes much more time and money, because while variation, ecological requirements, pollination, cytology, genetics and seedling biology of the species and relatives must be equally well investigated in detail, trees have a much longer life cycle than herbs.

The genetic base of a crop in cultivation is often very narrow. This means that there is not enough diversity among these plants to enable a percentage of them to survive a massive outbreak of pests or disease, whereupon an entire species may die for lack of a few resistant individuals. In the 1970s, the delicious jeruk Bali, *Citrus maxima* (Rutaceae) was wiped out in Indonesia by the 'phloem degeneration virus'. Timely hybridization with wild *Citrus* (present in the rain forests) might have prevented this calamity.

We remember that Malesia is the world's richest reservoir of fruit trees. Ashton (1976b) gives an excellent review of the genetic potential of the Malesian rain forests which he ascribes as partly due to the diversity of the soils on which they grow.

Besides the preservation of wild genes for fruit culture there are the purposes of forestry to consider. If tropical forestry is to have any future, in terms of centuries, use will have to be made of the entire spectrum of suitable tree species and of all their local races. This necessitates the special protection of those rain forests which contain many commercial tree species. The book, edited by Burley and Styles, *Tropical Trees/Variation, Breeding, Conservation* (1976) adduces many strong arguments for such protection, and warns against 'genetic erosion' which threatens populations subject to selective harvesting. This could easily result, for instance, from the technique of lifting fine logs out of the forest by helicopter to limit damage: only inferior trees are left behind as seed trees and inevitably a population of poorer quality is formed (p. 62). The importance of protecting forest genetic resources in situ was confirmed by a recent panel of experts convened by FAO/UNEP in 1981.

h) Shelter and Food Source of Animals and Hunting-Gathering Tribes

Chapter 12 (Plant-Animal Relations) dwelt on food consumption, pollination, dispersal and recycling. It also mentioned the high percentage of animal species living in rain forest that fully depend upon it for all their food and shelter needs.

For several years there has been a general awareness that trees in a rain forest do not exist in an ecological vacuum, but together with animals are part of a 'web of life'. The book edited by Burley and Styles (1976) gives much attention to pollination biology, particularly with a view to silviculture.

Forest animals are also, rightly or wrongly, an important research tool in modern medicine, for instance the tens of thousands of monkeys used in experiments. The chimpanzee yielded an anti-polio vaccine and Myers (1979, p. 96) mentions the 'golden lion marmoset' in Southeast Brazil as a potentially valuable species because of its resemblance to the 'cotton-topped marmoset', *Saguinus oedipus*: experiments using this species produced an anti-cancer drug (Laufs and Steinke 1975). *Saguinus oedipus* originally inhabited 6500 km^2 forest; for the last 600 individuals of this species now only 550 km^2 is left.

Chapter 12 also discussed the defence mechanisms of plants against animals, such as hairs and spines. The chemical weapons (arrow poisons and fish poisons) used by man against vertebrates are few in comparison with the array of chemical substances produced by plants in their defence against animals; the case of *Carica* was mentioned in Chapter 11 (Evolution). Just as recent research of feromones, the substances by which insect species attract members of the opposite sex, has opened up new fields of promise for biological control, a great variety of substances is awaiting discovery in the rain forests, with their many biological checks against parasite explosion.

Tribes of hunter-gatherers live in all the big rain forest regions, roaming large territories in small groups. The Punans in Borneo, the Kubus in Sumatra and the Pygmies in Central Africa are well known (see Richards 1970). Goodland and Irwin (1975) give an informative account of Indian tribes in Amazonia. All these people, who live entirely in and off the forest, are the only ones who have mastered the art of exploiting the rain forests on a really sustained basis, thanks to an enormous amount of practical knowledge about not over-withdrawing substances from this nutrient-poor system. But only such species-rich ecosystems with their innumerable and different products can provide for the complete spectrum of human needs. They know everything about food plants, medicinal species, edible insects and their larvae, and the collection of wild honey. With bow and arrow or a blowpipe they hunt the scarce animals of the rain forest to obtain protein. Most hunter-gatherers are extremely shy: the Tasaday in The Philippines were only discovered as late as 1975, and live (almost) entirely outside financial economies. Such utensils as bush knives and cooking pots may be obtained through cautious bartering using forest products, but for the most part everything they need they make themselves.

These tribes are subject to some persecution from those societies which they have so far shunned, but which now lay claim to their territories. Some countries make efforts to settle them peacefully; in others they are prosecuted, killed or forcibly acculturated. In either case their great store of knowledge is in danger of becoming lost. The organization 'Survival International' in London helps such people to defend their own traditional life-styles, to which they are just as much entitled as we are to ours.

i) Matrix of Evolution

To grasp the idea that the primary forests may continue to evolve for as many millions of years as there exists a biosphere on the planet Earth, is beyond the ability of most of us. It is easier to comprehend the four main achievements of the rain forests through evolution to date:

1. They have produced the largest number of plant and animal species, and still contain most of this diversity;
2. They have preserved many of the ancient forms of land flora still in existence, from which more recent forms outside the rain forests have derived;
3. They have built up the world's largest amount of biomass per hectare;
4. They have succeeded in making their structure independent of soil conditions to a very great extent, by evolving a very tight system of nutrient recycling.

As the outcome of a very long process of mutation, selection, isolation and adaptation, the rain forests are unique and cannot be reproduced. They function perfectly and no man can add to them or improve them, although we can extract crop plants from them which we can hope to 'improve' genetically, thus creating insignificant, short-lived, evolutionary side lines. But only if we handle rain forests extremely cautiously can we utilize them at no ecological cost, ad infinitum.

But that the rain forests should be allowed to continue their slow evolution, thereby perfecting existing forms of life, and producing new ones unconceived of by any human being, according to the principles of speciation and the durian theory – that is too bold a thought. If the loggers have their way, all the attainments of the dipterocarp forests in over 30 million years will have been annihilated in 30.

j) Source of Knowledge

The scientific study of plants and animals, through personal examination of the facts, was revived in the 16th century, after a virtual standstill of one and a half millennium. The oldest 'modern' herbal which started to provide authentic data is of 1530, the oldest botanical garden dates from 1543 and the herbarium technique originated during that same period. In the course of the next 4 centuries an impressive body of biological views and knowledge was accumulated, largely obtained from the study of the relatively poor plant and animal world of the temperate regions. Later work in the tropics resulted in large quantities of additional knowledge, but it was only in the period between 1930 and 1950 that the foundations were laid for a 'tropics-centred' botany. (In zoology, with its emphasis on the study of individual species matters were different.) These are some of the landmarks in its development:

1. The concept of the climax vegetation and the derivation of seres from it (Richards 1952);
2. The durian theory (Corner 1949; see Chap. 13);
3. The view on all life forms of the monocotyledons as variations on one theme (Holttum 1955);
4. A general view on the architecture of all (woody) plants (Hallé and Oldeman 1970);
5. An outlook on defences of plants against animals under tropical, ever-wet conditions (Janzen, various publications);
6. A first insight into the genetics of rain forest trees (Ashton 1969, and subsequent papers);
7. A full idea of the significance of animals as seed dispersers and their role in the survival of the rain forest (Corner 1964; Rijksen 1978);
8. The dependence of certain epiphytes on nutrients drawn from the host tree by fungi: epiphytosis (Ruinen 1953);
9. A first precise indication of the age of rain forests (Flenley 1979);
10. An understanding of species richness as a key factor in the functioning of the rain forests and in their relations with man.

This list represents a nice result of half a century. Further expectations are high; see the long lists of topics for research in *Tropical Forest Ecosystems* (UNESCO 1978).

All future scientific work on tropical ecology, vegetation dynamics and forestry in the widest sense will have a need for virgin (climax or undisturbed) forest as a standard, for from this stage all others are derived, through damaging or destructive factors. Such factors are numerous, as we shall see, and their effects must be determined with precision if we want to devise a fruitful way of silvicultural management.

k) Object of Respect for the Creation

Overt respect for nature is a rare ingredient in scientific biology. The two passages which I encountered while exploring the literature of my profession, are quoted in the Preface. There is not much more. In all deliberations among scientists and policy-makers this subject is usually avoided. Yet respect, as something to give and to receive, is an essential good in life, under all circumstances, and a condition for enrichment in the metaphorical sense. In the context of this book, it is immaterial whether the object of our respect is termed creation or evolution, when the forces that made the tropical rain forests are meant.

Respect can be stimulated and nurtured. To that end, I reproduce here this *Declaration of the Rights of Animal and Plant Life:*

i Each living creature on earth has the right to exist, independent of its usefulness to humans.
ii Every effort should be made to preserve all species of animal and plant life from premature extinction. Special protection should be afforded to those species whose survival is already threatened.
iii Suitable living conditions for animal and plant life should be guaranteed in order to safeguard their existence, taking into account the natural ecosystems and necessary minimum populations concerned.
iv The right to exist of animal and plant life may, in principle, be violated when necessary for the survival of humans. In every human act affecting this right, the need for violation should be carefully balanced against human interests.
v To guarantee a long-term existence of humans, animals and plant life, humans should regulate their own population growth and adapt their patterns of production and consumption with a view to the least possible violation of the right to exist of animal and plant life.
vi Should control of organisms noxious to humans and their livestock be necessary, nature should be used as an ally wherever possible. Control measures should be taken in such a way as not to affect other species.
vii Unnecessary suffering of animals caused by humans should be prevented in every possible way.
viii Wild animals and wild plant life should be left in their natural environment whenever possible. Living conditions of wild animals and wild plant life removed by humans from their natural environment, should correspond as closely as possible to their natural conditions of life. Agriculture, the farming of livestock and fishery should preferably be conducted in an ecologically responsible way.

The original text, formulated by Van Heijnsbergen (1977) has been republished in various journals (*Fl.Males.Bull.* 31:3048. 1978; *Habitat/Australia* 7, 4:31. August 1979; *BioIndonesia* 7:93. 1980). What matters most, in my opinion, is the point that no species must be allowed to become extinct through human action. Of all earth's ecosystems the rain forests contain the greatest numbers of species, each connected to the other in a Web of Life, and each of them a fellow creature of man. The human being who still feels no respect for the creatures of the rain forests and their evolution, should cease reading this book.

l) Medium for Education and Recreation

Still, many people do hesitate to recognize the right of other species to exist. They may have less difficulty, however, in subscribing to the right of every human to an education. And since biology in some form is an acknowledged part of every proper education, it would be irresponsible not to pay some attention to the richest biological ecosystems on earth.

Each of the many species in those ecosystems symbolizes a page in the Great Book of Nature. Scientists inscribe the pages for others to read, but all damage and destruction enhances the risk of species extinction. For each one lost, a page is then torn out of the Great Book, which might have opened, shaped or refined someone's ideas on nature. In every educational process ideas are transferred, nurtured and made meaningful; luckily, many universities now possess research facilities based in a number of different habitats, including the rain forests.

It occurs to me that the scandalous treatment suffered by the rain forests at the hands of man, however, stems from a lack of articulated ideas about what these forests are and what their values are, in spite of present educational efforts. Even educated people have yet to learn much about the rain forests, and clumsy misunderstandings about them are circulated. But ignorantly to do away with such potential material for education and research is to burn a book before it has been read – though it is better to burn books than forests, for books (except the Great Book of Nature) can be reprinted.

One of the newest functions of a rain forest, it seems, is to serve as a medium of recreation. The 1970s saw the first amateurs coming to field stations and national parks to seek their own personal acquaintance and relationship with a rain forest. These were resolved, enthusiastic people, both rich and poor, all driven by curiosity. In Ketambe in Sumatra I have witnessed a non-biologist falling in love with the rain forest in 3 day's time, soon walking barefooted and still feeling 'homesick' for it years later. Although such persons are, no doubt, exceptional, and although the rain forest may never attract crowds, it nonetheless has become the subject of many popular books, films, lectures, meetings and articles, to the extent that no one can be mistaken about what is going on: the tropical rain forest is, at last, and with hesitation, being incorporated into man's spiritual realm. It is a happy sign that man has reached adulthood as an inhabitant of this earth.

Conclusions

There is so much variety in the spectrum of values, that each of the 12 points actually represents a set of values on its own. They can be grouped in several different ways.

The first grouping sets off *the timber value against all others*, because through utilization of this value none of the others are left intact, as we will see in Chapter 17 (Damage and Destruction).

The second grouping is (b) retention of soil, (c) regulation of run-off and (d) stabilization of climate. These functions concern non-biotic factors and, being independent of composition, can be fulfilled by tall, secondary forest as well. They can be regarded as 'latent' values, because removal of the forest results in actual damage, by erosion, flooding and decreased air humidity, contrary to the others where the losses are unobtrusive.

The third combines (e) source of minor forest products, (f) pool of new useful plant species and (g) pool of useful genetic material. Here, the value is proportional to the numbers of species present, and *utilization* is possible *without damage to the forest structure*. Moreover, the values are highly specific: all these products and materials are unique to the rain forests in certain places, and irreplaceable.

The fourth grouping combines (h) shelter and food for animals and (i) matrix of evolution. In addition to species richness, *plant-animal relations* are involved in all of these, over a long evolutionary period, and they are an important component in the complexity and fragility of the forest. These values exist completely *independently of man*.

The fifth grouping combines the values (j) source of knowledge, (k) object of respect for the creation and (l) medium for education and recreation. These are *spiritual values*, depending on the *input of man*.

The sixth grouping, in contrast to the fifth, combines the values which can be expressed *directly in terms of monetary values*: (a) timber and (e) minor products, although they are to a large extent mutually exclusive.

The seventh grouping segregates the *more local values* like (b) soil retention, (c) regulation of run-off and (d) stabilization of climate in the region, and a large component of (d) source of minor products, (h) shelter and food for animals and the means of existence of hunting-gathering people, from the others which are appreciated in (much) wider areas, within the tropical belt as well as in the highest developed countries.

The eight grouping singles out those values which may be discovered *in the foreseeable future*, (f) new economic plant species, (g) useful genetic material, (j) source of knowledge, and the *indefinite future*: (i) matrix of evolution.

The ninth grouping emphasizes the values of the *tropical* rain forests, due to their *richness in species*. No other forest type (a) supplies such a variety of timber, (e) and of minor products, is so promising a pool of (f) new economic species and (g) useful genetic material, has such importance as (h) shelter and food for animals, (i) matrix of evolution and (j) a source of knowledge.

What a complex mixture of values! How poor is the vision of a forest as a mere mine of timber! And should we consider a city as simply a collection of bricks?

A Note on the Value of Secondary Forest

The existence of secondary forest cover can make the return of primary forest possible, if primary forest borders or surrounds it. In large areas where not enough primary forest is left, secondary growth will become the climax vegetational stage. Young secondary forest can also develop into mature secondary forest (see the differences in Budowski's Table) if such forest is not too far away. It is therefore important in land-use planning to save plots of mature secondary forest, lest the species assortment in a region becomes too impoverished. For our purposes here, we henceforth presume that both young and mature secondary forests are present.

The percentage of plant species useful to man seems to be approximately the same in primary and in secondary forest; the fact that the percentage seems higher in the secondary forest only reflects the fact that secondary species have been known longer, are generally to be found growing closer to human habitation and have been better investigated. But since the secondary forest is so much poorer in species than the primary, the number of useful species it contains is actually, of course, much smaller.

Secondary forest is mainly important for:

1. Rehabilitation of Land After Clearing. The pioneer and young secondary vegetation which develops on bare soil breaks the force of falling rain and thus protects the soil from erosion. Litter is produced which is transformed into humus; thereby the cycle of inorganic matter is set up again. After some years there is once more a canopy, with a favourable microclimate under it. After 10–20 years (depending on conditions) soil fertility has returned, enough to support one to three crops. The practice of shifting cultivation, to be discussed in the following Chapter (Damage and Destruction), is based on this mode of recovery.

2. Pool of Fast-Growing Trees with Soft Wood. The word 'soft' is here to be taken literally. There is an assortment of a few dozen species, tested for their usefulness at various altitudes, in a range of climates on different soil types. Von Meyenfeldt et al. (1978) have made a useful compilation. Their fast growth renders them suitable for control of light-loving weeds like alang-alang, and for reforestation in general. Their soft wood finds a broad spectrum of industrial application, for instance crates, pult, chipwood, extraction of chemicals and also local use as firewood.

Here are a few well-known genera, most of which contain more than one useful species: *Albizia* and *Leucaena* (Mimosaceae), *Anthocephalus* (Rubiaceae), *Cecropia* (Moraceae), *Eucalyptus* (Myrtaceae), *Gmelina* (Verbenaceae), *Macaranga* (Euphorbiaceae), *Ochroma* (Bombaceae), *Pinus* (Pinaceae), *Terminalia* (Combretaceae) and *Trema* (Ulmaceae). The bamboos (Gramineae), should also be included, but as a family they need much taxonomic work. Several of these genera are already being planted on a large scale, but further research and trial will no doubt reveal the rich new potentials of various regions which will hopefully enable a larger share of native species to be used in plantations and reafforestation projects.

Flora Brasiliensis, tabula XVI. On the *left*, a large *Ficus* (Moraceae) with buttresses. On the *right*, the forest is logged, presumably for shifting cultivation (note the shed and the people); the biomass, after a period of drying, has been put alight. Near St. Johannes Marcus, province Rio de Janeiro (Martius 1840–1869)

Secondary forest can no doubt be overexploited, as any ecosystem can be. It has, however, great capabilities for extension and renewal. In contrast to primary forest, it is poor in diversity but, if sufficient patches are conserved, almost inexhaustible. The management of secondary forests and the silviculture and utilization of the species belonging to their various stages, will be one of the main tasks of tropical forestry in the future.

17 Damage and Destruction

An Outline of the Factors Involved

The ways and patterns in which the world's tropical rain forests are degraded, are manifold and complex. A multitude of factors are involved; we will first review them, with an occasional word of explanation. We then will discuss some of the more important ones: collection of minor forest products, timber extraction, silvicultural measures, shifting cultivation, planned destruction; and then we will comment on consequences and causes. A comprehensive, well-illustrated chapter on human impact was given by Boerboom and Wiersum (1983).

1. Natural factors are:

a) 'Soil creep' on steep slopes subject to high rainfall; trees slant, their root systems suffer, growth is poor.
b) Landslides, small and large, whereby the entire vegetation comes down along a front; undercutting of forest by rivers changing their course or during floods.
c) Deposition of rocks, sand or mud, by water during floods, if in thin layers, trees will persist, but undergrowth will need time to recover.
d) Inundation, when natural drainage becomes impeded, or the sea level rises; the forest then dies as a whole, decomposing slowly.
e) 'Wind-throw' when cyclones or local gusts blow over a tract of forest, in which case young stems may grow more or less crooked; by this means wind damage can be detected long after the event (Fig. 3.5).
f) Emission of volcanic gases, causing branches and/or trees to die; as a result some species, more resistant than others, can come to dominate an area around a point of eruption, as for instance an *Eugenia* (Myrtaceae) in North Sumatra, found surrounding a sulphurous hot spring (Jochems 1930).
g) Volcanic ash and rock, which causes suffocation of root systems and breakage, or sometimes wholesale destruction.
h) Lava streams, whose effect is similar to that of landslides, but lava takes much longer to weather.
i) Mud flows, when a water barrier gives way, through volcanic or other geological activity; forest is swept away as if by a landslide, but these can occur on relatively level terrain.
j) Damage by such large herbivores as elephants, rhinos or tapirs. These animals tend to knock over trees (elephants like the soft substances in hollow parts, Corner 1978, p. 53); large animals also eat and trample on herbs and seedlings (ditto, p. 20), causing the extension of open spots.
k) Fire from lightning; although this is a rare phenomenon in rain forest, some cases are on record (see Whitmore 1975).
l) Increase of drought through slow climatic change, accepted for the Pleistocene Ice Age in parts of Africa and South America; under this influence former rain forests gradually gave way to deciduous forest types and savannas.

Fig. 17.1 Submerged trees in Lake Brokopondo, an artificial lake in Surinam, August 1973

m) Cooling of climate during the Ice Age; resulting in a narrowing of altitudinal zones and, in mountainous regions, a decrease in the size of the warmest areas (Fig. 13.12).
In addition, there is the natural replacement process in which old mature trees fall, forming *chablis* that are refilled by young growth.

2. *Anthropogenic, or man-induced factors are:*

a) Selective removal of animals, e.g. monkeys and birds, for trade or consumption.
b) Selective removal of a few plants or their parts, such as orchids or rattans, which are collected for trade.
c) Removal of trees by hand, for canoes, construction or conversion into charcoal.
d) Extraction of timber with the aid of machinery, also known as mechanical logging.
e) Silvicultural measures, to kill non-commercial species in favour of commercial ones.
f) Tapping of trees for latex, resin or gum; if this is not done expertly, the tree will die.
g) Browsing at ground level by domestic cattle with the result that seedlings are eaten and soil is trampled, thus prohibiting natural regeneration and leaving decrepit trees standing on bare soil.
h) Cultivation at ground level of shade-loving crops such as cardamom, *Elettaria cardamomum* (Zingiberaceae), for medicinal purposes, in southern India; the effect is the same as in (g).
i) Clearing of isolated patches for shifting cultivation, in which the biomass is burnt as far as possible to enrich the soil temporarily with minerals from the ash; very large trees are killed by specially applied fire, although exceptionally useful trees, i.e. durians, are often left standing.
j) Clearing of forest along a front, for agricultural enterprises or road construction.
k) Inundation by artificially created lakes; where the biomass is not first removed, it will decay, inducing weed growth in the water and acidity (Fig. 17.1).
l) (Nearly) complete removal of forest for 'total utilization' i.e. for paper and chipboard.

3. *The patterns of damage can be:*
a) Diffuse, ranging from extensive hunting and collection of minor products to intensive application of silvicultural measures, or logging.
b) Patchy, through degrading isolated plots of forest, while the forest as a whole remains recognizeable.
c) Along a wide front, as in large-scale clearing for plantation.
d) Along a narrow front, as in road construction; secondary fronts of destruction are often started sideways.

4. *The frequency of degradation can be:*
a) Once, as in a landslip, after which the forest can regenerate.
b) Continuous, which has an effect similar to the grazing by animals over long periods.
c) Intermittent, such as the periodic harvesting of timber.

5. *In time, degradation can be:*
a) Short, so as not to interfere with spontaneous regeneration.
b) Long, interfering with regeneration.
c) Terminated in the past, as for example when a human population has departed, permitting a new equilibrium to be reached.
d) Continuing to the present.

6. *Site factors influencing regeneration are:*
a) Altitude above sea level: the higher, the cooler, the slower growth proceeds.
b) Soil conditions: minerals added in sedimentation and through volcanic activity are lost by leaching and extraction of biomass; compression of soil by trampling or vehicles has a deteriorating effect on soil condition as also erosion.
c) Accessibility to the seeds and animals of the original forest: if no primary forest grows beside a clearing, there is no hope for its return to the site; this has been explained in Chapter 8 (Primary and Secondary Forest).
d) Occurrence of fire, which may penetrate the forest fringe, destroying seedlings and favouring grass.

We can surmise that the outcome which results from natural factors alone is a mosaic of great intricacy. Damage is followed by regeneration, unless setbacks occur, which may include human interference in one or another form, and the state of any tract of forest is the result of historical processes. But what is vital to know is if this state is one of equilibrium or not. If not, the forest will change in structure and/or composition, and it is important to know what sort of changes can be expected, as it may well be possible to take advantage of the changes, or to steer the changes in a desired direction.

Since changes in the future are the continuation of changes in the past, an understanding of the history of a forest will provide the key to its possible future, and the use that perhaps can be made of it. For this understanding, we need a standard of assessment. This standard is provided by the most stable, i.e. most mature, least-damaged parts of the forest under consideration, bearing in mind that most changes will have brought about some kind of impoverishment and have led to stages derived from the stable, mature state.

We, therefore, should first distinguish between stable, mature forest, on the one hand, and damaged, hence unstable, *modified* forest, on the other. In Chapter 8 (Primary and Secondary Forest) we presented a number of points relevant to this distinction, and others were mentioned with regard to composition (see Chap. 7); the above list of factors also helps. Careful, expert scrutiny of forest data will reveal much of the history of a tract. While this is a very difficult task, which involves considerable knowledge of geomorphology and skill in interpretating both aerial photography and satellite imagery, it is essential to make the attempt.

We will return to the distinction between virgin and modified forest later.

The Fragility of Rain Forest

Renewable forests are common in the temperate zones, and near places of habitation in the tropics. Under good management, these yield a predictable harvest. The crucial difference between a renewable forest and a primary tropical rain forest is the number of species per hectare: few in the former, many in the latter. The fewer species a forest contains, the more easily it is renewed. Take an oak out of an oak forest, and it will be replaced by another oak. In a rain forest where only 4–10% of all the tree species are commercial, the chance that a harvested tree will be replaced by another commercial tree is also about 4–10%. Although this percentage can be increased through silvicultural measures, this is only achieved by reducing the diversity; we will presently see how this is done.

But even though silvicultural measures may not have been applied, all rain forests are especially vulnerable to exploitation because of their complex composition, and their delicate economy of minerals. The more species per hectare, as we have seen, the fewer individuals of one species. The smaller, too, are the populations by which species are represented and by which each must ensure its survival. This means that the removal of even a few individuals has a large effect in terms of percentages. In Fig. 7.5 we noted 12 trees of jelutong, *Dyera costulata* (Apocynaceae) in nearly 3 ha. Removal or killing of a mere two of these trees per hectare; i.e. removing a mere 200 trees km^{-1}, reduces the population by 50%; the elimination of 3 trees ha^{-1} reduces it by 75%.

I do not know about the animal associations of *Dyera*, but we remember *Heritiera elata* (Sterculiaceae), of which Rijksen (1978, pp. 63, 71) in northern Sumatra found one large tree in his 150-ha plot. This tree bore fruit in the 1 month that *Ficus*, the staple food of the orang-utan, yielded next to nothing. During those weeks of scarcity, all orang-utans of Ketambe flocked around this one *Heritiera*, which tided them over that period of hardship. We may suppose that the loss of this one tree would have repercussions on the orang-utan population, even if we are unable to predict the details in the long run. Dispersal of seed may also suffer; we have already mentioned this aspect.

In view of the interdependence of plant species and animals through their food consumption, pollination and dispersal interrelationships, a decrease in the number of one species will inevitably lead to a decrease in the number of another species. Since rarity is the norm in a species-rich forest, declines in scarce items are soon felt. This is a well-known ecological law, but in a species-rich system the percentages of loss which result from even quite small removals are very high.

Another aspect of low population densities is the long time it takes to build up viable populations of individual species. Reproduction rates of many animals are low: for example the orang-utan and hornbill character of bearing only a single young. As for trees, once their seeds have been dispersed and have germinated in sufficient numbers, the seedlings and saplings run the risk first of exposure and then of suffocation by pioneers if the canopy overhead is opened. Kramer (1933) showed – probably in montane forest, but on fertile soil – that small gaps of 50 m^2 are closed without delay by seedlings of primary forest species. If the gap measured 50–100 m^2, some secondary growth developed, but this hampered very little the closing of the gap. However, if the gap were 100–300 m^2 or more, it would be first occupied by a very dense growth of pioneers, sometimes augmented by blanket-forming climbers, killing all primary seedlings and saplings, and causing a severe setback in the

Fig. 17.2 Stand of *Anthocephalus* (Rubiaceae), a species of the secondary forest, having established itself on soil that was damaged in logging operations. All trees have grown simultaneously. (Photograph Cockburn in Whitmore 1975)

regeneration. Gradually, he found, the primary species will win, but certain species such as *Anthocephalus* (Rubiaceae) which like places where the soil has been disturbed, often remain for 40 years (Fig. 17.2). In fact, all openings that occur in addition to the normal amount of natural chablis serve to diminish the area available for the primary species populations.

As for the second area of vulnerability, in the mineral economy of the forest, we have seen that recycling is almost complete in undisturbed forest. But this carefully balanced system can be disrupted in two ways: by exposure and by extraction. In places newly exposed to sunlight, humus oxidizes very rapidly, and the soil loses its capacity to adsorb and retain minerals. As the soil temperature increases under the full heat of the sun, the mycorrhizae are killed, which makes possible the growth of many tree species, dipterocarps among them, through the cycling of minerals. Losses will thus occur. This is also the case when quantities of biomass are removed in the form of logs. Brünig (1977) gives figures: one round of 'average' selective logging in Malaya costs the forest 30 dry matter per hectare, containing 240 kg Ca, 90 kg K, 30 kg Mg, 180 kg N, 12 kg P; that is, about 1 year's budget of minerals. Application of fertilizer to make up for this loss is expensive, and a risk to the structure of the soil, and may diminish or destroy the N-fixing bacteria in it.

The exact consequences of such mineral losses for the ecological processes in the remaining forest are not known. Foresters attach great value to good soil, and we may suppose that loss of minerals will inevitably result in poor growth and low bioproductivity in general, namely poor crops of flush, flowers and fruit. There will be less for the animals to eat; both animal and plant populations may decrease as a result of approaching closer to the critical minimum level.

Collection of Minor Forest Products

True hunter-gatherers generally do not remove much biomass from the forest. Only when they barter or sell products to the world outside, do their activities constitute a drain of materials produced by the forest ecosystem. What these materials may be has been listed under point (e) in the previous chapter. They are astounding in variety.

The damage done by minor product harvesting consists mainly in the killing of certain trees which contain latex, resin or gum by unskilled tapping. Rappard (1937) noted that in South Sumatra skills varied from area to area but were easily transferable, along with the proper tools. But many cases of large-scale depletion of certain species by tribesmen in Malaya are mentioned by Burkill (1935); these species can now be regarded as early examples of threatened plants. Rattan-pulling also causes some physical damage to both canopy and undergrowth. Yet of all exploitation methods, this ancient, traditional one is the least harmful, as very little biomass is taken from a great many species, all of which are well-adapted to growing in species-rich forests on generally poor soil. We will revert to this in connection with The Paradox of Logging (see later section).

Extraction of Timber

In the olden times of 'hand-logging', i.e. before the 1950s, a fine tall tree in the rain forest was carefully selected for felling, according to the available options for transporting it, which was often by water or on railways of various states of primitiveness. A scaffold (Fig. 2.2) was constructed if necessary. The tree was felled with axes, the bole sawn into logs by hand and the logs hauled towards the railway, using a series of wooden rollers or greased beams and a great amount of animal or human muscular effort (Fig. 17.3). In the old literature one often comes across photographs of a group of labourers with their foreman, posing by some huge tree trunk at the railhead, the place of their triumph.

Awe for the achievement of modern foresters is of a different kind. Mechanical logging using heavy equipment, still handled, no doubt, by nice, hardworking people, transforms the scene before one's eyes. The starting point may be a road or a riverbank or even the seashore: in coastal regions, equipment loaded on flat-topped freighters is put ashore in any suitable

Fig. 17.3 Transport with elephants and railroad lorries. Dipterocarp forest near Ban Bao, in the southernmost part of Thailand (Photograph Rappard 1946)

Fig. 17.4 The loggers have gone. From this location a road was made inland, the equipment was landed by way of this landing stage. The road has fallen into disuse and falls prey to erosion (Photograph MJ)

Fig. 17.5 Haulage of logs with tractor-bulldozer, with rope. New Guinea, from *Nothofagus* forest. (Photograph MJ)

place (Fig. 17.4). Bulldozers can thrust roads through the forest for 10, 20 km or more. Far from the noisy bulldozer battlefield one hears the heavy growling of diesel engines and the screeching of portable chainsaws; the air is blue with oily fumes. The trees, selected in advance, come down in mighty crashes. Once the branches are cut off, the bole is attached by hawser onto a bulldozer, or clasped in the fangs of a 'skidder', a specially designed vehicle whose front and rear parts are jointed. Such vehicles pull the logs to the road unhindered by the deep muddy ruts created by the passage of heavy vehicles over friable forest soils, which continues, rain or no rain (Figs. 17.5 and 17.6). Soil that has been compacted under the heavy machinery and in the skid trails where trees were extracted, is inhospitable to mycorrhiza growth. For 10 years or longer such tracts are revealed by the poor regeneration of trees.

In very steep country, the freshly felled trunks are divided into chunks 6–8 m long to avoid breakage. A tall bole is left standing as 'spar tree'. A pulley is fixed to its top to support a cable which pulls the logs out of the forest to the road, track or river (Figs. 17.7 and 17.8). In the case of land transport, the logs are loaded onto strong, slow-moving trucks or bullock carts which carry or drag them away, eventually to reach the coast and a seagoing freighter, or directly to a sawmill. From places near a long river, logs can be rafted. Those heavier than water, known as 'sinkers', are tied to 'floaters' or less dense logs, often cut for the purpose. In swampy areas, ditches are often created with the aid of explosives, then dredged so that logs may be floated away in the artificially created drainage canals. The damage this causes to swamp forest and its specialized vegetation can be imagined.

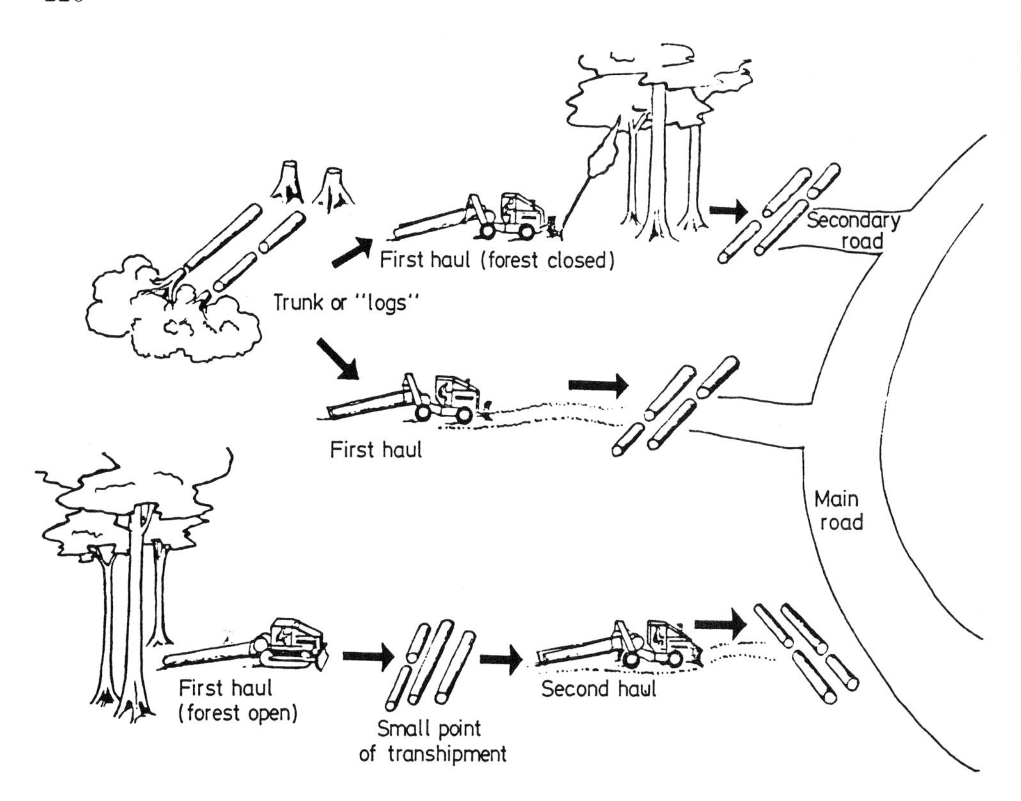

Fig. 17.6 Outline of mechanical logging (*Bois et Forêts des Tropiques* 103: 77, 1970)

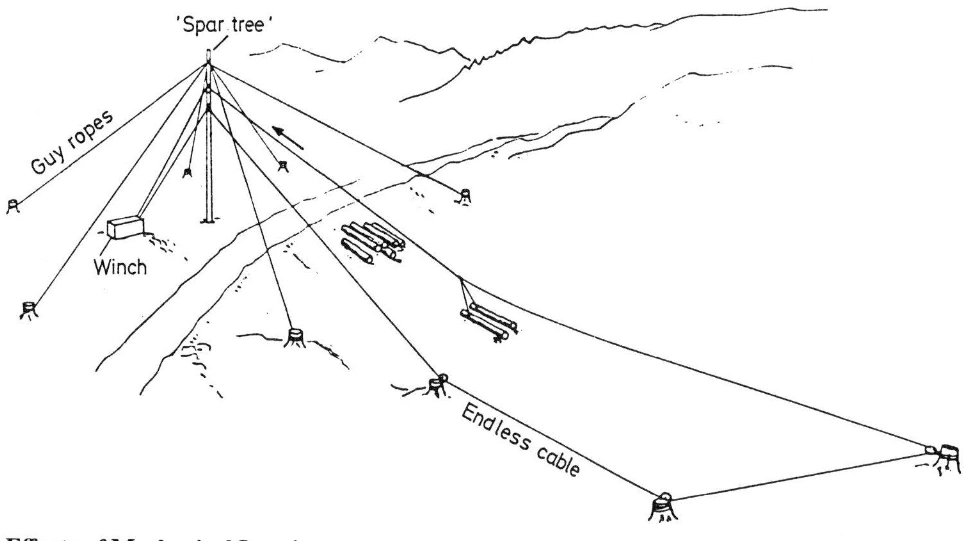

Fig. 17.7 Transport of logs in hilly terrain, with a cable (*Bois et Forêts des Tropiques* 144: 71, 1972)

Effects of Mechanical Logging

Most of the detrimental effects of mechanical timber extraction are due to the construction of logging roads. Since bulldozers, skidders and trucks can manage slopes of 30%, little trouble is taken to align roads in moderately hilly country. The roads are usually ca. 3.5 m wide and on either side some 20 m of forest is removed to enable the sun to dry the surface.

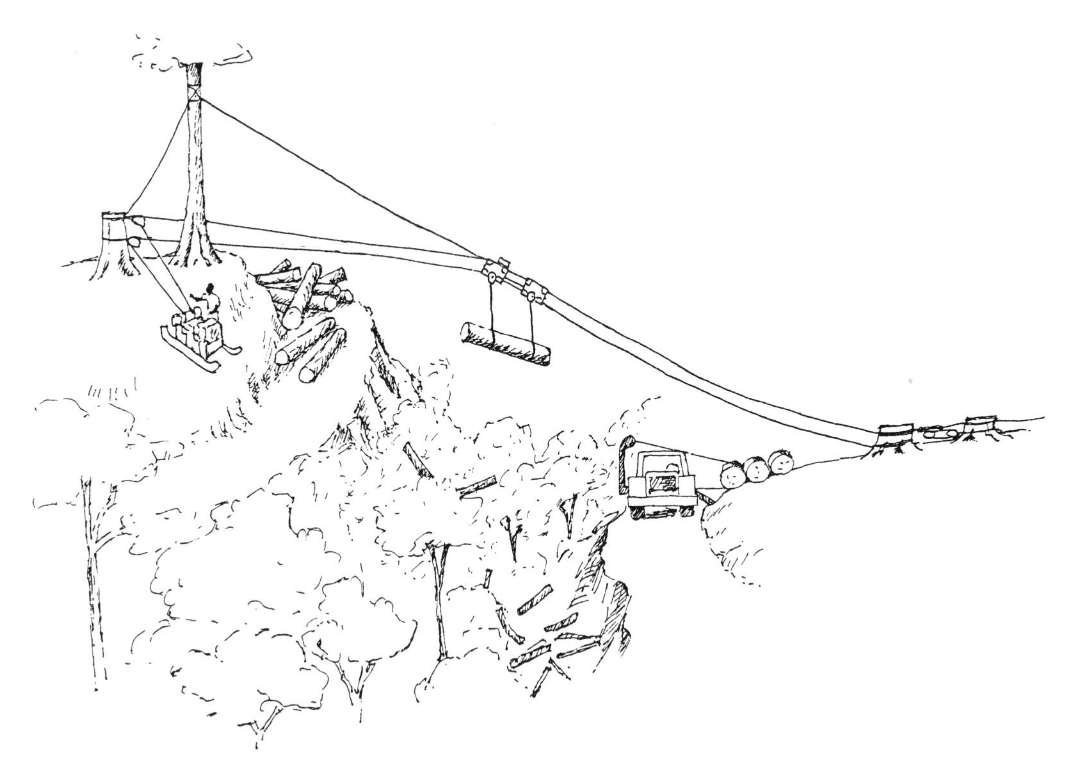

Fig. 17.8 Loading trunks onto a truck. Same as 17.7

Neither asphalt nor gravel is applied, which would increase the cost. For the same reason, embankments are insufficiently sloped or reinforced, and no more culverts are placed for water drainage than are strictly necessary.

When rain comes, undrained water collects in any natural or more often man-made roadside depressions, causing trees to die (Fig. 17.9), and creating ideal places for mosquitos to breed, so that malaria expands. Furthermore, bare soil is a quick prey to erosion. Again using a bulldozer, the worst spots can be filled up, but before long, potholes and gullies will threaten the road. Timber extraction often becomes a race against road deterioration. In addition, the exposed soil erodes and is washed into streams, filling the riverbeds further down, spoiling the water for human consumption and silting up artificial lakes and irrigation dams. At the coast, an excess of silt penetrates the mangrove belt, ruining the spawning grounds of fish and crustaceans, or smothering coral reefs and thus harming both coastal and pelagic fisheries.

While actual logging is in progress, many animals flee the area. Burgess (1971) estimated that 48% of the mammals do so. Rijksen (1978, p. 363) agreed, and referred to an investigation revealing that in logged-over forest no more than 40% of the original mammal fauna remains; there are similar figures for birds. Where do the animals go? To neighbouring forests, if available, where the displaced animals disrupt the social structure of those already there? I do not know.

When trees are felled, they crash upon others which break their fall, but which are broken in turn. From the logger's point of view, this is desirable since it reduces the risk that the large boles will split or break on hitting the ground. Burgess' shocking findings (1971) in an 'average' forest in Malaya have been widely quoted: when 10% of the standing stock of trees were harvested (expressed in 'basal area'), road construction and breakage resulted in the loss

Fig. 17.9 The effect of road construction, where for the sake of economy not enough culverts have been used to drain all rainwater. Local inundation causes the death of trees. The stagnant water at the same time provides a good breeding place for mosquitos. North of Manaus, Brazil, March 1982 (Photograph MJ)

of another 55%, so a mere 35% of the trees were left (Fig. 7.10). In Sarawak, Proud (1979) produced even lower figures: after the extraction of 6–8 trees ha^{-1}, he estimated that only 21% of the forest was still intact.

Along the logging of roads and in the newly formed gaps, where full sunlight reaches the soil, the microclimate will become hotter and drier; Proud has quoted data on this subject as well. As the forest floor is now open in many places, the general worsening of the microclimate will have unknown effects on the delicate recycling process. Machinery will have compressed the soil in some places, churned it up in others and tracks will remain visible for years. Seedlings have been destroyed in masses and larger gaps will be filled with secondary growth, thus suffocating the seedlings and saplings there, as has already been mentioned.

A logged forest can aptly be compared to a bombed city. A number of functions have been eliminated; only some are restored, others are not, but no aspect of life remains untouched. The forest equilibrium with its many biological checks and balances has been disrupted. Patches of secondary growth will occupy space in it for many years to come. Some species will expand temporarily, then give way to others. The ecosystem will have become destabilized. In biological terms, this means that the surviving plants and animals are

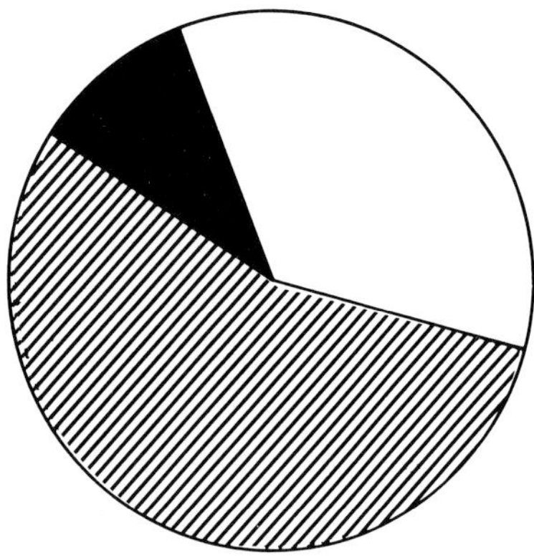

Fig. 17.10 The effect of mechanical logging on the remaining forest, in terms of the net area of the trunks in diameter ('basal area'). *Black*: Harvested trunks, 10%; *grey*: destructed as a result of road construction and cutting, 55%; *white*: remaining trees, 35% (Burgess 1971)

subjected to strongly altered criteria of selection. Some of the remaining primary forest species will fail this sudden survival test and they will vanish. More species will disappear from the depleted forest if the logging is repeated, or if it is combined with silvicultural interference and the once wide range of forest functions and values will dwindle away to zero.

Very heavy logging leaves little more than a landscape of isolated, broken stumps, eventually to be overgrown with climbers and the low vegetation of secondary scrub. Such lands often give the impression of being stuck in this succession stage. It is not even clear when and how recovery will follow.

Silvicultural Measures and Systems

A logged-over forest, as explained, is in a state of imbalance. In a kaleidoscopic way, one stage of succession in a gap gives way to another. This irregular, almost capricious mode of regeneration after selective cutting, eloquently described by Fox (1976), has challenged foresters to search for ways to ameliorate this process, to obtain more wanted trees and fewer unwanted ones. Methods of 'improving' mixed species-rich forests to promote growth of commercial species while suppressing others are called by foresters silvicultural measures. They are:

1. Cutting the stems of climbers and stranglers, plants which take away light, tieing trees together and enmeshing trunks, making them unsuitable for harvesting;
2. Cutting unwanted smaller trees;
3. Eliminating undesired adult trees, including poorly shaped ones, or very large ones; since felling the latter would cause damage, they are killed by the 'poison-girdle' method, i.e. by pouring a solution of mostly sodium arsenite into a groove made around in the bark; ring-barking is often unsuccessful in the rain forest. In either case, the tree dies and gradually comes down in parts;
4. Removing of palms, bamboos and other 'weeds' in the undergrowth;
5. 'Liberation' of young commercial trees, by removing neighbours which are too close;
6. Working the soil, to improve it as a seedbed, sometimes adding fertilizer;
7. 'Enrichment planting' with commercial seedlings or saplings.

Silvicultural measures, which have been described extensively by Wyatt-Smith (1963–64), and in shorter form by Whitmore (1975, pp. 89–93) and by Tran (1974, pp. 204–207), can be combined with well-timed logging operations, to become silvicultural systems. There are polycyclical and monocyclical systems. Under the former, every 5, 10, 15, 20 or more years the mature trees are selected and removed from the forest, which thus as a whole theoretically remains more or less intact. This system is commonly applied to temperate forests.

Under monocyclical systems, harvests are less frequent, but more abundant. The silvicultural preparations may begin as much as 5 years in advance, so that during the time of harvesting a large crop of the desired seedlings is standing. When the canopy is opened, light stimulates their growth; thus cutting will give the commercial species the upper hand, if they are all of the same age and grow up simultaneously, to a more or less equal diameter, they can be taken in one big harvest, while superior trees can be saved as seed producers.

In this way, foresters in Malaya learned to take advantage of the irregular fruiting of the dipterocarps. During decades of ingenious work on suitable soils, the 'Malayan Uniform System' was elaborated. This system was grafted onto a long tradition of forest management to which foresters like D. Brandis and R.S. Troup contributed, and which in various countries, e.g. Burma, India, Nigeria, The Philippines, Puerto Rico and Queensland was elaborated according to local possibilities. Neither in Amazonia nor in Indonesia, however, were the forests worked so intensively, as this system required far more skilled manpower than there was available. Exploitation in Amazonia and Indonesia did not go much beyond selective logging.

Effects of Silvicultural Measures

When silvicultural measures have been applied for some time, the result is a much more 'orderly' forest. No very big stems occur, and hardly any climbers or stranglers; the canopy is rather even. What the ecological effects are of the above measures, which together are reassuringly known as 'wise management' is not easy to quantify. Much depends on the composition of the forest in question, on the thoroughness, skill, frequency and duration of the measures' application. But a number of effects can be indicated as likely:

1. Decimating lianas, which represent 8–10% of the plant species, will curtail the mobility of many animals, which in turn may contribute to seed dispersal. By removing medicinally valuable species of climbers, particularly those of the Apocynaceae, Connaraceae and Menispermaceae families for instance, the medicinal resources of the forest will be greatly diminished;
2. Controlling of strangling figs threatens the main supply of food for orang-utans and other larger animals which disperse many seeds;
3. Killing the largest, oldest trees threatens the epiphyte flora which on their branches attains its maximum development. Among this flora are many orchids;
4. Eliminating large, old trees with hollows removes from the forest all their accumulated organic matter as well as the minerals contained therein; nor will hornbills, important seed dispersers, be able to use them as nesting places;
5. Inflicting harm on seed-dispersing animals will reduce their vital role in the forest web of life. Less genetic material will be transported, individual plants will become more isolated; eventually this will lead to a decrease in the adaptability of populations to change and in the long run, enhancing their risk of extinction. The whole process can be characterized as 'genetic erosion';
6. Controlling the growth of non-timber species may eliminate the wild relatives of fruit trees and other plants supplying minor products, and will lead to a serious decrease of the genetic capital of the forest; many wild fruits which otherwise could, on the contrary, be brought into cultivation, in ways we will indicate later.

For all the reasons mentioned, we may conclude that 'wise management' must have a severe long-term impact on the natural functioning of a rain forest, drastically reducing its potential. It is, in fact, better not to confuse silviculturally managed vegetation with undisturbed rain forest at all, but to call it – in line with R.A.A. Oldeman – a 'forest-derived system'.

The Paradox of Logging

Shifting cultivation cannot really be called a method of forest exploitation, as only the land is taken, while little or no use is made of the biomass. In genuine forest utilization, there is a great contrast between minor product collecting and the harvesting of timber. Logging does three things to the forest ecosystem which minor product collecting does not, (1) it damages the canopy, (2) it damages the soil and (3) it results in the removal of large quantities of minerals. Only in species-poor ecosystems on good soils, as in temperate forests, does logging appear to be an ecologically sound form of exploitation.

In the species-rich, non-renewable forests, where trees of one species are widely spaced, the situation is very different. Such a finely-tuned ecosystem, slow to adjust, but capable of yielding hundreds of exquisite and often irreplaceable products with the most economical use of scarce minerals, is suddenly ripped open by bulldozers and chainsaws. For only a small fraction, all the others are, in principle, sacrificed. Even for timber, the present extraction methods are conceded to be notoriously wasteful. The whole potential of this ecosystem for converting scarce minerals into fruits, latex, medicine, spices, stimulants and other precious commodities is squandered. It is almost too clumsy an approach to spend any more words on; yet, not only does it continue but, even worse, rain forests are often completely removed to be replaced by (often ephemeral) agricultural crops of a few species at best. Future generations will find it hard to understand how we can condone the destruction of such diversity for so small a return.

It is especially the long history of minor forest product exploitation which places modern rain forest exploitation in such an unflattering light. It reveals that the present-day fixation on timber is of very recent date. The early trade in rain forest products consisted, for instance, of oil and wood from the camphor tree, *Dryobalanops aromatica* (Dipterocarpaceae), cinnamon, the bark of *Cinnamomum* (Lauraceae), benzoin, the resin of *Styrax benzoin* (Styracaceae), incense wood of garu, *Aquilaria malaccensis* (Thymelaeaceae) and a variety of others, but not much timber. The book by Dunn, *Rain Forest Collectors and Traders* (1975), dealing with modern and ancient Malaya, is one of the few comprehensive works on the subject of early utilization. Timber extraction as an organized means of utilization only began in the course of the 19th century. As late as 1938, the value of the trade from Indonesia amounted to 16 million Dutch guilders for timber and 13 million for minor products (Cohen 1939). At present, the proportion has shifted, both in Indonesia and Malaya, to about 95% vs 5% of the trade value respectively.

Various factors have caused this shift. First, a wish for ‘profit maximization’: situations where ‘economy’ is synonymous with consumption of capital favour large-scale timber exploitation and exportation. Second, the collecting and marketing of minor products is usually small-scale and in private hands, which makes it difficult for a government to collect revenues, and the sums involved are modest in any case. Third, logging permits the extension of government power over remote lands, which minor product collecting scarcely does. Fourth, small-scale exploitation by its nature does not attract large investments involving an ‘attractive’ flow of capital, enabling aid agencies to dispose of the large sums they are committed to spend. Fifth, the local people who do the collecting have no power base from which to protect their immemorial source of income. The middlemen to whom they sell their products are far more flexible: if rattan supplies diminish, they can always sell plastic instead.

Yet isn’t the harvesting of minor products a far more responsible way of exploitation than logging? In a species-rich ecosystem on poor soil, the logical thing to do is to take a little of many species rather than all of one or two. If the present rate of forest destruction continues, in 10–50 years (depending on the region) there will be no natural forest left to log; and simultaneously *all* rain forest produce will by then have been exhausted. And this is in spite of the fact that the need for minor products is considerable, in tropical as well as in temperate countries, as Myers has argued in *The Sinking Ark* (1979).

Where ‘sustainable yield’ is a genuine and honest motive in forest policy, and decisions are to be made about the destiny of forest tracts that for some unfortunate reason cannot be saved from exploitation, the prospects for minor product utilization should be weighed heavily against logging. While it cannot be admitted in strict reserves, as collecting also entails ecological change, forests exploited in this manner are certain to last much longer.

Fig. 17.11 A Papua 'garden', eastern New Guinea. Light forest has been selected, cut down as far as possible and the area is fenced to keep out wild pigs. At the *right*, a tree with buttresses (Photograph MJ)

Shifting Cultivation

The original, traditional method of agriculture is similar in all rain forest regions. Gourou (1958, Chaps. 4 and 5) wrote an outstanding account of it, so did Bartlett (1956). Indeed, there is voluminous literature on the subject. Shifting cultivation is known as *caingin* in The Philippines, *chena* in Ceylon, *conuco* in Venezuela, *ladang* in Malaysia and Indonesia, *milpa* in Central America and *roça* in Brazil, to name but a few terms. Throughout the tropics the methods used are essentially the same.

A plot of forest, about a 0.5 ha^{-1} in extent, is expertly selected. The undergrowth is cleared with machetes, the trees are felled at the onset of the dry season, during which the biomass is burnt. The very biggest trees are killed by setting fires around their boles; shade is not wanted. Many of the dead trees stand for a while as skeletons, but are eventually demolished by rain and wind.

The resulting wood ash fertilizes the soil for a short time. At the beginning of the wet season, a mixture of crops is planted, such as peanut, *Arachis hypogaea* (Papilionaceae), sweet potato, *Ipomoea batatas* (Convolvulaceae), cassava or manioc, *Manihot esculentus* (Euphorbiaceae), tobacco, *Nicotiana tabacum* (Solanaceae), rice, *Oryza sativa* (Gramineae), maize, *Zea mays* (Gramineae) and dozens of others; able shifting cultivators known precisely what combinations to use. After a couple of harvests, the humus that was left in the forest soil has been oxidized, and the minerals have disappeared. Another plot of forest is then selected, while the first is left fallow. On the original plot, secondary growth develops (Figs. 17.11,

Fig. 17.12 Shifting cultivation, probably an old 'garden' used again after a period of rotation. *Right of the middle* a pawpaw, *Carica papaya* (Caricaceae), at the *extreme right* a banana, *Musa* (Musaceae). Eastern New Guinea (Photograph MJ)

17.12, 17.13). After a certain 'rotation period' of between approximately 15 and 30 years, depending on conditions, sufficient humus has been produced for another couple of harvests. The secondary growth is then recleared, which is much easier than felling primary forest, and the whole cycle starts afresh. One family thus operates ca. 15–30 plots in all stages of succession. According to one estimate, this method of agriculture takes about 4 months of human labour annually, working half days.

Shifting cultivation makes use of the natural processes of regeneration discussed in Chapter 8. Light-loving, fast-growing crops are cultivated in the bare soil; thereafter the pioneers and seedlings of secondary species are allowed to take over. Most rain forest soils are too poor to support such systems as irrigated rice cultivation, which are usually only possible where the *sawah* (paddy fields) receive minerals from the weathering of young volcanic products as they do in Java and Bali.

For a man to support a family of five by shifting cultivation, he needs about 6 ha on good soil, 20 on poor. If about half the land area in a region is suitable, 1 km^2 can thus support a population of 3–12 persons. But when the population increases, the balance that existed between the areas of virgin forest and those under cultivation, breaks down. There are only two ways to meet the increased demand. The first way is to shorten the fallow period: even though the agriculturalist realizes that the secondary growth has not had time enough to improve the soil, it is again cleared for another crop. Two sequences of this degrade the soil so much that afterwards it will only support low-quality pioneer vegetation. In Malesia this

Fig. 17.13 Shifting cultivation. In *front* probably *Xanthosoma* (Araceae), a tuber grown for starch. In the *background* young secondary growth, the area was probably abandoned a few years ago. Middle area a blanket of lianas of the pioneer and early secondary phase (Photograph MJ)

consists mostly of the notorious alang-alang, the lalang or cogon grass *Imperata cylindrica* (Gramineae), which in Indonesia alone is thought to colonize a further 2000 km^2 $year^{-1}$. Cogon grass forms a network of rootstocks highly resistant to fire, flood or drought. Even after a fire, in which all other vegetation is killed, cogon roots are able to sprout an abundance of fresh culms.

The second way is to extend the area under shifting cultivation by felling more forest. Population growth often coincides, however, with a decrease in the considerable skills required for the successful practice of shifting cultivation. The rotation system, known as *'swidden'*, which affects small tracts of forest, then gives way to 'slash and burn', whereby the forest is cut along a front (Fig. 17.14). As the front always forms the limit of civilization, well away from roads, there is no way to dispose of any valuable timber other than by burning it. Thus, a large area of forest some 45 m tall, the product of millions of years of evolution, is sacrificed for perhaps eight harvests in all; whereupon the hard to eradicate alang-alang invades the whole front and eventually takes it over. Meijer (1975, p. 60) estimates that over 20 years of this sort of 'agricultural development' to accommodate migrants from Java, in Lampung Province in southern Sumatra exportable timber valued at $ 60–100 million were lost, and three times that amount lost to local markets.

Politically, shifting cultivators are popular scapegoats on which to blame forest destruction, but their real share in it, worldwide, is hard to determine. Differences in population density, population increase and government control all play a part. Myers (1980, p. 114)

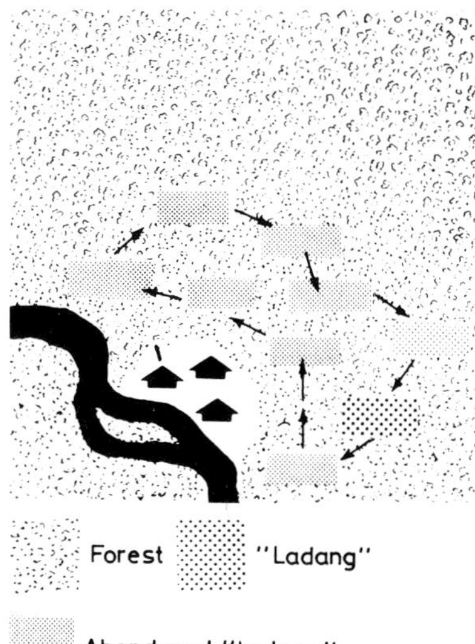

Fig. 17.14 The transition of rotating, shifting cultivation ('swidden') to the replacement of forest by agricultural land over a wide front ('slash and burn'), under the pressure of rising population. The shortening of the rotation period impoverishes the soil within a few years, thereby making the growth of secondary forest impossible, enabling only alang-alang, *Imperata cylindrica* (Gramineae) to establish itself (Rijksen 1978)

summarizes the situation: "All in all, these forest farmers have been estimated in the mid-1970's to total at least 140 million persons, occupying some 2 million km^2 (or over one-fifth) of all tropical forests. Of these 140 million, approximately 50 million are believed to occupy at least 640,000 km^2 in primary forests, and another 90 million to occupy twice as large an area. According to preliminary reckonings, they are thought to burn at least 100,000 km^2 of forest each year. The greatest loss occurs in Southeast Asia, where farmers clear a minimum of 85,000 km^2 each year (some of which are allowed to regenerate), adding 1.2 million km^2 of formerly forested crop-lands in the region. Tropical Africa is believed to have lost 1 million km^2 of moist forest to these cultivators before the arrival of modern development patterns in the last quarter century; of Africa's present tropical forests, as much as 400,000 km^2 may now be under this form of agriculture, with a current loss of forest estimated at 40,000 km^2 per year. A similar story applies to Latin America, though fewer details are available; all forms of expanding agriculture in Latin America, of which slash-and-burn cultivation is a major type, are considered to be accounting for 50,000 km^2 per year". This is a low estimate. The actual area destroyed, Myers continues, may even be twice that much.

When a road towards or through a rain forest has been made, migrant settlers move in. Generally, because of population growth and/or the mismanagement of land resources in their native region, they no longer are able to find the means to exist there and society is incapable of absorbing them; explanations are supplied by Brown et al. in *Twenty-two Dimensions of the Population Problem* (1976). Swarming out on the road, they build ramshackle houses and cut down the forest to grow crops.

As a consequence of this combination of population growth, lack of facilities and unwise road construction, planners in tropical countries often see their calculations frustrated. Tracts of forest they had counted on to exploit for watershed protection are occupied by migrants and cleared before the state can be mobilized to protect them. Reserves, too, are threatened. Nobody knows what to do with so many spontaneous migrants, and those who drift to the cities give rise to a different set of intractable problems there.

Planned Destruction of Forest

Whether intended or not, road construction marks the beginning of virtually all forest destruction. It is the single most disastrous factor in the vanishing of rain forest, and plays a crucial role in the environmental degradation of all tropical countries. Vehicular access means the immediate export of wood and import of settlers bringing with them the tools for further destruction. Side roads and trails are soon made off the main road, enabling human influence to penetrate far into the forests on both sides.

Any road is, of course, the outcome of some sort of planning under the aegis of government. The large initial investment often comes from development agencies, from which contractors benefit, but afterwards the country itself is saddled with the costs of maintenance, which run high in a wet climate. If the region fails to produce enough to meet the cost of maintenance, traffic is reduced and hope for prosperity in remote areas served by the road dies. In addition, after a few years, all forest along the road is gone. The result may well be a region of degraded land, where poor farmers try to make a living by minimally shifting cultivation under the most backward conditions. The opening up of the forest in such cases has been a complete waste of investment and resources.

Another planned form of forest destruction is forest clearing over extensive areas for such large scale agricultural enterprises such as oil palm plantations, rubber or other tree plantations, cattle ranching, resettlement projects and hydroelectric dams. In Malaya, for example, the government decided in the 1960s to convert many of its well-managed forests

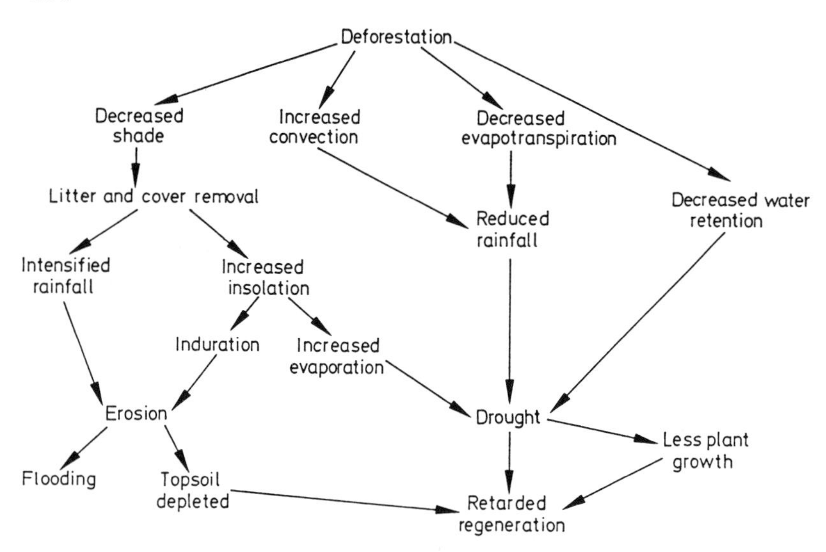

Fig. 17.15 Diagram showing the consequences of deforestation for the environment of the area (Goodland and Irwin 1975)

into oil palm plantations. The Jarí project, which subsidized clearing for cattle ranching and road construction in Amazonia, has already been mentioned in Chapter 9 (Tropical America).

On the effects of large-scale deforestation we can be brief: the forest as a resource ceases to exist. All its valuable contributions to human welfare are totally annihilated in the stricken area. The consequences are shown in two well-known diagrams by Goodland and Irwin (Fig. 17.15 and 19.4). On aspect, however, deserves special discussion, i.e. the loss of species.

Species Threatened with Extinction

All forms of rain forest destruction lead, separately or combined, to a reduction in the numbers of individuals of such species as are bound to the primary rain forests. If the number is reduced beyond a certain critical level, a population cannot function properly any more. That is, it cannot maintain the genetic diversity which allows it to respond to (small, gradual) changes in conditions which favour some individuals over others. The population comes to suffer from 'genetic drift', a process whereby the elements of diversity, the genes, become fixed or lost by chance; for a further explanation, see Ashton (1969, p. 157). The narrower its genetic basis of diversity, the easier a species is wiped out. A small population may still be able to maintain itself under certain conditions, but it is unable to colonize new ground, or in other words, to occupy new ecological niches. Its disappearance is then only a matter of time.

When a population of one species vanishes, this affects populations of other species which depend on it for food consumption, pollination or dispersal. Extinction thus has a domino effect: one case of extinction after some time causes another. If the one species is a seed-dispersing animal and the other a long-living tree, centuries may elapse before the effect manifests itself. A famous example of this is the small remnant population of *Calvaria major* (Sapotaceae) in Mauritius, Indian Ocean. In 1973, 13 trees were still standing, but although they were producing plenty of fruit, no seedlings were ever seen. Then, in a swamp in Mauritius, fossil *Calvaria* seeds were found together with skeletal remains of a dodo, the big bird of which the last individual was killed in 1681. *Calvaria* seeds have an extremely hard layer surrounded by fleshy pulp, and while they are eaten by several extant birds and bats, these consume the pulp without damaging the hard layer enclosing the seed. Stanly A. Temple (1977) suspected a relation between *Calvaria* and the dodo. Would the probably very muscular stomach of the big bird have been strong enough to crack the hard layer without also crushing the seed? And would this have been the prerequisite to its germination? To find the answer, he fed *Calvaria* fruits to turkeys. We know the approximate size of the dodo, and so its weight could be estimated at about 12 kg, the size of a big male turkey. Any lighter bird would also have lacked sufficient muscular strength in its stomach. Temple experimented. Of 17 fruits ingested, 7 were destroyed by the turkey's intestine, but 10 left the body in a recognizable state, and 3 of them were brought to germination. *Calvaria major* brought back from the brink of extinction, after 3 centuries of lone survival!

Similar delays in extinction can be expected in the rain forest after certain animals have gone. Data are scarce, however, for actual cases of extinction. *Stemonoporus* (Dipterocarpaceae) is endemic in Ceylon. Kostermans, the botanist who made the largest collections on the island, studied the genus (1980). Of the 20 species he distinguished, 4 had not been found since 1854, a loss of 20% in one and one-third century. Of *Diospyros atrata* (Ebenaceae) all he could locate was one old tree in the Peradeniya Botanical Garden. If, as is scheduled to happen in Malaya, virtually all forest below 300 m is destroyed apart from that in the 625 km^2 of reserves (which represents only 0.5% of the land area), we may expect a result similar to that in Brunei (Fig. 7.9), where 26.5% of the dipterocarp species were wiped out, and similar percentages are likely to be similarly affected in other low altitude plant families.

The term 'species' is used here for the 'smallest unit of biological diversity' which can actually be a subspecies, a variety, a race or even a small population. The extinction process is the nibbling away at such entities by whatever means. The rambutan, *Nephelium lappaceum* (Sapindaceae) occurs in Borneo in several dozen local races, all different in minute characters. While forest destruction does not therefore directly threaten the species as such, it will narrow its genetic base. The same can be said for the majority of polymorphic, widespread species in the lowland rain forests. For homogeneous plant or animal species in small areas, the future is much bleaker. The extent to which a considerable part of the normal territory of one of the two subspecies of orang-utan has been lost is shown in Fig. 17.16.

Loss of forest area, however, is not paralleled by loss of flora and fauna in exactly the same proportion: the curves run differently towards zero. In half the number of forested hectares much more than half the total number of species is present. Some patterns of destruction, however, namely the patchy ones, lead to a decrease of space available to the primary species because gaps become filled with secondary growth. Newly exposed forest edges, too, are normally invaded by secondary species, and animals will stray out along these long perimeters. This is why saving strips of primary forest rarely if ever helps conservation.

That natural factors have exerted enormous influence on the proliferation (or otherwise) of species can be observed in Malesia. Geologically fractured land areas are poor in comparison with others. Volcanic Sumatra, 480,000 km^2 in area, has 95 dipterocarps and 17 endemic genera; non-volcanic Malaya, with an area of 133,000 km^2, has 155 species of Dipterocarpaceae and 41 endemic genera of seed plants, even though these two parts of Sundaland have a long common history, and were often connected. Climatically seasonal areas also tend to support poorer flora.

Species extinction is, however, a natural process, but one which is normally much slower than species differentiation. It is man who has speeded up extinction rates mainly by damaging and destroying the primary lowland tropical rain forests, where most of the diversity is located. The following important forest functions suffer (see p. 197): minor forest products, stock of new useful plant species, gene pool, food and shelter for the animals, matrix of evolution, source of knowledge, place of education and recreation.

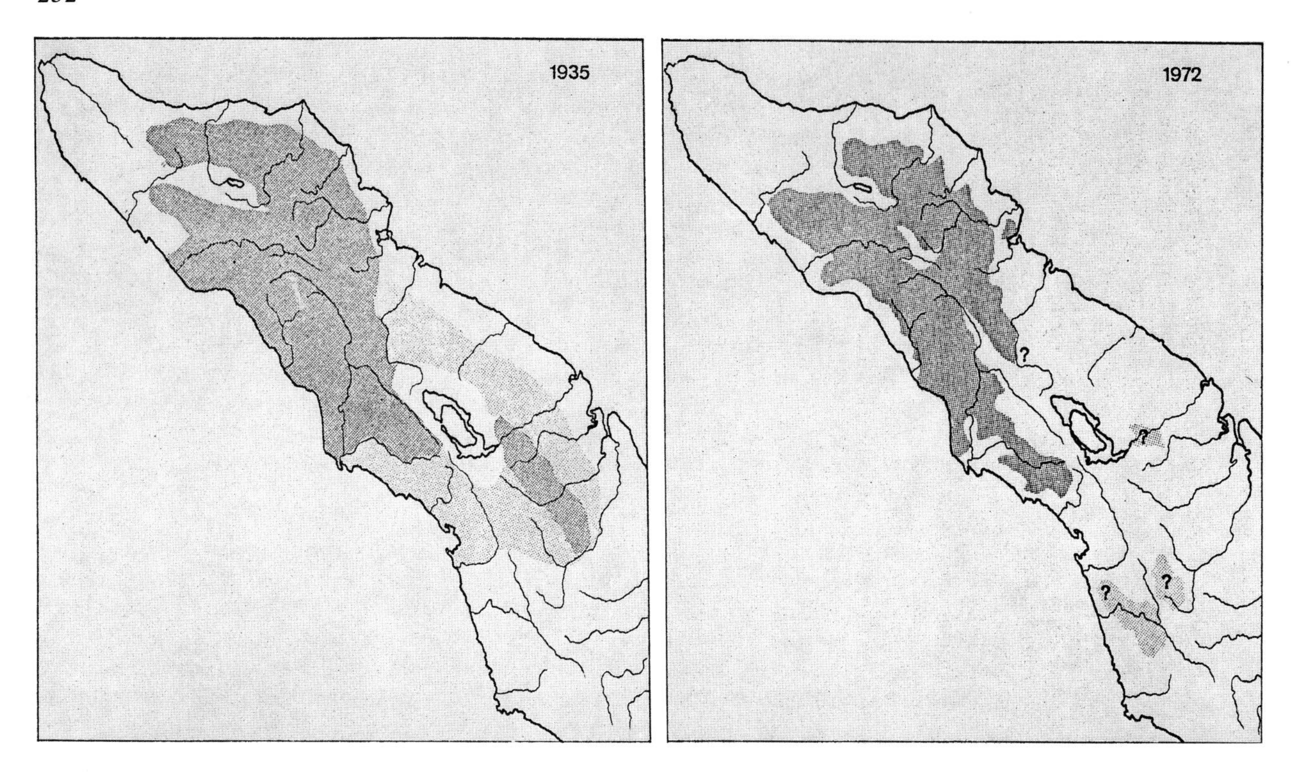

Fig. 17.16 Reduction of the area of the orang-utan in North Sumatra, over a single (human) generation. The remaining area is actually smaller than indicated, as the area covers much high ground, unsuitable to the orang-utan. Approximately halfway between Lake Toba to the south and Lake Tawar to the north, the study area Ketambe is situated (Rijksen 1978)

The problem has been explained in compelling terms in a small book by Eckholm, *Disappearing Species: the Social Challenge* (1978), and in a larger one by Myers, *The Sinking Ark* (1979). It is aggravated by a common aversion to, or even fear of, a variety in wild plants and animals. Many biologists have even expressed disdain for the study of this variety, taxonomy, trying to suppress it in academic curricula. Their efforts to block knowledge and understanding of nature's diversity have contributed, indirectly, to its endangerment.

Causes of Forest Destruction

The present rate of rain forest destruction is ca. 150,000 km^2 $year^{-1}$. At least 40%, maybe half, of all the rain forest that ever existed is no longer present. These forests with their many functions, after tens of millions of years of evolution are at the 'mercy' of man, being destroyed as a matter of course, and no small part of it on purpose. How is this possible? I see seven causes.

1. Systematic discrimination by Man Against Everything that is not Man. Every human being enjoys a status which entitles him to recognition, protection and assistance, if he can obtain them. With a few exceptions, enforced by recent legislation, man is free to help himself to everything in nature, as if the earth were made only for him. There is no widely accepted ethical system or principle which explicitly holds that man exists to maintain all life on earth and to treat his planet respectfully.

2. Growth of Population. Before world population growth set in, shifting cultivation was not a problem. Growth was even welcomed at first, as a source of prosperity and power, by many nations. Many Third World leaders still cling to this idea. But it has long since become evident that population growth has pervasive and disastrous effects on both human societies and the environment. Those who still wish to close their eyes to it[5] should read the booklet by Brown et al. (1976), and *The Global 2000 Report to the President* of the United States (Barney 1980).

3. The Rights of Man. Isn't the shifting cultivator, the poorest of the poor, entitled to a living, just as we are, if only by virtue of his needs? If his needs involve land, who are we to deny it him, even if the soil is poor and the finest forest grows on it?

4. Progress. A better life for everyone! In the tropics a huge effort to that end has manifested itself under the banner of *development*. Demands are created, promises formulated and expectations aroused in the so-called less-developed countries. To fulfill them, natural resources are to be tapped. As no one wishes to halt 'progress', any reminders of any limits to natural resources will be silenced for a variety of reasons.

5. The Free Market. Under a capitalist system, adjustments are made on a voluntary basis, through the flow of money. The main factors regulating this flow are supply and demand. Through their stimulus, production is generated, competition leads to the best possible quality of commodities at the lowest possible price, to everybody's benefit. In this system, a pivotal role is fulfilled by private property and free enterprise. These furnish the money for investments to tap new natural resources. Success or failure will follow. To learn, experiment is necessary. This requires a no man's land, a 'frontier', where initiatives are put to the test. Experiments by private enterprises deserve a reward, or incentives, for instance in the form of tax rebates which means that part of the risk is taken by the public treasury. The rain forest is often considered one such no man's land.

6. Technology. The bulldozer, developed during World War II, is only one beginning of the highly sophisticated machines now used to harvest timber, which is then further processed in sawmills by the most modern equipment. Technology has given man the upper hand over the forests that have from time immemorial stood in his way.

7. Economic Growth. To provide for the costs of 'progress', markets must be sought and found. We see the results around us, in the consumer society with its rapid turnover of products. Timber which the Japanese cut in Indonesia, for instance, is exchanged for a multitude of industrial products. Even in the smallest villages the placebo of 'pop' music is available to all via resplendent portables.

These causes may not be the only ones. Nobody should underestimate the age-old hate the people of tropical countries have for the formerly omnipresent forest. Add to this a widespread indifference towards the future. Add again the greed for money and the corruption that are a common plague in many countries, rain forest nations no exception. Yet most of them seem less specific, and can be classified under point 1. Hate against the rain forest has in common with many other points this curious feature, that it was born in the past and that its present validity is questionable.

This whole syndrome of causes explains, in my opinion, the demise of the tropical rain forests. A complex of values, opinions and ideals, widely hailed and promoted, has led to one of the most tragic and far-reaching events of the 20th century. Yet to these same developments the highly developed countries owe their prosperity and poor countries their autonomy. Nor can we blame western expansion with regard to the devastations. The governments of rain forest nations must take the responsibility for opening up their own forests to depletion for money, or even for subsidizing deforestation. By doing so, they have affirmed their sove-

[5] Plumwood and Routley (1982) bagatellize population growth as a major cause of deforestation. Rather, they blame certain governments and multi-national corporations, who open up and destroy forest for their own political and financial purposes, which are opposed to, or which negate the rights of the peasants to justice and equal distribution of land. The rain forests thus are enlisted for the political arguments of Plumwood and Routley.

reignty, contributed to progress or development, stimulated economic growth and made population growth possible. In the name of these goals and aided by the most modern equipment, man exerts his rights over the forest that resisted him so long.

Obsolescent Values

But the curious feature of these goals is their root in the 18th century and their obviously obsolescent value now. While discrimination against nature is much older (it will be discussed in the last chapter), the upward trend in human numbers which has persisted to this day started about 1740, in western Europe. So did nationalism as a value: in that same year, the anthem *Rule Britannia* was written; *God Save the King* dates from 1745. Van den Berg (1977) pointed out these historical facts, identifying nationalism as the main cause of World Wars I and II. In addition, that seminal idea, the 'rights of man' were first formulated in the *Déclaration des Droits de l'Homme et du Citoyen*, in 1789, the logical outcome of the new philosophy of, among others, Jean-Jeacques Rousseau, that man is good by nature, an outright and irreligious reversal of the belief that man is essentially wicked and consequently only had duties (e.g. the Ten Commandments) and only the King had 'rights'. Nowadays, we all have rights; the only question is what will be left of them when the earth is populated by some 8 to 10 billions of us right-bearers. However that may be, it remains very hard, nay, inhuman, to refuse anything to someone who is good by nature if he needs it. It is disturbing to realize that this concept of man's innate goodness has led to so much devastation among man's fellow creatures.

The term *progrès* (in French) was coined as early as 1687, but it only really caught on in the 19th century. The problem with 'progress' is a lack of formulated purpose and direction to which everyone agrees. There is no built-in end, no finish line except the limit of the resources themselves. Schumacher, in *Small is Beautiful* (1973) explains the dangers of this shortcoming. The ravages caused by 'development' are beginning to show everywhere (Fig. 17.17). Curry-Lindahl has even posed the question *Is aid for developing countries destroying their environment?* (1979), and answered, yes.

The free market as a system is another 18th century concept, set forth by Adam Smith, in *An Inquiry into the Wealth of Nations* (1776). In a free market, the worth of everything is expressed in money. Labour, too, can be sold as a commodity. It does not matter if a commodity is renewable (like wheat) or non-renewable (like fossil fuels), though the neglect of this vital distinction has had terrible consequences, which are also discussed by Schumacher. The exchange of non-renewable timber for renewable transistors is a case in point. As the free market system brings wealth to some, and ruin to others, many regulations have sprung up to limit the dangers of the system that have gradually revealed themselves. A book by Gunnar Myrdal (1968) bears the subtitle *An Enquiry into the Poverty of Nations.* In the wake of its criticism of the system, efforts are being made to establish a New International Economic Order.

The development of modern technology can directly be traced to the invention, in 1733, by John Kay, called the flying shuttle. This was a device for weaving at unprecedented speed. It marked the beginning of the Industrial Revolution in Britain. In 1763, James Watt began a series of improvements on the steam engine. Thereafter came MacAdam's new method of road surfacing (1783–1798), which greatly increased the speed of travel for the first time in history, then the railroad, the internal combustion engine, the motorcar, the chainsaw, the bulldozer, the nuclear bomb and the demonstrations against the nuclear bomb. The book by Farvar and Milton, *The Careless Technology* (1972) for the first time collected evidence of the effects of modern machinery on the environment.

Fig. 17.17 Progress, free enterprise, technology and economic growth symbolized in a single advertisement in *Bois et Forêts des Tropiques*, around 1970

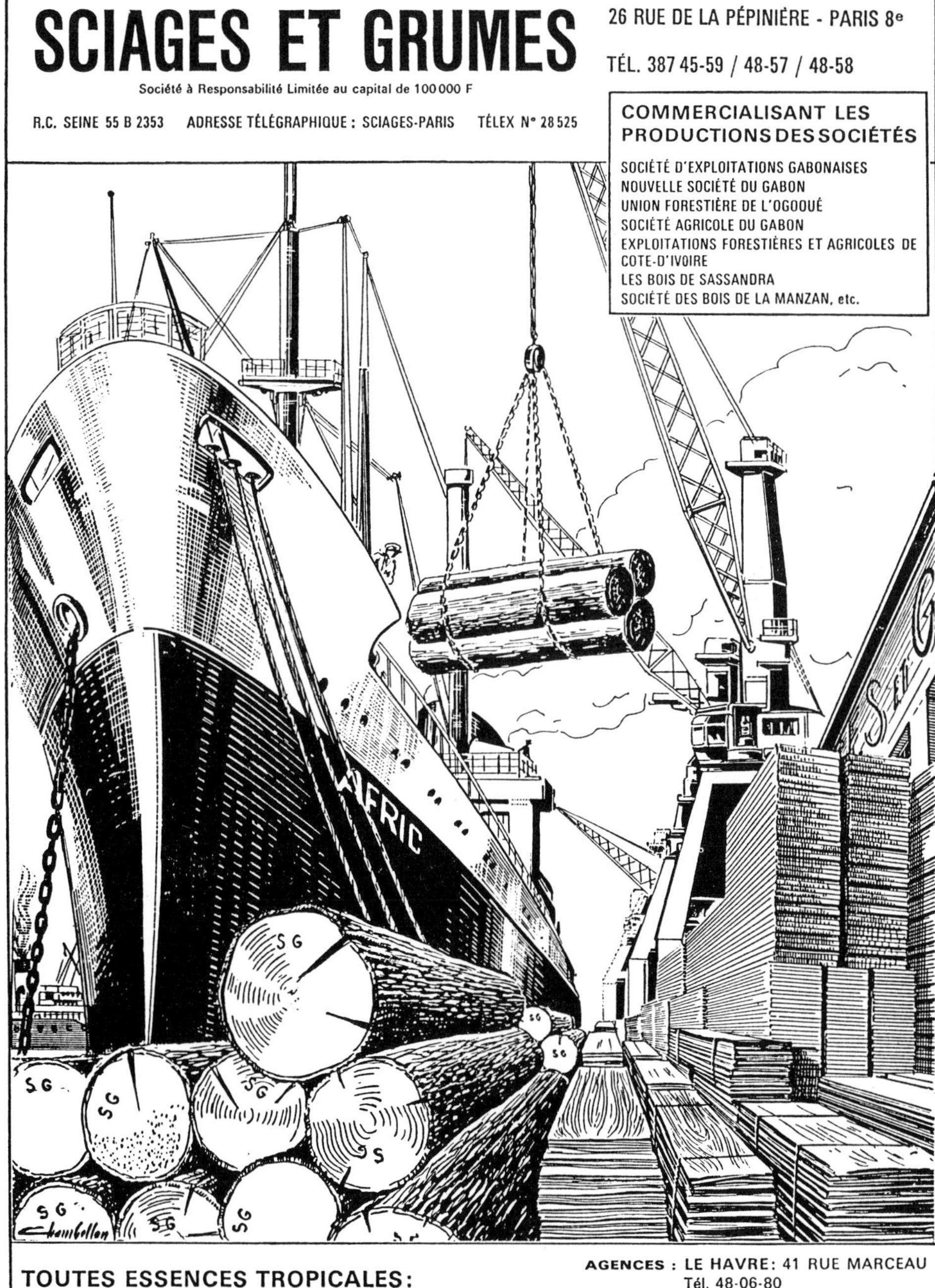

Economic growth, too, originated in the 18th century. The flying shuttle produced more cloth, so more yarn was needed. Spinning machines were invented; then the engineers raced against each other to create even more improvements. Division of labour, proposed by Adam Smith, led to production increases undreamed of. Markets had to be expanded, demands created, sales promoted with a variety of tricks which were not lost on developing countries. If economic growth lags, this is proclaimed to be disastrous. On the other hand, movements criticizing *The Waste Makers* (Packard 1961), consumerism and economic growth itself (again Schumacher), are gaining ground.

What I have tried to outline, in the briefest of terms, is the context of ideals, values and imperatives in which the destruction of tropical rain forests is allowed to proceed. They originated some 2 centuries ago, were taken up and have been carried into the present time, refined and expanded along the way. They are part of our 'heritage'; we have reaped the benefits that they generated. Most people genuinely believe them, and model their everyday actions accordingly. It is unsettling to discover that the same ideals, values and imperatives, when carried far enough, lead to the destruction of so much life on earth, epitomized in the fate of the tropical rain forest. This makes it perhaps understandable why people sell half-truths or lies about the rain forest, and why others look away from the scandal. Not only the fate of the forests is at stake, but some of our deep convictions. And to most people it is easier to kill a forest with a stone axe than to examine critically their ideals and expectations, and to check the underlying premises. To begin with, that one about man's innate, essential goodness.

18 Protection

The Place of Conservation

In its early years, the conservation movement concentrated on the establishment of what the Dutch call 'nature monuments': tracts of land on which something beautiful grew or bred that was to be kept unharmed. From this period dates the view, still favoured by advocates of Progress, that conservation has an element of luxury, in that it results in the useless locking up of natural resources. This view is, however, obsolete. Nowadays we regard conservation as an indispensable part of the wide field also known as 'the wise utilization of natural resources', aiming at utilization *ad infinitum*. In the *World Conservation Strategy* (IUCN 1980) this is summed up in three points:

1. To maintain essential ecological processes and life-support systems.
2. To preserve genetic diversity.
3. To ensure the sustainable utilization of species and ecosystems.

As one can see, the broad concept of *conservation* consists of two elements: *preservation* (points 1 and 2) and *management* (points 1 and 3). Both are needed. Preservation, through protection, protects parts of the earth's surface of land or water as well as species of plants and animals. There are many degrees of protection: in some cases, populations of protected plants or animals can be cropped on a quota system and will still maintain themselves; that is why point 1 actually contains both elements.

Nowadays conservation is also a recognized academic discipline, even though sometimes known by different names like 'biomanagement' or 'environmental management'. To a high degree, it is an applied science. It includes much general ecology, knowledge of animals and their way of life, vegetation science, the study of major ecosystems (terrestrial, aquatic and marine), some human geography, sociology and environmental legislation. A number of fine textbooks are available, though they mostly concentrate on temperate ecosystems, and the concerns of pollution, urbanization and the pressures to which a modern national park or nature reserve is subjected in a developed society.

Botanical science has of course always played a role in conservation, but until recently this role was small in the humid tropics. Although tropical animals were hunted and traded, their habitat and source of food, tropical vegetation, remained essentially undisturbed. This was abruptly changed by the advent of the bulldozer and the chainsaw and multi-national interest in 'cheap' Third World supplies of tropical hardwoods. Today, botanists, who can estimate the conservation value of certain areas, have access to information on useful plants and are familiar with vegetation maps, are much in demand. According to Küchler, vegetation maps make an excellent instrument for planning both conservation and 'development'.

Protection Versus Management

Protection implies safeguarding, keeping inviolate; management implies action. What does the difference mean with regard to the rain forest? Tropical rain forest is, as we have seen, a climax vegetation. As a whole, it does not change when left to itself; it perpetuates itself in 'cyclical regeneration'. Ecologically, it is stable, with the maximum number of species possible under the circumstances. Any influence from outside, through silvicultural measures, for instance, not to mention timber cutting, can only result in impoverishment.[6] There is nothing to 'manage' in such a system. The only way to preserve it is to protect it from interference of any kind.

Climax vegetations have become a rarity. In Europe, the Middle East, dry Africa and North America they were degraded long ago or replaced by other, unstable ecosystems. Balances have been disrupted; the new ecosystems, if left to themselves, change continuously, through series of successional stages. Management consists of perpetuating man's influence in order to attain or maintain a known desired stage. This human role was explored in the section on the savanna (Chap. 15): unless burning is applied at fixed intervals, the (unstable, short-lived) biological equilibrium will always be succeeded by light forest, of less nutritional value to 'game'.

Since all forest surrounding human settlements has been degraded, destroyed or replaced by artificial and ecologically unstable forests, most people began to believe that *all* forests are in need of management. Few people these days even see genuine primary forest, or if they do, they approach it, as many professional foresters do, as something in need of managing or improving. Their silvicultural measures then render it ecologically unstable, thus generating a need for further management.

Ecologically unstable units do need to be managed, and this is also true for single populations of grazers or predators in unstable systems: a surplus of foxes can endanger small game and bird populations, a surplus of deer can destroy a generation of young trees. For ecologically stable systems the word is: hands off. Unstable systems, however, in need of management, dominate many parts of the world. This, and the fact that management is more labour-intensive than strict preservation, explains the emphasis on management in the study of conservation.

From the botanical viewpoint protection of rain forest is easy. All one has to do is to establish reserves, and to keep all exploitation out, and management, in the context of rain forest conservation, is, literally, a marginal activity, confined to adjoining tracts of depleted forest and secondary forest; the buffer zone of a reserve.

The problems of rain forest conservation are quite different to those of management and conservation problems are the subject of this chapter. There also exist indirect methods of conservation, for instance by reducing exploitation pressure.

Intentions Towards Protection

All conservation starts with the formulation of intentions, so phrased and reasoned that governmental policies will welcome them, adopt them and want to make them into law. A conservation law generally consists of a general framework supplemented by ordinances dealing with the specific problems. The legislative aims will be:

[6]Some Australian foresters have suggested that logging results in greater species richness per hectare. Numerically this may be so as secondary species invade the logged-over forest. Ecologically, it is a misleading theory. If the disturbance is slight, these secondary species are temporary. In case of large-scale disturbances, the secondary species will suffocate a high percentage of the seedlings and saplings of the primary forest, and cause a setback at best, or at worst contribute to the (gradual) disappearance of primary forest.

Fig. 18.1 Illustration from *Conservation Indonesia*, bimonthly journal of the WWF Indonesia Programme (Wendy Blower)

1. The establishment of protected areas, enjoying various degrees of protection, according to their legal status;
2. The regulation of hunting;
3. The regulation of trade in protected species.

All of this requires a well-manned and well-trained government department for guarding, inspection, prosecution and confiscation.

Once the legislation exists, the conservationists, who, of course, have nursed it through all its stages, set out to realise these intentions. The reserves-on-paper need to be staked out on the ground, and to be protected in reality. Hunting and trade must really be controlled, and punishment meted out to trespassers and transgressors. When gaps in the legislation become apparent, conservationists must propose improvements. Conservation in practice is, therefore, like politics, the art of the possible, first to see that intentions are officially declared, then to make them into reality.

It is thus evident that conservation is a public affair to a very large extent (Fig. 18.1). A conservationist will always try to get public opinion on his side. He knows too well that only in this manner the (weak) position of conservation in the national structure can be strengthened. Organizations of persons who give moral and financial support, who at crucial moments will come forward to defend the good cause, are essential to him. The public nature of his activities, however, means that in almost no rain forest country can a conservationist move a limb without the government's knowledge and approval; and if he is a citizen of such a country, social pressure can easily be exerted upon him, while if he is an 'expatriate', he runs the risk of expulsion or refusal of another visa.

Evaluation of a Forest

Wherever a possibility exists for direct, strict protection, it is this course which should be pursued. There are a fair number of points to be observed so that the maximum number of species will remain safe in their existence ad infinitum, and conservationists must be ready to produce alternatives if necessary, and to argue the merits of various protection proposals with the authorities, by explaining the conservation value of the forest in question.

Of course, the state of the vegetation and of the fauna will have been thoroughly surveyed in the field. The all-important distinction between intact and degraded or man-made forest largely determines its conservation value. The presence of large mammals, birds, reptiles and amphibians also counts heavily in the evaluation.

As for the birds, Schodde's idea, worked out in eastern New Guinea, of using avian species richness as a standard of botanical conservation value, seems worth further testing and application. It is based on the strong preference of birds for primary vegetation, and on the fact that an able ornithologist can make his assessments much faster than an able botanist can because bird species are fewer in number than plant species and much better known, yet highly specific to certain biotopes (Schodde 1973); for more discussion on this subject, see the data under Plant-Animal Relations (Chap. 12).

The flora is difficult to evaluate. There are so many species, the plants are often so large in size, and so rarely fertile, that a proper assessment in the forest can only be made by a very specialized and experienced botanist, if one is available. Failing this, two things can be done, as next-best yet responsible simplifications.

First, a start can be made by counting the *variety of life forms*, e.g. (half) parasites, rattans, Schopfbäume and herbs; the general richness in epiphytes may also be a good criterion. They can always be recognized and listed on a standard form; the one designed by Webb et al. (1976) is a good model. The analysis can further be pursued by distinguishing as many species as possible within each category more or less at a glance, and noting their number in a given plot. Terrain data should also be noted. By these means rather detailed comparisons can be made between plots which can show obvious differences in diversity.

Second, an evaluation can be based on a (more or less complete) *list of species*; for such an evaluation, a *well-stocked Herbarium annex library* is the best place. From this list, the families and genera for which recent high-quality taxonomic revisions exist are selected. Using these families and genera, we look up what species are on the list for our area, then see whether they belong to primary or to secondary forest, what their range of distribution is, and whether they are rare or common within this range. We consult the literature on useful plants, to see what the potential conservation value is in that sector. All such data contribute to forming an idea of the botanical value of a tract of forest, partial but well-documented. A sample of 100 species, which may take a herbarium taxonomist a couple of weeks to deal with, will give a reliable impression.

The species lists needed for this purpose already exist for many forests, in some form or another, if only in the up-to-date notebook of a major collector. In the absence of a list, however, collections must be made and identified. The more complete the list is, the better, but an incomplete list will also help, because an actual evaluation is always a kind of sampling, given our partial taxonomic knowledge.

Points to Note when Evaluating a Forest for Protection

These methods of evaluation fit into the more general considerations which come up wherever the selection of forests for various degrees of protection is possible. They follow largely the principles already explained. In view of the responsibility of such a task – isn't it, in fact, a matter of life and death? – the most important criteria are listed here by way of summary:

1. The less a tract of forest has been subjected to damage (both natural and anthropogenic), the closer the number of species in it will approach the maximum possible number. The presence of all diameter classes of all tree species is an important criterion of integrity. Many trees, however, are so rare that only one individual can be found in each hectare. Assessing the presence of all diameter classes then is a very difficult, if not an impossible task.
2. The more species an area contains, the greater its conservation value, i.e. the greater the reason to protect it; we refer to the '12 significant values' (Chap. 16). Endemics have particular value.
3. The longer (in geological terms) a forest has grown on a site under stable geological and climatic conditions, the more delicately balanced it will be as an ecosystem, the more specific its plant-soil and plant-animal relations, the greater its fragility and the more it will be in need of absolute protection.
4. The larger the original continuum, of which a tract of rain forest forms a part, the greater, generally, the number of species in it. In Malesia this means that a square kilometre of forest in the larger islands has more conservation value than one in the smaller.
5. Proven refugia, where during the Pleistocene Ice Age rain forests survived, like islands surrounded by drought, are likely to harbour concentrations of species.
6. Riverbeds and -banks in a rain forest probably serve as traffic routes for migration and dispersal. As such, they should be included in forest conservation efforts.
7. The greater the diversity in topography and in soil, the greater the number of species is likely to be.

Fig. 18.2 *Rafflesia tuan-mudae* (Rafflesiaceae). The survival of *Rafflesia* depends on the effective protection of large mammals (p. 73) (Photograph Ding Hou)

8. Due to the difficult and slow dispersal of many rain forest species, caused by their large seeds and woody habit, such geographic barriers such as mountain ridges, straits and wide rivers will always hold back a large percentage of them. This makes it probable that each river basin will have its own characteristic rain forest flora and fauna, with its share of endemics. At least one rain forest reserve per river basin is therefore needed.
9. Forests known for their wealth of commercial species deserve special protection, to safeguard genetic material of superior quality forestry will be needed in the future.
10. Forests known for their potential yield of such 'minor forest products' such as rattans, comestible fruits and their wild relatives, tanning agents, essential oils and medicinal plants also deserve special protection.
11. While each altitudinal zone will have its own range of species, the greatest species richness is found below 300–500 m. For this reason, the lowland forests deserve special protection.
12. Of each animal or plant species a sufficient number must be present in a reserve to ensure its survival forever. Seasonally migratory animals must be protected over a sufficiently large area (Fig. 18.2).
13. The more species there are in each hectare, the fewer the individuals of one species this hectare can accomodate and the larger a reserve in species-rich forest will have to be. We shall come back to this vital question of minimum size later.
14. In calculating areas to be protected, unequal biotopes must be segregated according to river basin, type of soil, state of forest and altitudinal zone. To put it bluntly: protection of a forest on limestone will not save a forest on alluvium, and the latter is usually much more vulnerable to exploitation than the former.
15. Forest types whose species have been recruited from other types have less value than the originals do. Swamp forests and kerangas, spectacular ecosystems though they are, have less conservation value than the adjoining 'normal' forest. If lost, they could (theoretically) be repopulated by the adaptation and evolution of normal forest species,

but not the other way around. Such 'matrix' forests therefore are to be protected along with the forests of special type, which have been derived from them. Such forest may deserve protection not only because it contains or adjoins a spectacular ecosystem, but also because it has an environmental value, for example to the hydrology of the region.

16. A tract of lowland forest is more effectively protected if the watershed and catchment areas above it are also protected.
17. The longer the boundary of a reserve, the greater the dangers of infringement. Elongated reserves are more vulnerable than circular-shaped ones. A 400 km^2 reserve has better prospects than four of 100 km^2.
18. Small reserves amidst depleted or 'managed' forest (like the Virgin Jungle Reserves; Putz 1978) are vulnerable and must be regarded as a risk, unless exceptional conditions ensure their survival.
19. An obligation on the part of logging firms to save part of each concession is desirable – insofar as the preservation of the entire forest is desirable – but such forest cannot be regarded as protected if it is in the hands of an agency whose primary interest is exploitation.
20. There is no point in saving strips of primary forest between plots of agricultural land. They will be invaded by secondary species which hamper regeneration; and their resident animal life will move out. Such forests will slowly die in a matter of decades.
21. Lianas deserve special protection, because of their ecological role in the forest, their great medicinal potential and because they are generally persecuted by foresters.
22. Feasibility of protection is always a weighty point to consider. Natural, patrollable boundaries are the ideal, while forest bordering human habitations or routes of communication should be designated as buffer zones (a concept to be elaborated below.)
23. The value of a forest reserve for scientific research and education should always be estimated, and the mutually reinforcing effect of research and conservation efforts remembered and stimulated. The presence of a field station in a reserve can add immensely to the knowledge about the reserve and to its protection.
24. Those the conservationist must influence to get the reserve declared must be made aware of all the greater contexts to which the reserve belongs:
    a) The extent to which it carries part of our world heritage;
    b) The vital place it occupies in the network of reserves of the nation and of the rain forest region it lies in;
    c) The place it takes in the hydrology and economy of the region;
    d) Its value as a location for scientific and educational work;
    e) The potential for the future benefits that it contains.

All these points deserve consideration and explanation in the proposal to declare a reserve. After the proposal's favourable reception, a 'management plan' can be drawn up. Such a plan will contain a description of the area, with arguments for its protection (with references to scientific documentation), maps with the projected boundaries, plans for a system of guarding and for the attraction of visitors and a budget for several years of operation. Only if something similar to a management plan is produced can serious negotiations be conducted with the local authorities, especially on the subjects of the boundaries and the traditional rights of the local population. Once an agreement has been reached, an official decree can be signed by which the reserve becomes legally established, on paper, that is. Then boundary marking can begin, guards can be appointed, trained, equipped, paid and encouraged to do their duty in spite of conflicting allegiance. An infrastructure of trails (with identified and numbered trees) and of observation facilities adds greatly to the value of the place; so does a network of camp sites and a good map. The journal *Parks* (established in 1976) contains many ideas on this line.

Setbacks

The effective protection of lowland rain forest is one of the most difficult tasks. Many people, officials as well as private, have great financial interest in exploiting it. Logging companies are often willing to pay corrupt officials for a concession to extract trees, or to pay guards not to control or patrol a given area within a nature reserve. High timber prices for certain species may result in the illegal extraction of trees from a reserve or even a national park. In this way, the original value of a reserve can be so adversely affected that it is eventually abandoned. Thus, the best remedy may well be continuous documentation and publication by independent authors in the media, as is already happening in some countries. Political instability, however, often has a negative effect because the authorities are afraid of taking any action against trespass in the reserves which may further alienate the local population (Figs. 18.3 and 18.4).

Fig. 18.3 Sign at the border of Sekundur, a reserve in North Sumatra, where 10,000 ha forest was logged by way of 'experiment'. The text runs as follows: Protect animals. Save animals. It is illegal to: enter this area without permission, to carry firearms and to hunt (Photograph MJ 1977)

Fig. 18.4 Bulldozer road in the Sekundur Reserve, North Sumatra, 1977 (Photograph MJ)

Conservation and Research

The combination of conservation and research is a good working concept. One good example is at Cibodas, near the volcanoes of Mts. Gede and Pangrango in West Java. The present National Park, 15,000 ha in extent and ranging in altitude from 1500 m to 3019 m, incorporates the oldest reserve in Indonesia, declared in 1889. In 1891, Melchior Treub founded at Cibodas a small field laboratory with facilities for expatriate researchers, and to date, there have been 246 important botanical publications based on work done there (*Fl.Mal.Bull.* 1953, pp. 312–351). In this way the reserve gained fame and attracted the attention of the Indonesian authorities. It is now a well-known park, popular with young Indonesian nature lovers and scientists alike. Another example is the effect of the Royal Geographic Society's expedition to Gunung Mulu in Sarawak in northern Borneo. For 15 months, 115 researchers worked there, a third of them Malaysians. This made the park famous throughout the world which has hopefully conferred on it considerable immunity from exploitation or changes in government policy.

More modest, but very effective work can be done by the single researcher who, often with a WWF grant, is stationed for a few years in a reserve and who can give part of his time to nature conservation. He visits neighbouring authorities, employs local labour, works together with young local scientists, receives distinguished guests. He may publish material, not only in scientific papers, but also in newspapers and magazines. When possible, he encourages and accompanies TV crews making films in the reserve. Gradually, public awareness increases and local authorities become involved, with the result that the reserve acquires importance and is likely to receive better protection in the future. From the conservation point of view, the money spent on such WWF grants is a very good investment.

Buffer Zones

The local population usually considers the measures taken to conserve a forest as an encroachment on their rights to hunt in the forest, or to extract timber and other products from it. Since clearing the forest for exploitation is also out of the question, the conservationists are clearly curtailing their way of living. Furthermore, the unfortunate conservationist is regarded as a representative of the central government, which they may not hold in great respect. What can the conservationist do to refute such strongly held and often justified opinions?

1. From the beginning he can try explaining to local people what he is doing and why. He can try to convince them of his opinions on proper soil use, and point out to them the important role of the forest in soil and water conservation. He can try to convince them that destruction or impoverishment of the forest will adversely affect both their lives and the lives of their children. He can involve them in establishing boundaries which they will protect their own interests; tasks which all require much patience and insight.

2. To protect the core area, a buffer zone should be created where human interference is unavoidable. Sometimes this applies to the entire circumference, sometimes only to specific places. The establishment of buffer zones, where necessary, should be included in the management plan, and if they are to survive in the long term, they must fit into existing ecological, social and economic conditions. Land-use planning should be based on the possibilities and limitations of the landscape, the support of conservation efforts by both the local government and the local population, the local measures already devised to meet (short-term) economic needs, and so on. Such constraints mean that every buffer zone project must be geared to a specific place. In areas where the population pressure is low, it will be relatively easy to design the buffer zone around the outside of a reserve, and it should be given

immediate legal status before the population grows. Where the population pressure is already high, there is often no suitable land to be found outside the reserve; in this case, part of the reserve itself will have to be rescheduled as the buffer zone. Buffer zone size depends on the circumstances. It must be large enough to reduce the negative impacts of human activity on the reserve. The cutting and collection of timber, fuel wood and minor forest products which otherwise would have been taken from the reserve, can take place in the buffer zone, and therefore some calculation of profits and monetary needs will play a part in determining the buffer zone size.

In most cases the buffer zone itself is zoned. The least disturbed area, where minor forest products can be collected, lies nearest the reserve. Further from the reserve, other zones can be created for rattan, fuel wood and timber production and at the periphery the buffer zone gradually merges, perhaps by means of agroforestry projects, with the villagers' agricultural land. Often the buffer zones consist of already affected (for example, selectively logged) forest, and only careful management under the supervision of an ecologically trained forester will bring about the desired structure and productivity.

Minimum Size of Rain Forest Reserves

What is the minimum size a rain forest reserve must be to survive in the long run? As yet there is no answer to this question. Although some research has been done on the subject in Amazonia, reliable data will not be available for decades. To estimate minimum size, a series of protected plots of varying size must be marked out and the populations of plants and animals inside them estimated and identified, then monitored as they gradually decrease beyond recovery, and this decrease must be directly related to plot size for all the different species. Where, as is often the case for some species of forest trees, only one specimen occurs in a hectare, how large must the area be to maintain a viable population of this one species? Also, some tree species 'migrate' through the forest, the old individuals dying off on one side of an area, while the young ones extend the species' range on to the other side. In this way a species can eventually migrate over an entire continent. But how large is an area, large enough for such species to survive? Extinction is a slow process, particularly in the case of trees. Long after the biological relations of pollination or dispersal have been disrupted, a tree will stand, apparently healthy, yet without any means of reproducing itself. When the Singapore Botanical Garden was founded in 1859, a small part of it was virgin jungle of which 4 ha still exists. But it is "dying on its feet", as Whitmore has expressed it, because when the old trees die there are no saplings to replace them.

In determining minimum size, the main factor would seem to be the retention of the animals in the forest ecosystem while the plants must be present in a sufficient quantity to maintain heterogeneity and thus their adaptability to possible change or disease. The best available estimate (Marshall in Whitmore 1977, p. 145) is that of some 1000 to 25,000 individuals, depending on the species, with about 5000 as a likely average. When this is related to a population density, true for many species, of less than one tree/ha, often one tree/3–5 ha, we can calculate that to maintain the viability of an individual tree species an area of 150–250 km^2 is needed, providing that all the land is suitable and the population is evenly distributed, both of which are unlikely. Whatever the estimate, we have to think in terms of hundreds of square kilometres. A 100 km^2 means a block of 10×10 km. For mobile animals like ungulates, monkeys, hornbills and other large birds, however, this is not a great distance to cover, and consequently such animals can stray out of the reserve and may or may not return. If they leave, their 'work', the dispersal of the larger seeds, comes to a standstill. If there are no orang-utans, who will drag fruiting *Heritiera* branches 150 m or more? The vital process of exchange of genetic material will then break down, slowly, imperceptibly. As for hornbills, Medway and Wells (1971) estimated that a viable population of 5000 individuals need an area of 10,000 km^2. In fact, the reserve's area needs to exceed the range of the most mobile animal which plays a vital role in the ecosystem. If a reserve is too small to contain all its original inhabitants, various processes will set in, i.e. invasion of secondary species, on the one hand, and extinction, on the other. The forest as it existed in its original form will be replaced by something quite different, less varied and therefore less useful for mankind, at least in the long run.

How Can the Potential of Nature Reserves Be Put to Use?

Norman Myers wrote an excellent book on this subject (1984) entitled *The Primary Source, Tropical Forests and Our Future.* 'Minor products', newly discovered useful plants and useful genetic material are all grouped together for this purpose. Commercial exploitation of plants or products direct from the forest will certainly lead to the extinction of the plants involved, and it is of great importance that they remain in the forest in sufficient numbers. The only thing we may do is take seeds, saplings or cuttings and propagate them for commercial purposes. However, if such material is sold, it should happen far from the forest where such plants grow naturally, or local entrepreneurs will remove them from the forest and sell them. What has to be done regarding (potentially) useful products is to identify the species, to study their ecology in order to find out the best way to cultivate them (*Cinchona*, the source of quinine, appeared to grow rather poorly in the rain forest and does much better on special plantations); and finally to cultivate them on a commercial scale for a definite market.

These ideas are not, of course, new. The story of the discovery, application, propagation and selection of rubber, *Hevea brasiliensis* (Euphorbiaceae), is only one example. Other important rain forest products are resins, for instance the damar, a product of planted dipterocarp trees (*Shorea javanica*), and certain rattans planted under near-rain forest conditions by the Dayak people of Central Kalimantan. The number of species and the diversity of products already used, however, is very small in relation to the potential of the rain forest. This is understandable. The finding and developing of products costs money which usually has to be paid for by private enterprises. But if money is available for tropical forestry and forest management, and the latest initiatives by FAO and the report *"Tropical Forests: A Call for Action"*[7] make this hopeful, it would be unwise, even irresponsible, not to develop useful rain forest products.

In this connection, further research in the following fields is most important:

1. Cultivation of forest trees, to broaden the number of hardwood timber species that can be grown on tree plantations;
2. Cultivation of plants (trees) that yield edible fruits, for genetic selection both to increase the viability and adaptability of existing species and to introduce new species;
3. Medicinal plants;
4. Cultivation of more rattan species, and protection of species threatened with extinction (Manau, for instance) to guarantee a sustained yield;
5. Research into resin and allied products; already well-known species are being over-exploited in the wild; assessment of the possibilities of tengkawan or singkawan nut cultivation (Dipterocarpaceae);

[7]By the World Resources Institute, the World Bank and the United Nations Development Programme (WRI, World Resources Institute 1985).

6. Spices, as also for fruits;
7. Other 'minor products' according to the area. Local people still have a wealth of knowledge which could be made more widely available.

This is a plea for the diversification of the use of the rain forests by man, through indirect harvesting (seeds, cuttings, etc.) and the best place to start is with research on the products currently harvested directly from the forests, and to conserve rain forests. It is also important to make inventories of the wild relatives of trees with comestible fruits. Forests in which many of these species grow should be given special status. If the institutes which do substantial work in this field would broaden their programs, much would be achieved in a short time. There is no way other than 'indirectly' that permanent use can be made of the primary rain forest.

A Ban on the Import of Tropical Hardwoods?

If the western countries stopped importing tropical timber, would then the remaining rain forest be saved? Large-scale exploitation only started after World War II. From a mere 4 million m^3 in 1950, the volume has grown to 70 million m^3 and an increase by more than half as much again is projected by the year 2000. The exporting countries pull in $ 8 billion a year now, and the value grows at a far faster rate than the trade value of all other forest products. Tropical timber ranks among the leading exports of the Third World, earning as much as cotton, twice as much as rubber and almost three times as much as cocoa. In 1980, timber earned Indonesia $ 2.2 billion in foreign revenues, second only to oil. But although timber exports from the tropics are increasing, tropical timber still amounts to only a small share of the worldwide trade in forest products. Zaïre and Finland have land areas and forest estates of roughly the same size, yet the value of Finland's forest exports is about 50 times greater than that of Zaïre. When weighing commercial importance of tropical hardwoods, many people in the producing countries believe this resource can fuel their drive for development, and not surprisingly, look forward to the day when commercial logging will contribute much more to the public welfare.

Considering the impoverished state of most citizens in these countries, it is difficult to contest this goal. But what the policy-makers overlook is that they are cutting their forests on a once and for all basis, and now, as a result of over-harvesting, several countries, Thailand and Nigeria for example, have become net importers of timber. At the current rate of cut this will also be the case in Indonesia by the end of the century, and before then in The Philippines and Malaysia. As the world's largest single consumer of tropical hardwoods, Japan has played an essential role in this over-exploitation. Among Japan's imports, wood ranks a strong second to oil, and the hardwoods come entirely from the nearby tropical forests. Unprocessed raw materials attract no import levies in Japan, and the processing and re-export of hardwood products like plywood and veneer to the United States and Europe plays a vital role in the Japanese economy. Most of this processed timber goes to the United States, the world's second largest importer of tropical hardwood, where demand has been growing faster than the country's growth rates for population and GNP even though the United States could easily be self-sufficient in hardwood timber. The main reason for buying tropical hardwood is that it is cheap. Moreover, while the tropical hardwood represents only a fraction of the US timber consumption, it represents a sizeable share of the harvest, or over-harvest, of hardwood currently being extracted from Southeast Asian countries. Nor does the European community escape censure in this matter, with her timber imports second only to oil.

The rain forests are also being increasingly exploited for paper pulp and chip wood, and where the existing degree of timber extraction is already disastrous for the forest, allowing clear-cutting for pulp and chip wood production may lead directly to the creation of an ecological desert. Both for the forest and for the economies of the exporting countries, it is a good thing that the Southeast Asian Lumber Producers Association is trying to form a cartel, drastically cutting back log exports in order to retain as much of the lucrative processing industry as their infrastructure can cope with.

As for the importing countries, it would be difficult, if not impossible, to ban the importation of tropical hardwood. It would be better to reduce its use. Pine wood, when treated in the right way, is as good as or sometimes superior to tropical hardwoods, since these are often only used because they have become fashionable. Here, I am talking about planks and boards, i.e. sawn timber, rather than plywood or veneer, which is popular because it is cheap. Furthermore, many tropical hardwoods are already cultivated or come from secondary forest. In that case a ban on hardwood imports should not apply to those hardwoods cultivated on plantations or of secondary forest origin. Western countries would do better to help tropical countries to manage their forests in a sustainable way, leaving substantial areas of primary forest undisturbed as reserves. International aid should be directed to projects which benefit the forest, not projects which tend to deplete them, and an economically harmful ban on importing tropical timber could thus be avoided.

Organizations

Some organizations active in protecting the rain forest are:

1. CATIE, Centro Agronomico Tropical de Investigacion y Ensenanza, Turrialba, Costa Rica. Important centre for research on and the protection of rain forest in Central America.
2. FAO, Food and Agricultural Organization of the United Nation (Forestry), Via delle Terme de Caracalla, 00100 Roma, Italia. Is engaged primarily in forest exploitation, but also in long-term conservation. FAO organizes a World Forestry Congress every 6 years, the last in Mexico in 1985. Supports, together with UNEP (United Nations Environment Programme) the Tropical Forest Cover Monitoring Project to monitor the decrease in forest area. Publishes an excellent journal, *Unasylva*. Finances aid programs in the Third World.
3. ICSO, International Council of Scientific Unions, 51 Boulevard de Montmorency, 75016 Paris, France. Overall organization of scientific organizations. Supports SCOPE, Special Committee on Problems of the Environment, with the institute MARC, Monitoring and Assessment Research Centre, which collects data on forest areas.
4. IUCN, International Union for Conservation of Nature and Natural Resources, Avenue du Mont Blanc, CH-1196 Gland, Suisse. Founded in 1948. Membership open to governments and NGO's. Publishes a quarterly journal, the annual subscription for which is US$ 30.00 for non-members. The largest organization of nature conservation experts in the world Published *Ecological Guidelines for Development in Tropical Rain Forests* (Poore 1976b) and the *World Conservation Strategy* (IUCN 1980). Is active in helping member countries in designing their own *Conservation Strategy*.
5. NRDC, Natural Resources Defense Council, 917 15th Street, NW. Washington, D.C. 20005, USA. A private organization active on the interests of the environment. *The Tropical Moist Forest Bulletin 1* (NRDC 1978) gives a survey of organizations and programs dealing with rain forests.
6. Sierra Club, 777 United Nations Plaza, New York, N.Y. 10017 USA. Large American society for nature conservation only marginally interested in rain forests, but has published a study on Venezuela (Hamilton 1976).
7. UNEP, United Nations Environment Programme, Box 30552, Nairobi, Kenya. Founded in 1972 by the Stockholm Conference. Organized the first 'Experts Meeting on

Flora Brasiliensis, tabula XL. Banks of the Parana. In the foreground stems of lianas (Martius 1840–1869)

Tropical Forests' in 1980, with a broad and excellent *Overview Document* (UNEP 1980).

8. UNESCO, United Nations Education and Scientific Cooperation Organization, 7 Place de Fontenoy, 75700 Paris, France. Published books on *Natural Resources of Humid Tropical Asia* (UNESCO 1974) and *Tropical Forest Ecosystems* (UNESCO 1978). Supports the program MAB (Man and the Biosphere), section one of which is devoted to the influence of man on the tropical rain forest in several countries.
9. WRI, World Resources Institute. A centre for policy research. 1735 New York Avenue, NW. Washington D.C. 20006 USA. Published *Tropical Forests: A Call for Action* (1985). Discussion document for governments, development banks, etc. on the ways to solve the problems of deforestation.
10. WWF, World Wildlife Fund, Avenue du Mont Blanc CH-1196 Gland, Suisse. Founded in 1961. The main organization for collecting funds from private people for nature conservation projects, education included. In many countries WWF has suborganizations, the most active probably being those in the USA, Switzerland and The Netherlands. A *World Wildlife Fund Yearbook* is published showing how the money is spent and describing funded projects.

19 Forest and Man

Was the Earth Made only for Us?

"With regard likewise to wild animals, all mankind had by the original grant of the Creator a right to pursue and take away any fowl or insect in the air, any fish or inhabitant of the waters, and any beast or reptile of the field: and this natural right still continues for every individual, except where it is restrained by the civil laws of the country. And when a man has once so seized them, they become, while living, his qualified property, or if dead, are absolutely his own".

According to R. & V. Routley (1980, p. 109), this passage was quoted with approval as late as 1967. The idea behind it is biblical or older, and explains why man regarded the degradation of nature as his privilege. And this is still the prevalent attitude of man towards nature. One biological species has thus turned against all others. In this attitude, man supports himself by legal systems which have all in common the feature of embedding man in a network of rights and obligations towards other human beings. Within this net, human beings enjoy protection; what lies outside it, is literally outlawed. Remarkable in scale are the ecological ruins left behind by civilizations with elaborately codified laws.

In recent times this unproven, yet universally adopted principle that only man has rights over nature, has come under scrutiny. R. & V. Routley (1980, p. 97) suggest that the tenet of "everything for one species", is in fact analogous to systems of discrimination based on nation, race or sex, or even, as in the mafia, to systems used to obtain money.

Through tacit yet systematic and sustained discrimination, man is able to claim power over nature. Until recently, nobody seriously contested such claims, because the exertion and growth of man's power over nature was generally regarded as sound and necessary. As for power over the rain forests, this is universally made legitimate by references to the plight of humans who need land and money.

And then there is a silence. It seems as if this very principle justifies all the half-truths and lies that are told and written these days about rain forests.

The silence deserves to be broken. The ruinous results of man's abuse of his power over nature have become too conspicuous, witness the books *Man's Role in Changing the Face of the Earth* (Thomas 1956) and *The Careless Technology* (Farvar and Milton 1972). Moreover, on what grounds does this strange discrimination rest, which isolates one species in hostility from all others on which it is nonetheless dependent? The possession of culture and intelligence seems scant justification for exploiting all living things said not to possess them. Man's alleged superiority over animals seems untenable in view of the many senile, retarded, invalid or otherwise handicapped people. Animals can withstand many hardships better, are aware of their environment and, in contact with rivals, also of themselves. Their forms of mutual communication seem perfectly suited to their purpose. Animals also possess a certain body of basic knowledge – do not orang-utans locate unerringly the fruit trees in their forest? – although knowledge is not the source of virtue, as Socrates already observed. And if it comes to feeling and intention, even rats and rabbits cannot be denied these.

The Routleys conclude "... that it is not possible to provide criteria which would *justify* distinguishing, in the sharp way standard Western ethics do, between humans and certain nonhuman creatures, and particularly those creatures which have preferences or preferred states. For such criteria appear to depend upon the mistaken assumption that moral respect for other creatures is due only when they can be shown to measure up some rather *arbitrarily-determined* and *loaded* tests for membership of a privileged class (essentially an elitist view), instead of, upon, say, respect for the preferences of other creatures. Accordingly *the sharp moral distinction*, commonly accepted in ethics by philosophers and others alike, between all humans and all other animal species, lacks a satisfactory coherent basis" (1980, p. 103).

The Situation in Indonesia

Indonesia is second only to Brazil as the nation which holds the world's largest area of rain forest. It seems therefore appropriate to add some specific data to the Regional Notes with regard to exploitation. From the 19th century onwards, forestry in Indonesia operated in two main fields: the plantations of teak, *Tectona grandis* (Verbenaceae), in the drier parts of Java, and the natural forests in the other islands. Huge tracts of forested land were explored and administered by a relatively small staff, who reported on their activities in the venerable journal *Tectona*. When in 1955 this was suspended, it filled 3.15 m of shelf. Already in the 1920s it carried papers on problems caused by population growth, shifting cultivation and lack of firewood.

Indonesia was a colony of The Netherlands when early in 1942 the Japanese occupied the country and interned all Dutch. This fact marked the end of Dutch influence as an expression of a coherent policy backed by physical power. The surrender of the Japanese in September 1945 was followed by a long period of confusion and upheaval. Late in 1949 Indonesia became formally independent, except western New Guinea which was kept by the Dutch for the time being. In the ensuing diplomatic dispute over that province, Indonesia expelled all Dutch experts and tradesmen from her territory, early in 1958. This resulted in severe breakdowns in the Indonesian economy. In 1962, western New Guinea was transferred to Indonesia all the same. In 1967, a change in national policy led to a welcome of foreign investments and an opening up of the natural forests for large-scale exploitation.

In May 1977, Chandrasekharan submitted to FAO *A Report on the Forestry Situation in Indonesia*. It gives a large amount of data, in a spare and objective tone, without comments or conclusions. Yet, no reader can fail to be moved by the heartbreaking series of facts. By the time western New Guinea, now called Irian Jaya, had hardly entered the picture. The report concentrates on the dipterocarp forests of Sumatra and in the Indonesian part of Borneo, which is now called Kalimantan. We reproduce a few figures:

Island	Area (km^2)	'Forest land' (km^2)	Population
Sumatra	473,600	260,000	20.18 million
Java, Madura	132,200	30,800	76.04
Kalimantan	539,500	419,000	5.15
Celebes	189,200	113,900	8.53
Irian Jaya	421,900	310,000	0.84
Others	148,200	80,400	7.00

As can be extrapolated from these figures, 63.5% of the surface area of Indonesia, or 1,200,000 km^2 is regarded as 'forest land'. This figure already occurs in pre-war statistics, but did not mean even then that all of these lands were actually under forest. On the contrary, forests damaged and/or cleared by shifting cultivators now amount to something between 200,000 and 370,000 km^2, while 126,000 km^2 is bare land, of which 85,600 is still designated as forested land.

In Indonesia, about 11.5 million people are engaged in shifting cultivation. They have affected 110,000 km^2 forest in Sumatra, 100,000 km^2 in Kalimantan, 120,000 km^2 in Celebes, 40,000 km^2 in other islands. Most of these lands are now under alang-alang, *Imperata cylindrica* (Gramineae); and the extent of alang-alang increases by about 2000 km^2 year^{-1}.

Land under actual forest includes: mangrove forests, which occupy an estimated 10,000 km^2, swamp forests, 130,000 km^2 or 11% of the 'forest area', coastal forest, 10,000 km^2, peat-bog (i.e. probably peat swamp forest), 12,400 km^2, rain forests 890,000 or 73%, seasonal forest (including the plantations of teak and conifers in Java and Sumatra) 10,000 km^2, secondary forest 152,000 km^2. But of the 4000 or so species of trees, only some 60 are commercially exploited.

Stock estimates start at an average of 100 m^3 ha^{-1}, of which ca. 60 are commercial, but in the dipterocarp forests of Borneo this may reach 270 m^3 ha^{-1}, or 50–80 in commercially sawn logs. In the East Sumatran freshwater swamp forests, the estimate is 58–125 m^3.

'Potential growing stock' volume, which is an estimate of what will be accessible for logging during the next 15–20 years, is expected to be some 119,113,000 m^3 timber in plantations, on a total of 12,030 km^2 land, and in natural forests, on 289,200 km^2 land, to some 2,524,780,000 m^3.

Concessions in 1977 covered 202,350 km^2 of Kalimantan, 90,022 km^2 of Sumatra, 1,967 km^2 of Celebes and 35,460 km^2 elsewhere. By the end of 1975, there were 267 firms operating on 202,040 km^2, with 379 more waiting for 391,560. Philippine firms held 24% of the total area under concession, Japanese 21%, those from the United States 17%, from Hong Kong 8%, South Korean 5%, Singaporean 4%. Concession size varies between 400 and 5000 km^2; the average duration, 20 years. Management plans have to be made, and a sawmill must be established in 3–5 years, with plywood and veneer factories to follow, but in an unspecified way and only if it is considered economically feasible. "And only few companies seem to be simultaneously involved in reforestation following logging. A tendency to make quick returns on minimum investment and a fast withdrawal possibility is pre-eminent" (p. 45). "The granting of utilization rights has probably progressed so fast as to get out of control. This problem is not easy to solve as the forests have been committed for a period of 20 years" (p. 46).

Chandrasekharan spells out what this means. Concessions are assessed by a 0.2% sampling for trees 35 cm diameter or thicker. By 1982 all land will have been thus inventoried, with foreign assistance. The 'allowable cut' is set at ca. 1 m^3 ha^{-1} year^{-1} reckoned over 35 years. The 504,581 km^2 of lands declared to be production forest, must therefore yield 45 million m^3 annually, with ca. 15–18 million projected for export. Thus, the forest has to meet the targets set by a handful of planners. "But, concession agreements are usually set for 20 years, a situation which will make sustained operators in the following 20 years a dubious prospect" (p. 37). Extraction is therefore done as cheaply as possible, and with a minimum of care. Breakage is common, and during the 3–3.5 months between felling and delivery, owing to various delays, borers attack. The result is an estimated overall waste of 40–60% of the stock volume.

"In spite of the conservative assumptions and the uncertainties of the estimates, the yield of the first extractions from unit areas of the natural forests has so far been good. But the future potential would depend on the quality of the working area and growth expectations. The expectations can be achieved only with greatly improved silvicultural knowledge and application greatly improved marketability of the tree species and fuller utilization of harvested trees. The control on logging excised is by volume. And sometimes annual cuts can be achieved only by logging more than 1/35th of the area. Since in case of some species, for instance those of which the wood sinks in water, trees of larger size than 50 cm dbh are left because they are non-merchantable and/or their extraction is unprofitable, the structure of the forests will undergo considerable change; and the second cut will not contain the assumed proportion of desirable species" (p. 39).

Such is the story of plunder in Indonesia. In 1975, Japan took 7.5 million m^3 timber, South Korea 2.7, Taiwan 2.2, Singapore 0.7, Italy 0.2. Nearly all this wood was exported in the form of logs; only 4.5 million m^3 was sawn inside the country, 24% of this was exported. Logs were shipped to Japan, 20% under Indonesian flag, 80% under Japanese flag. Singapore planned to import more Indonesian logs (mostly from Sumatra) to fill the gap in its timber requirements caused by a recent ban on the export of logs thicker than 40.5 cm from Malaysia to Singapore. For 1 m^3 wood, valued at US$ 40 Free On Board, Indonesia collected a mere $ 8.20.

Production of paper in 1975 was 20% of consumption; 18,950 t paper pulp and 241,850 t of paper were imported. In view of the lack of reforestation, which between 1969 and 1975 amounted to only 6950 km^2, there is no reason for surprise at this figure.

The need of firewood for cooking is 4 m^3 $year^{-1}$ for a family of five, but no more than 1.7 m^3 is available. In Java, people deforest the mountain slopes for fuel, and erosion sets in, to such an extent that in some places bitumen is applied to prevent further slope wash.

So far Chandrasekharan; meanwhile, the population in Java is growing by 5500 a day. One remedy is transmigration. Since 1905 people have been moved, with government assistance, to South Sumatra, which in 1883 was fertilized by volcanic ash from the Krakatau eruption (see map in Verstappen 1956, p. 10). In later years, other, less fertile provinces have also been colonized. In 1975, some 12,500 families moved, and the target for the next 5 years was 25,000, then 50,000. This does little to relieve the over-population in Java (12,500 families of 5 equal 11–12 days of increase), but it takes a lot of forest, as areas are cleared for transmigration: 30,000 km^2 was earmarked in Sumatra, with 20,000 to follow. And some people have already had to be re-transmigrated, because since their arrival the soil has been too degraded (and catchment areas were not properly protected from deforestation in the first place).

In 1975, the botanist Meijer wrote a booklet on *Indonesian Forests and Land Use Planning* which every interested person should read. Meijer knows his business, and went everywhere to look for himself at situations that arise from lack of planning, hurry and ineptitude. In Sumatra's southernmost province of Lampung alone, during the preceding 20 years of 'rural development', about 20,000 km^2 of lowland dipterocarp forest with 60–100 million m^3 wood was cleared and burned on the spot to make room for (semi)permanent agriculture: all wasted.

The Indonesian Forest Service was heavily understaffed, and unable to provide any direction or control. As for forest exploitation, a mere 0.5% of the revenues gained were re-invested in the Service, vs 10% in Sabah (formerly North Borneo), which is part of Malaysia. Knowledge of forest botany is inadequate, as is the training in proper forest conservation methods. No trees thinner than 60 cm in diameter must be cut, but this is ignored everywhere, and small wonder: the salaries of foresters are such that they can't refuse a tip from the logging firm for looking the other way. A young graduate forester, with a family, has a salary which equals US$ 50–60 a month, i.e. an annual income of about the price of a large tree. Bribery and corruption exist at high levels as well, however: on pp. 68–69 of Meijer's booklet a satellite photograph annex map is reproduced, of a logging area inside a reserve in Southwest Sumatra, for the benefit of some high naval officers.

It looks as if all human mistakes and failures in Indonesia are ultimately revenged on the rain forest! This may be excusable in a weakly organized society that was unprepared for a

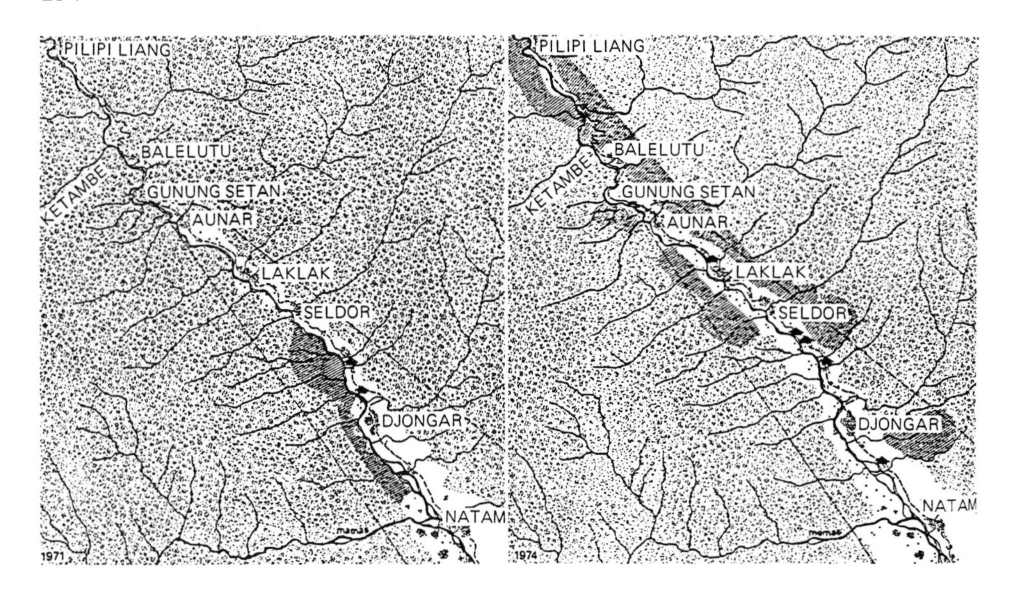

Fig. 19.1 Degradation (*hatched, dark*) of forest (*lighter*) in the valley of the Alas, North Sumn the *upper left,* the study area Ketambe; the Alas runs towards the southeast. Year: 1971 (Rijksen 1978)

Fig. 19.2 Degradation in the same area, 3 years later. The forest belongs to the Gunung Leuser reserves. Bataks from the Karo Heights, north of Lake Toba (to the southeast of this area) have depleted the soil there, and now encroach upon the Alas valley. Because they can dispose of more money than the native population they can buy plots of land; the Alas people then move upstream to convert forest into agricultural land. The buffer zone that was originally added to the reserve has already been used in the same way (Rijksen 1978)

boom in timber and in 'development', but such human considerations are immaterial to the rain forest *as such*. To the forest it makes no difference whether it is sacrificed for the rich or for the poor, for the political right or left, after deliberation or just carelessly. Every hectare less is a hectare less, and as the habitat area of each species of plant and animal, which together form the forest, shrinks so the risk of extinction is increased (Figs. 19.1 and 19.2).

Since the mid-1970s, much has happened. A minister for development supervision and the environment has been appointed. The government has been working towards its goal of setting aside 10% of the Indonesian land area for conservation purposes, and a team of professional conservationists is working towards realization of that goal. A popular environmental movement has sprung up, the *Yayasan Indonesia Hijau*, or the "Green Indonesia Foundation". A series of projects to upgrade the Forest Service, with assistance from The Netherlands, has started. Family planning is gaining ground. To be sure, destruction of forest and growth of population proceed. But a front has been formed, and counterforces do exert influence. Although she has lost a great deal, Indonesia seems unwilling to have her environment completely ruined.

Corruption

No society is free from corruption, and in many tropical countries it is accepted as a way of life. Some estimate that about one-third of the money flow fails to reach its originally intended destination in certain countries. In tropical rain forest exploitation, as in so many other fields, the rather sudden lifting of limits to man's power over nature has led to disastrous effects. While corruption always had been a means to grease the wheels of a gritty bureaucracy and to keep ties with one's friends in good repair, nations in tropical countries now find that large-scale corruption threatens the biotic substratum on which the whole society depends. The *Far Eastern Economic Review* of 30 November 1979 carried a substantial article on this aspect of rain forest exploitation. It reported that on Pulau Maya, south of Pontianak in southwestern Borneo, in 1 month 20,000 m^3 of ramin wood, *Gonystylus bancanus* (Thymelaceae), was stolen by about 100 logging gangs, and that this operation was directed by the very man who had been hired to end the practice, and it was estimated that an amount equal to US$ 100,000 a month was being channelled through the regional military command.

Corruption, however, like other phenomena in human society, can sometimes be regulated. It is the fact that corrupt exploiters are these days equipped with modern machinery for forest destruction that makes such a difference and leads to a disruption of ecological balances and processes very fast indeed. The moment has come for society to react in accordance with the new reality. This calls for a new criterion in the regulation of society: corruption must stop where it contributes to the degradation of unrenewable resources. The dangers to the quality of life are too great for tolerance in this respect. The unrenewable resources that come to mind first are primary vegetation and topsoil.

The UNESCO Book

Late in 1978 UNESCO published a book weighing 2.4 kg with 683 pages of two-column print called *Tropical forest ecosystems* / A state-of-knowledge report. It was announced as a tool for action, a synthesis of the work of the last 2 decades.

There are 21 general chapters arranged under (1) description, functioning and evolution; these deal with: inventory; statistics on tropical forests; paleography and climatology*; composition; forest structure*; animal distribution; animal populations; biology and regeneration*; secondary successions; primary production; secondary production; water balance and soils (but hardly anything on erosion); nutrient cycles; pests and diseases*; human population; nutrition; health and epidemiology*; human fitness; population density and civilization; types of utilization; and conservation and development. (2) Man and the patterns of use; there are also eight 'regional case studies'. The asterisks indicate oustanding papers.

The absence of an all-round introduction, summaries and indices, however, makes the information in the book well-nigh inaccessible except for the most determined reader. Although several contributions are excellent [they have been reviewed extensively in the *Fl. Males. Bull.* (32) 3256–3365 (1979)] and long lists of references are given – altogether 3515 in the general part alone – the omissions are noteworthy. The subject of minor forest products, for instance, which greatly affects all considerations about utilization and value, has been ignored. Evolution, and in general the time factor, which alters so many perspectives if given due attention, is virtually absent in the text. The all-important subject of environmental education, too, is lacking, a strange shortcoming in a book published by the world's largest organization for education. A clear, comprehensive vision of the future of the rain forest is missing as well.

There is a vagueness about the value of rain forests, and exactly what would be lost with them, besides "a psychic loss to some people resulting from the disappearance of some specialty timbers", is anybody's guess. "The case for retaining the moist tropical forests, if they could be managed as such, depends therefore on doubts about the future that are too speculative to match the urgency of demands on them. The uncertainty may point to a need for caution, but it is hardly strong enough to justify a halt" (all on p. 542b). Yet on these forests "it can be confidently stated that it is possible to carry out the various stages of their development" (p. 547b).

Besides such lack of vision, embodied in a plethora of United Nations prose, on the one hand, and a large display of knowledge, on the other, there are astounding lists of 'research needs and priorities'. These are summed up at the end of almost every chapter; together they cover 23 of those large pages in dense print. Taken together, these well-intentioned items amount to one big mockery. The grandest funding would be insufficient to finance all those programs. What a sad combination of abundant knowledge and failing wisdom. No book better reflects the general lack of clarity in thinking about the rain forests and their fate. If one common feature common to all the chapters can be pointed out, it is that something is amiss.

What the UNESCO book reveals indirectly is a sad disarray in the relations between man and forest. No word of wisdom guides man by telling him what to do and where to stop in his dealings with nature.

Renewability, a Complex Matter

As for the future of the rain forest in the hands of *Homo sapiens*, much depends on his judgment on the question of to what extent these forests are a renewable resource, or whether they are renewable at all. A renewable resource is one that can be 'cropped' in certain quantities, and continuously cropped if conditions do not change. Until recent years, the North Sea herring, a well-known delicacy in Holland has been a good example of such a resource. When catches become too small due to over-fishing, limitations are imposed for a couple of seasons, until recovery has been achieved. On the other hand, oil, as everyone knows, is a non-renewable resource: the total reserve can only grow smaller. Tropical rain forest is often termed renewable: for instance in the *World Conservation Strategy* (IUCN 1980, Chap. 16, first sentence). In the first chapter of the present book, I have contested this, and it seems to me a point on which there should be no misunderstanding.

To elucidate, we must return to the distinction between *vegetation* and *flora*. The question of whether or not the rain forest is renewable in its entirety, i.e. as vegetation, has been discussed in Chapter 8 (Primary and Secondary Forest), in which Budowski's Table showed four succession stages lasting from a few years to a few centuries, with true primary forest returning only on sites adjacent to a stand of similar intact forest which enabled even the biggest seeds to be spread and to germinate. The early Budowskian stages of 'pioneers' and 'young secondary forest' are easily renewable; so is, of course, plantation forest. But only if it is true that *all* 'tropical forest' goes through the same succession stages can one suggest that *all* such forests are renewable.

Let us now consider 'careful' or 'enlightened' exploitation, in which so many people seem to believe. We remember how easily a population of scattered trees in species-rich forest is decimated even if only a few trees in 1 km^2 are extracted. Well, that is but one population. Let us say that there are 50 truly commercial species of trees and lianas in 1 km^2. Now kill off (whether through harvesting or through silvicultural measures, it makes no difference ecologically) the population of two of them, beyond the point of no recovery. Forty-eight species will now remain – for the time being. Suppose that no further damage is done in the form of broken biological relations, are we then justified in calling this an intact forest? If the answer is yes, we can go further in our exploitation: 45 species left in functional numbers, 40, 35, 25 . . . who shall set the limit?

The exploiter will be glad to. He declares: "all silvicultural measures can be executed in such a manner that leaves the ecosystem virtually unharmed" (as stated in a Dutch leaflet in 1980). Thus, in one sentence he reassures the public and for himself creates a margin, i.e. his 'virtually', in which he can operate in his own terms. Justification can be found in a 'demand' that he wants to satisfy in the widely uncontested, accepted right to exploit nature that he wants to exercise, and in the scantiness of the yield obtainable from rain forest, so majestic and forbidding.

So what Amount of Exploitation Is Possible?

Nowhere have I found a well-considered and specified answer to this question. The *Ecological Guidelines* (Poore 1976b) everywhere imply responsible planning, caution, carefulness, high standards, but that is all (Fig. 19.3). However, if we agree that no species should become extinct through man's doing, and that exploitation is inadmissible unless it can be perpetuated ad infinitum, what possibilities remain?

Fig. 19.3 Relationships between shifting cultivation and other land uses, in each of the four categories (Poore 1976b)

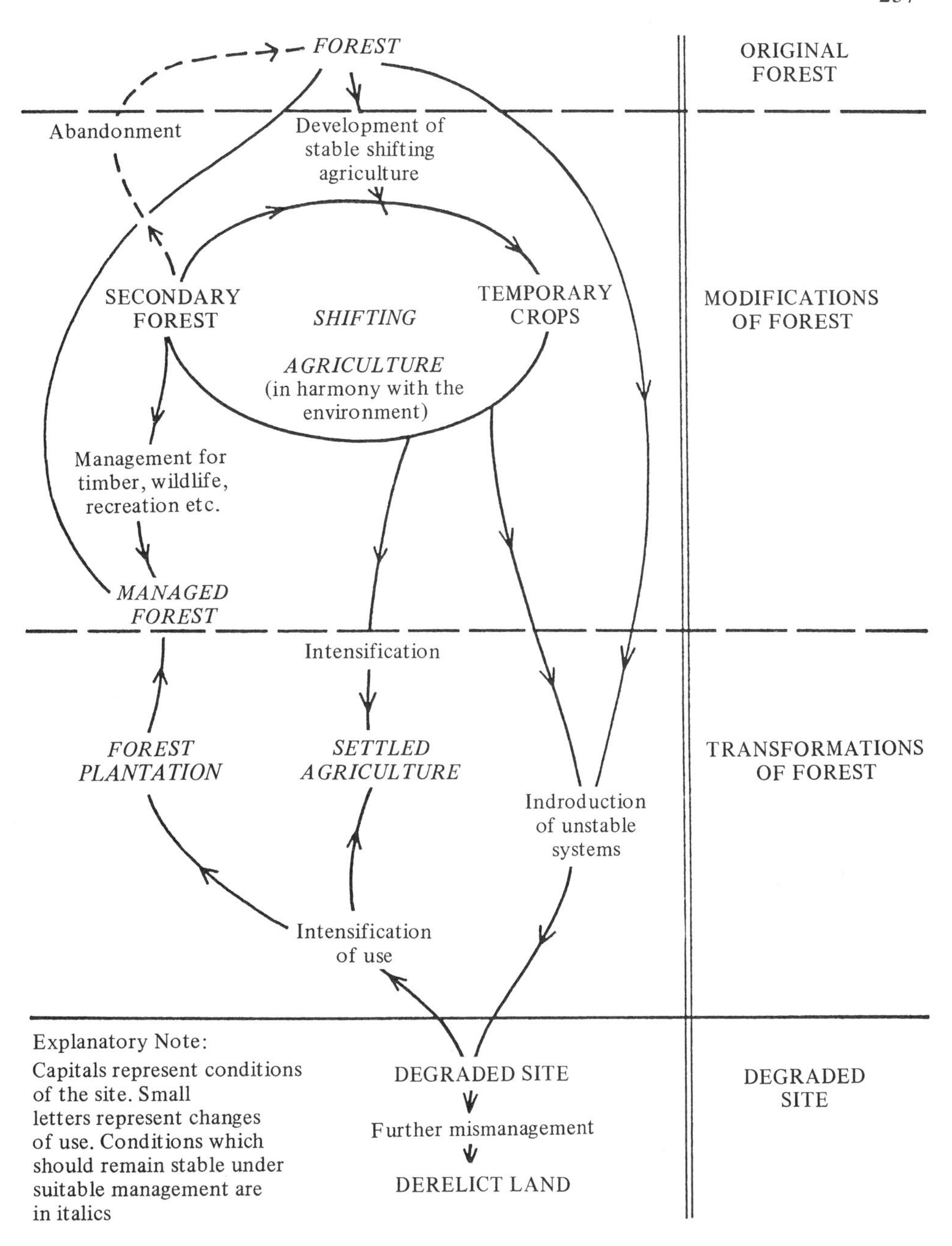

From previous chapters a number of rules can be distilled which should be obeyed. They are as follows:

1. No population of plants or animals must be diminished through human influence beyond the minimum size needed to ensure its continued existence.
2. All food chains and other biological relations, like those of pollination and dispersal, must be left intact.

3. The genetic composition of a population must not be altered. Selective harvesting, through which commercially inferior specimens are left behind is dangerous, as is any destruction of part of the genetic variability: the remaining population may be less adaptable to changing circumstances.
4. No one age class must be completely taken out of a population. The composition of the ecosystem would otherwise be influenced over very long periods, endangering the patterns of its biological relations.
5. The peculiar rhythm of building and breakdown of the canopy must not be disturbed, for the same reason as in point 4. Changes in light quantity on the forest floor can easily endanger primary regrowth.
6. No more inorganic matter must be withdrawn from the mineral cycle than can be resupplied through natural courses.

This is quite a package of limitations, from which not one can be omitted. They have been derived from the total body of our knowledge about rain forests, and there has so far been no evidence to prove them invalid.

Large Capital, Small Interest

The six conditions above leave very little room for manoeuvre. On the basis of permanency, which means without *any* loss of species in the long run, only very small quantities can ever be harvested. However, the *quality* of the harvest is unsurpassed, because of its variety of unique, precious material: rattan, cacao, rubber, curare, an array of anti-cancer substances, bird feathers, etc. as has been enumerated in "The Significance of the Forest" (see Chap. 16).

Hunter-gatherers come closest to sustained utilization, if they do not carry anything out of the forest, i.e. if they live outside a cash economy. Even so, Richards (1973b) suspects that over ten thousands of years they have caused considerable impoverishment of the African rain forests.

Nearly always, forests fail to fulfill human expectations. If one believes that all biological riches exist on earth merely for the benefit of man, then one can also believe that this most diverse, most majestic type of vegetation has only obstructed, throughout history, the movements of man, and that it has only been lack of technical equipment that has prevented man from its full utilization. But now that we have the technical equipment, are such restrictions to be imposed on the utilization of this enormous capital that we must now regard the rain forest as *un*exploitable? To find out what we want from the forest can be realized only under the conditions set by the forest: is this our reward for having finally gained power over nature?

This attitude is extreme, but one must take a critical look at one's own set of expectations or preconceptions – which may now need changing – and at beliefs shared by many botanists. It is hard to let down colleagues, friends, subordinates, superiors: one feels compelled to declare that the rain forest is indeed an important renewable resource. The assumption that forests will replenish themselves naturally (provided we manage them well) is a pleasant prospect to everyone. To discuss the difficult distinction between renewable and unrenewable is painful, and one can always try to believe that the forested tracts on earth are still vast (as Lanly and Clement of FAO themselves have stated), and development must at no price be stopped or even slowed down. In any case, who are we to contradict someone silly enough to sell out his riches cheaply? From the beginning, human weakness has obscured the problems of exploitability.

So many different products can still be fetched from the forest, each of them in quantities that seem tiny and yet may lead to the destruction of a plant population, that even the imposition of a theoretical limit on a harvest is out of the question. Who is to deny the neighbours of a forest some benefits to alleviate their penury? Not Norman Myers, who in *The Sinking Ark* (1979, p. 200) seems to favour "some very light and careful logging".

This is a most unfortunate statement, for two reasons. First, because logging is the most common, most damaging and most wasteful sort of utilization. Second, because here a kind of 'blanket utilization' is advocated from which no forest is to be spared – with untold consequences in the long run in terms of slow deterioration.

If we had the means to determine how any of the several dozens of useful species could be cropped and in what quantity, and also to implement these views, there would be nothing against opening reserved forests to such wise and limited exploitation. But the means are lacking. Nobody knows the limits of safety and nobody can regulate a harvest of forest products and their flow into a cash economy.

Any permission to exploit primary rain forest in whatever form, except to collect seeds for cultivation elsewhere, will therefore result in a process of impoverishment that no one can stop or revert. Some populations will decrease below the vital minimum that ensures their survival. Through severance of their biological relations, other populations will be affected. Disappearance goes on spot-wise and very slowly, hence imperceptably in most cases, but without the possibility of recovery. Whether this multiple process might come to a standstill at a certain point is anybody's guess.

Since this fate awaits all tropical rain forest under the influence of man, *decisions must be made now on which tracts of forest are to be protected completely and which are to be exploited in some form or other*. Any form of exploitation will result in disruptions of the biological relations, and some loss of species. Any forest opened up for exploitation will be exposed to the processes of 'ecological simplification' which will result in the eventual disappearance of part of the species in them. Through wise and careful management, these processes can be slowed down, their effects delayed, but such a forest will progressively become stunted in many of its functions.

The priorities involved in determining what tracts shall be given absolute protection, and how the remainder shall best be managed has already been given much attention in Chapter 18 (Protection). But it has to be considered afresh, in the wider context of integrated land-use planning.

From Intact Forest to Degraded Land: A Four-Point Scale

The dark green blanket that once covered the earth in all regions where natural conditions allowed the growth of tropical rain forest has now, under the influence of man, been replaced by a patchwork quilt. On a clear day, a ride in a helicopter offers a fine display of the variety in colour and surface of the different patches. Some order can be brought in this variety if we organize our perceptions according to a scale which has at one end intact forest, and at the other, well, no forest at all. The simplest subdivision of the scale results in four categories already distinguished in the *Ecological Guidelines*. We will discuss them briefly, with some elaboration of category 2.

1. Intact Forest: the original undamaged forest which is the subject of this book. It is full of diversity, in life forms and species, hence complex, hence fragile. The diversity diminishes, however, with increase in altitude above sea level. Its full regeneration even from small disturbances is very slow; see the Hallé-Oldeman circle in Fig. 8.6. A number of natural gaps always occur; in them a succession proceeds, along the stages given in Budowski's Table (Fig. 8.1). Expansion is possible only in a continuum.

2. Modified Forest has been influenced by man in such a way that the original structure of the vegetation can still be recognized; in practice, this is the case where at least a few big trees are still standing, with long slender boles, some buttressed, and half-globose crowns. Chapter 17 (Damage) began with a list of destructive factors from which we now briefly summarize some main types:

a) Through selective hand-logging, decades ago, some age classes of one or several species are missing. The canopy closed afterwards.
b) Through moderate mechanical logging in more recent times, considerable damage has been done to the canopy and to the soil and regrowth by road-making and tracks. Trees of the various succession stages may have filled the gaps, depending on conditions.
c) Through exhaustive mechanical logging, several broken stumps and only an occasional non-commercial tree still stands, often overgrown by climbers. The canopy has been destroyed.
d) Through silvicultural measures, the stand has been 'homogenized': the very large trees have been killed, climbers and stranglers are few and far between, of some commercial species certain age classes are abundant. The canopy is rather level.
e) Through the collection of minor forest products, the canopy and undergrowth has only been occasionally damaged.
f) Through the browsing domestic animals, or the cultivation of such crops as cardamom, *Amomum cardamon* (Zingiberaceae) on the forest floor, as is done in southern India, the original canopy has been left intact, but seedlings, saplings and further undergrowth has been destroyed.
g) Through clearing for shifting cultivation in former times, secondary forest is surviving in isolated spots in various stages of succession, depending on age and accessibility.
h) Through continuous shifting cultivation, only forest remnants have been left and the fringes of these are now being invaded by pioneers which suffocate the original regrowth. Often such forest pockets are occasionally exploited further for timber and minor products, and their shape varies according to the prevailing social conditions.
i) Through exploitation of the borderlands between rain forest and savanna, only the gallery forest along streams remains, but is subject to exploitation as in point h.
j) Through exploitation on all level tracts of land and lowland generally, no forest is left, although on the steeper slopes and in the higher hills and mountains, protection forest is maintained, either for water catchment and soil conservation, or because no one has yet cut it.
k) Through exploitation a forest has been opened up completely, and when logging is finished, people enter and clear plots for agriculture. If some big trees are left, they are killed by fires lit around their trunks after which they slowly decay.

To diagnose all these modifications is very difficult. It requires a great deal of practical knowledge, and data obtained on the ground as well as from the air. An excellent paper by Sicco Smit (1978) demonstrates a number of cases from 11 pairs of stereo-photos, one of which has been reproduced in Fig. 19.5.

3. Agricultural Land is all terrain where forest has almost completely been replaced by man-made vegetation. The *Ecological Guidelines* (Poore 1976b) speak of 'transformations from forest', and include urbanized land. The 'almost' refers to certain useful trees that are saved during the clearing, like sugar palm, *Arenga pinnata* (Palmae), fruit trees such as durian or kempas, *Koompassia excelsa* (Caesalpiniaceae), in which wild honey bees like to construct their huge nests. All plantations belong to this category. Some, like coffee, *Coffea* (Rubiaceae), are shaded by tall trees overhead; others, such as rubber, *Hevea brasiliensis* (Euphorbiaceae) and oil palm, *Elaeis guineensis* (Palmae), are planted in the open, and soon reach 20 m. An account of the main tropical crops is given by Purseglove (1968, 1972).

Timber plantations or man-made forests are monocultures of fast-growing pioneer or savanna species, which have been selected for ecological versatility and bioproduction. Among the best known are *Anthocephalus chinensis* (Rubiaceae), a pioneer of the Malesian rain forest region; *Eucalyptus deglupta* (Myrtaceae), wild in the eastern seasonal parts of Malesia; *Gmelina arborea* (Verbenaceae), another Malesian rain forest pioneer; *Pinus caribaea* (Pinaceae), a pine from Central America. Expeditions are sent to their native lands to collect 'germ plasm', i.e. seeds for cultivation to improve the stock, which is kept and exchanged by tropical forestry institutes the world over. They can be harvested 15, 20, 30 years after planting. They are used for light timber, paper pulp and firewood, and they characteristically remove relatively small amounts of minerals from the ground.

Various systems of agroforestry, to be discussed later, are also composed of this sort of vegetation, as are various types of homestead gardens or groves planted in the vicinity of houses: pawpaw, *Carica papaya* (Caricaceae), bananas, *Musa* (Musaceae), kapok, *Ceiba pentandra* (Bombacaceae), breadfruit, *Artocarpus communis* (Moraceae), frangipani, *Plumeria* (Apocynaceae), betel nut, *Areca catechu* (Palmae), petai, *Parkia speciosa* (Mimosaceae), *Bougainvillea* (Nyctaginaceae), many kinds of bamboo (Gramineae-Bambusoideae) and several other species which contribute to the superficial impression of vegetable luxuriance in the humid tropics, for the same few species proliferate everywhere.

4. Degraded Land. This name is given to lands that once were under (modified) forest or agriculture, but which have lost their vegetation cover and were then abandoned as the soil became exhausted. In Malesia enormous tracts of such land are covered with alang-alang (lalang or cogon grass), *Imperata cylindrica* (Gramineae), which benefits from recurrent fires. There are also vast areas of fern and scrub: secondary forest in its poorest form. Land in an advanced state of degradation, however, is almost denuded because the soil is of laterite, or has become a prey to erosion. In many places, such lands stretch to the horizon, often where no more than a few decades ago there was all primary forest!

A great deal of such land could be brought under modest cultivation again, but this requires investment and considerable administrative and social effort. Alang-alang can be controlled by preventing fires for several years, and seeding with a succession of legumes; the trees then have time to grow and shade out the grass. Eventually, a suitable kind of (agro-)forestry can be embarked on.

In recent times, the exploitation of forested land has often followed this sequence: intact forest (1) is logged over (2b), then occupied by people (2k), who cultivate it as long as possible (3), until the soil becomes exhausted and is left to its fate (4), constituting a localized ecological disaster. The area of intact forest thus steadily dwindles, while the area of degraded land increases. In Indonesia in 1978 the area of degraded land amounted to 126,000 km^2. This transformation is always one way: degraded land will never regain its original forest cover, if only because of its distance from a source of primary forest seeds.

Another important point to consider is the impossibility of determining exactly and in detail over large tracts of forest where intact forest still exists and where and to what extent the forest has been modified. The interpretation of air photographs for this purpose is both difficult and risky; see again Fig. 19.5. Surveys on the ground are needed to back up remote sensing, but at the same time loggers and shifting cultivators are still busy. Furthermore, if a forest is left in peace for some years, the damage lessens to the untrained eye, and can only be diagnosed by a qualified expert. For these reasons no one can venture to say exactly how much intact forest has been left anywhere. If one receives a detailed estimate, it always concerns a small area, and even in such cases a survey usually reveals less intact forest than had been estimated.

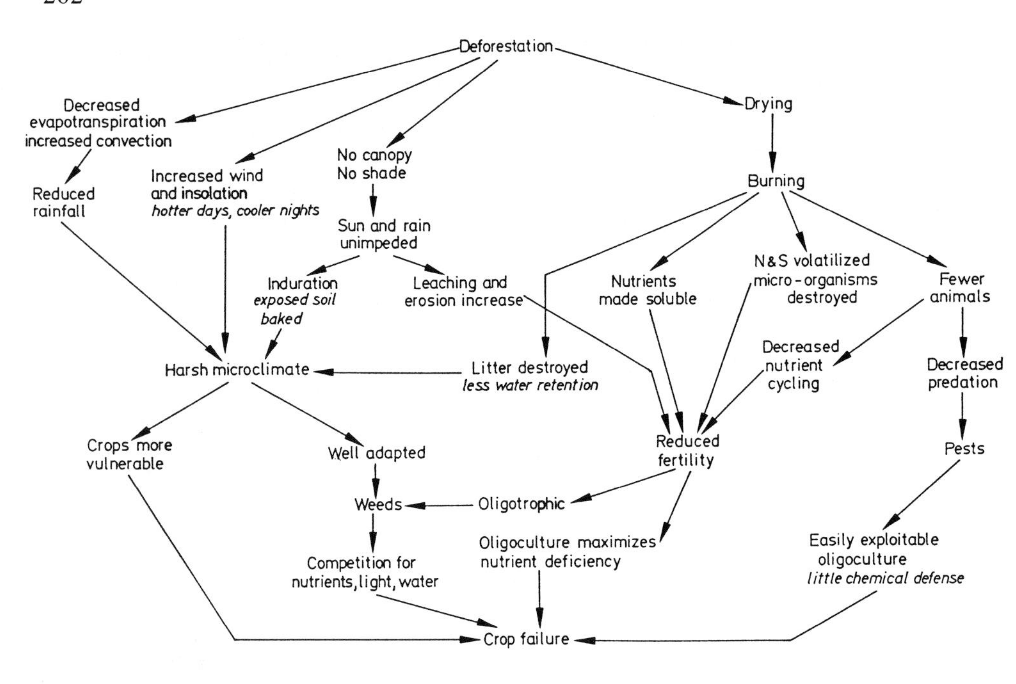

Fig. 19.4 Relationship between deforestation and crop failure (Goodland and Irwin 1975)

Fig. 19.5 Aerial photograph of an area of degradation in East Kalimantan, Borneo. In a valley (*1*), among disused rice fields (*white*), attempts are made to establish permanent cultivation, strips with a width of 20–30 m lie along a central road or canal. In the hills (*2*) shifting cultivation is practiced, resulting in various stages of destruction and secondary growth. After a few years it becomes difficult to distinguish primary and secondary forest from this height. On the ridge (*3*) the forest is intact, on the other side (*4*) it has been cleared by way of a different system of shifting cultivation (not visible on this photograph). As houses and huts are not visible, it is impossible to estimate the density of the population. Without visiting the area it is impossible to predict its future. Possibly the area of permanent cultivation will be extended, it is also possible that the area will be abandoned following depletion of the soil; that depends entirely on the quality of the soil (Sicco Smit 1978)

Nevertheless, classification of land into the above four categories is a prerequisite to making any decisions and to forming any management plans, and if all rain forest is not to be sacrificed to uncontrollable degradation, the borderlines must be drawn clearly.

Intact forests must be accorded full protection to retain the full spectrum of the significance they still possess, while modified forests, agricultural and degraded lands are to be brought under good management, to make the best of what is possible on each site.

Ecological Guidelines

The question of what 'good management' can and should be has already received much attention, although by no means enough (Fig. 19.4). Two booklets in this context deserve perusal. The first, by Dasmann et al. (1973), *Ecological Principles for Economic Development*, is a masterpiece of composition and clarity, and was written for a wide public.

Dasmann and his co-authors maintain that the goals of economics and ecology are, in fact, identical and inseparable, provided they are considered in a very long-term context, for what they both aim at is the attainment of the highest possible standard of living for mankind, ad infinitum. Those efforts towards economic 'development' that did not reckon with the biological substratum of man's existence often turned out to have highly harmful ecological side effects, convertible into large liabilities on the cost/benefit account, for years and years to come. An impressive collection of such failures is given in the massive book by Farvar and Milton, *The Careless Technology* (1972). Another collection, more specifically dealing with land use and deforestation, is the book by Eckholm, *Losing Ground* (1976).

As it is, any act of 'development' means that some change is imposed on an ecosystem of vegetation + animals + soil, and for this reason is subjected to limiting conditions. Trespass of the limits results in damage, like erosion, siltation, desertification, loss of biological diversity through disappearance of species, domination by noxious or inferior species, and so on. The damage can be expressed in terms of loss of bioproductivity. Repair, if possible, requires time,

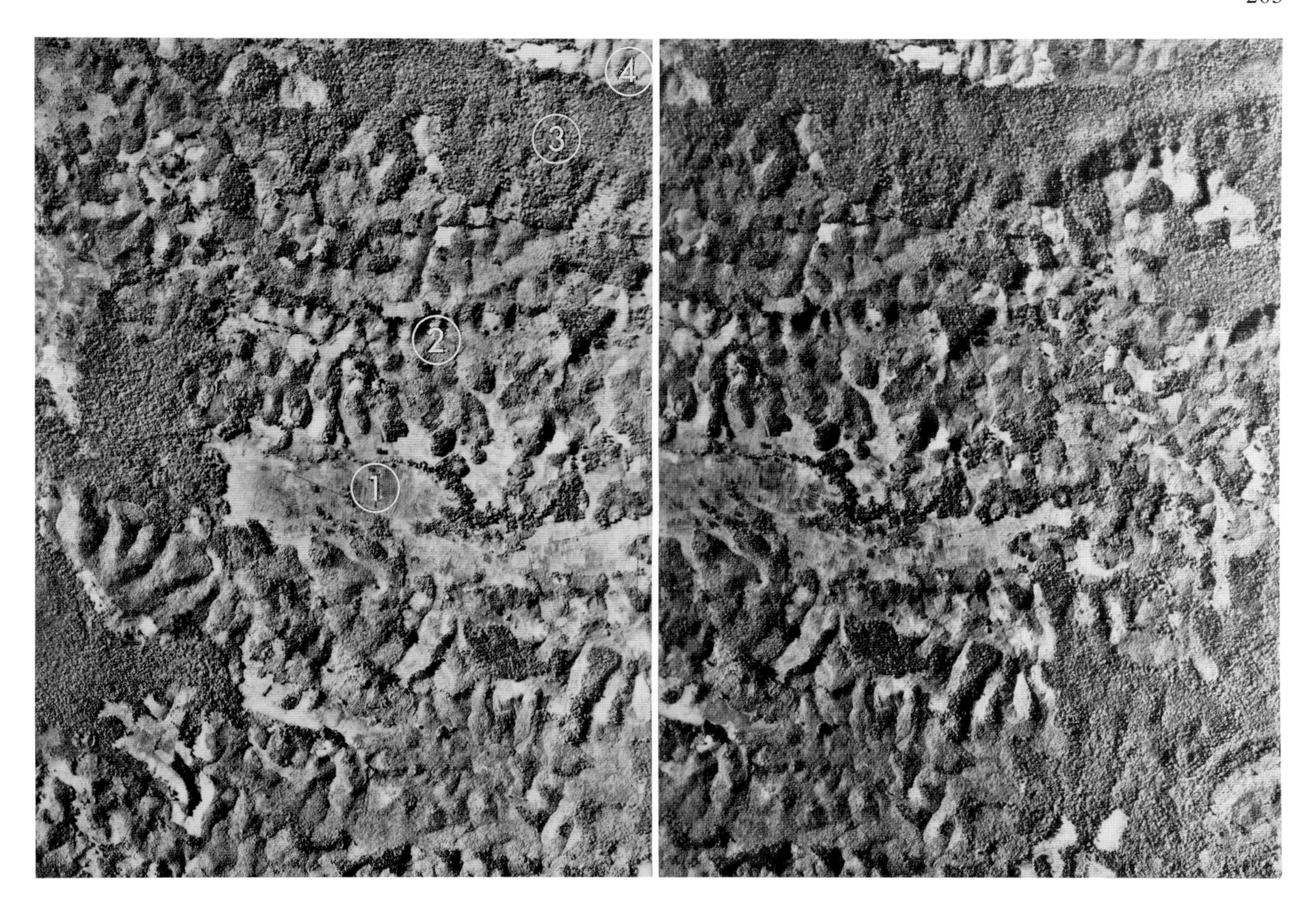

during which heavy restrictions must be imposed. These may drastically alter the pattern of expectations formerly aroused.

As Dasmann et al. (1973, pp. 22–23) explain, for any virgin tract of land, there are various options: (1) it can be left in its natural state; (2) it can be developed as a national park, open to the public; (3) it can be left largely intact while providing a limited harvest of wild vegetable or animal produce; (4) the natural vegetation and fauna can be managed for intensive harvesting; (5) part, most or all of the natural vegetation can be replaced by agricultural crops; (6) the vegetation can be completely replaced by pavement for urban, industrial or transportation purposes. One can see that the options diminish as the degree of exploitation increases: more exploitation/destruction is possible, but less is rarely so. From this emerges Dasmann's principle: "in making a decision to develop hitherto untouched land, the need to keep a range of resource use options available to future generations should be a major consideration" (p. 24). Each escalation up the exploitation scale therefore deserves careful scrutiny as to why and how the change in land use is to be imposed.

After their explanation of such general ecological principles, the authors elaborate them and supplement them for humid tropical lands, for pastoral lands in the drier regions, for tourism, for agricultural projects and for river basin projects.

The second booklet tries to make this philosophy applicable in practice. Duncan Poore, under IUCN aegis, compiled sets of *Ecological Guidelines*, for some of the world's main ecosystems. One covers the American Humid Tropics (IUCN 1975), another deals with the Malesian rain forest region (Poore 1976b).

The nucleus of both Guidelines is a list of 70–80 numbered criteria, grouped into sections. Each section contains explanations in lucid, eminently intelligible wording, as befits its author's broad background in the classics. Three cardinal rules underlie the Guidelines: wise, primary allocation to various uses, high standards in changing from one use to another, high standards of management. The titles of the sections give an idea of the subjects covered:

1. Land-use policy and allocation of land to various uses.
2. Allocation, conflicts and multiple use.
3. Preservation of natural ecosystems and genetic resources.
4. Protection forests.
5. Modification of natural forest for timber production.
6. Indigenous communities and shifting agriculture.
7. Water resources.
8. Transformation of natural forest into field and plantation crops.
9. Management of fisheries in river systems.
10. Pest control.
11. Settlement, engineering works and industry.
12. Categories of land use in relation to the conservation of flora, fauna and ecosystems.

As an example, the following text is given under "Protection forest" (Poore 1976b):

There are certain areas where the forest acts as a guardian of soil fertility, prevents erosion, regulates the run-off of water and possibly has a moderating influence on climate. It is very important to maintain these characteristics.

Guideline

Where slopes are so steep or unstable that disturbance of them would lead to soil erosion and accelerated run-off of water, the protection of these must be the primary aim of management.

Explanation:

The purpose of this is to maintain the quality of the soil in the catchment and to regulate the quantity and quality of the water delivered from it, by preventing erosion, siltation and excessive fluctuations in water flow.

Derived rule:

22. Areas must be designated whose primary function shall be catchment protection. Such areas may be used for a harvest of forest produce or other use provided this does not interfere with their primary function of protection.

Those who read this booklet must become impressed by the size and complexity of the 'development' field, and by all the possible causes of failure. No wonder Duncan Poore makes an unending plea for the use of well-balanced reason in the long-term plans of all parties concerned. But once again the slopes of Luzon come to my mind, burning while the inhabitants look on inertly, and once again I fear that ecological discipline needs an iron hand for its implementation.

Again: Minor Forest Products

If we review present-day tropical rain forest exploitation, we see a strong fixation on timber. This is a recent and curious development. Until well into the 19th century, the prevalent forest use was not for the extraction of timber, but of minor forest products. An interesting reconstruction of the activities of rain forest collectors and traders in Malaya from ca. 20,000 years ago was made by Dunn (1975). Almost throughout history, utilization of the forest itself touched it lightly and over a broad spectrum. Conscious efforts by tropical foresters to extract timber and to change the forest composition are scarcely older than a century, while mechanical logging which has so markedly speeded up this process, is even more recent.

These recent developments seem anomalous because the rain forests are so evidently unsuited to timber extraction. As we have seen in the chapter on Degradation (see Chap. 17), it is hard to think of a manner of utilization more wasteful for the present and more destructive for the future. A finely tuned, slow-motion ecosystem, a 'factory' capable of yielding hundreds of exquisite products, some of which are highly valued in international trade, is ripped open by bulldozers that pull out the very pillars supporting its roof, while quantities of scarce minerals that the ecosystem might convert into fruits, latex, medicine, spices, stimulants and other precious articles, are lost in the process. To obtain a few useful species, all others are left to their fate. It seems almost too clumsy an approach to spend any more words on . . . not to mention the waste of removing such a multi-purpose ecosystem completely to replace it by (often ephemeral) agricultural crops.

If the present type of utilization is continued, in at most half a century's time the remaining 'pillars', too, will have been extracted, the roof of our living factory will have caved in and the factory itself demolished altogether over certain areas. How is it possible that people have invoked 'economic need' to justify such stupidity?

The answer is supplied by Schumacher in *Small is Beautiful* (1973), a book which contains much food for thought and some inevitable observations. He explains that modern concepts of economy fail to distinguish between renewable and non-renewable resources. Consequently, 'economic' is any action that generates profit, whether or not the resource is renewable. Exploitation of non-renewable resources amounts to destruction of capital: which is what is happening now to the tropical rain forests. Anyone who does not choose to be blind can see that the present ways of handling them are self-defeating. In this chapter, we have sought to define the limits of exploitation, and on biological grounds have concluded that these are exceedingly tight. Since it is only in strictly protected reserves that we can hope to retain the full potential of tropical rain forests, the accessible remainder will, in fact, be given up for degradation (in biological terms) by man. But in a book like this, it still seems appropriate to outline what sort of exploitation is the most responsible towards the resource, on the strength of the same biological considerations which have guided us throughout. If exploitation is inevitable but options exist as to the manner, what is the least destructive way to follow?

The two chief constraints have been discussed already: poverty of the soil, and wide spacing of individual species, despite overall species richness. Failure to observe these constraints results in heavy, irreversible damage to the forest capital. This means that we should withdraw only those quantities of biomass whose mineral content can be replenished on a basis of ready availability. It also means that we should withdraw a little from many species, rather than draw much from a few. This points to the harvesting of minor products as an exploitation form that is much better geared to both the weaknesses and the strengths of rain forest. The weaknesses have just been mentioned as 'constraints'; the strengths consist of the unique variety of minor products which are available nowhere else.

The present style of logging and conversion seems not only disastrous, but futile: in half a century no tropical timber will have been left in the rain forests and the forest as a source of minor products will then have been destroyed along with it. New sources of timber, in the form of artificially managed stands, as well as replacements in the form of other materials, will have to be found in any case.

There are many sources of timber outside the tropical rain forests, though not of the same quality since many kinds of unique and highly valued properties only occur in tropical tree species. But a great deal of the timber demand can be met otherwise than directly from the rain forests. With minor products this will not be so easy. How are rattans, chewing gum, tropical fruits, cocoa, curare, quinine to be replaced? Growing them on plantations can be done, no doubt, if their stock is kept 'safe' in 'genetic reserves'. But there is no way to substitute most of them. So even from an 'economic' point of view, i.e. a replacement possibility, it seems wise to phase out logging in favour of the collection of minor products. Of course, the latter too can be destructive, and should never be done in fully protected forest. There can be no doubt, however, that forests exploited only for minor products will last much longer than those exploited for timber.

Apart from the ecological merit of this idea, if no attention is paid to the sources of supply of minor products, and paid quickly, many of these products will vanish totally from world markets. Essential substances required by local and western medicine and technology will cease to be available. What the effect will be, there is little way of telling. As people so far have relied on the supply, it seems unlikely that instant replacements will be available, if they can be found at all. Norman Myers in *The Sinking Ark* (1979) has worked out a few examples, but concern about minor forest products has certainly not grown in recent years. Yet, if any form of rain forest utilization is to have a future, this must change.

Agroforestry

If the soil after cultivation is exhausted and the plot is abandoned to recovery by itself, a mixture of light-loving plants develops which eventually is removed and burned. From an agricultural point of view, this mixture is usually a haphazard one, but man can influence its composition. While the annual crop is still on the land, certain tree species can be planted to take over when the harvest is over. During the fallow period, these planted species can give crops of firewood, forage, fruits, green manure and so on in many combinations, provided only that the species planted belong to this particular stage of succession.

In Burma this system has been known for a century, under the name *taungya*. In Java it is called *tumpangsari*. Many variations have been worked out, in combination with homestead gardens. The repertoire of combinations of agriculture and forestry was publicized afresh during the 1970s whereby King (1979) played an important role. An institute has been set up in Nairobi, the International Council for Research in Agro Forestry (ICRAF).

Agroforestry thrives on the hope that the long fallow period vital to successful traditional shifting cultivation can now be made productive. One method is to plant, under local management, certain fast-growing useful tree species adapted to the generally poor soils in rain forest regions. Small domestic animals can also be incorporated in the system and a recycling of minerals attempted. If solutions exist for the horrendous problem of shifting cultivation, they will be along these lines, but they require knowledge of the local possibilities and of the species that suit them, nurseries for planting material, funds to prepare the soil and to encourage local people to take up this work and knowledge of the stability of infrastructure, administration, economy and population size. Finally, a clear view of the future and responsibility towards that view must be accepted and believed by the people concerned.

Forestry and Rural Development

In virtually all tropical countries, forestry has been a centralized affair from the (colonial) beginning, and national autonomy has kept it that way. Forest policy is decided by the government in the nation's capital; from there it filters down to the provinces, the districts, the villages; no one should be surprised that the general public is basically indifferent to the fate of the forest in its vicinity. Only in recent years have pleas been heard for small-scale, local participation, for devolving responsibility for abiding by or even creating forestry codes of practice onto the people who live near the forests. FAO has recently issued two important publications in this line, one technical paper, *Forestry for local Community Development* (1978), and the other a more 'popular' booklet called *Forestry for Rural Communities* (1979). Eric Eckholm also wrote a splendid little book, *Planting for the Future*, in 1979. Thus, a discipline is emerging that could be called 'social forestry'. The aim is the involvement of the local people, so that through participation human society will become closely integrated with its semi-natural environment, thus lessening the risk of (further) degradation which would threaten both. Successful 'social forestry' is economic good sense. Degradation precludes self-reliance and people in a degraded environment can only become a burden on others.

One of the most urgent concerns of social forestry in the tropics is the supply of firewood needed for cooking at the rate of about 1 m^3 per person annually. In the humid tropics this should be no problem if the population is sparse and planting is done regularly. Fast-growing species exist that can be harvested in 10 years; yet a family of five needs a hectare of such trees. This determines to a great extent the carrying capacity of the land. When pressures mount, the people collect firewood from the wild, in Java on the volcano slopes, with enormous erosion ensuing. In rain forest, wood is converted into charcoal, for the same purpose; this was already discussed under the Values of the Forest (see Chap. 16).

Firewood problems received ample attention in the forestry literature as early as the 1920s, but before the publication of Eckholm's compelling book, *Losing Ground*, in 1976, few governments took an active interest in the matter.

Besides agroforestry and the provision of firewood, the adequate control of erosion, the supply of water for agricultural and domestic use, the supply of timber and minor forest products and the rehabilitation of degraded land, are all regarded as the proprietary concerns of any forest department. No biologist should therefore be surprised if little time for discussion is made available to him by foresters unless his points are well made.

Development Assistance

Nearly all rain forests are situated in countries that nowadays receive development assistance, technical cooperation or overseas aid; there are many terms. It is therefore an absurdity when at the same time such countries allow their irreplaceable resources to be depleted. But false expectations have been aroused, and immense financial and political interests have been formed for whose miscalculations the environment generally pays. Curry-Lindahl, who is a highly respected expert, has even asked the question "Is aid for developing countries destroying their environment?" (1979), and on a solid basis of documentation, has answered "yes".

This was to be expected. 'Development' has come to mean a tapping of natural resources for an immediate increase in profit. The other course theoretically open, namely to put a check on population growth first of all, and to streamline existing production processes so that waste can be avoided, has nowhere been made. In most development agencies, ecological constraints are considered grudgingly and with suspicion. Even in The Netherlands, it took more than 5 years of lobbying before an advisory committee of ecologists to the Minister of International Development Cooperation was established, early in 1982. Ecology will have to take a more prominent place, however, if only from sheer and painful necessity, and this is now already perceived in several tropical countries though the mood needs strengthening. I well remember a conversation with a businessman from Japan with whom I shared a taxi, in Port Moresby, in 1973. Asked for his view on the ecological future of Malesia, he answered: "But we need these raw materials!" I have often wondered since, if Indonesia should not deliberately avoid being too attractive a resource.

The whole forestry sector needs strengthening, in nearly all tropical countries, with talented, incorruptible, and dedicated people. Only by an enormous effort will it be possible to stop the avalanche of environmental and social disasters that are already striking many rain forest areas, of which population growth and the increase of degraded land are only the most visible, and most easily quantifiable.

One eloquent document that makes a case for the strengthening of forestry is a *Sector Policy Paper* issued by the World Bank in 1978 which expresses the intention of giving more support to ecologically responsible forestry projects. The manifold economic significance of forests in general is explained concisely and with clarity, and the main points of many current misunderstandings about utilization are covered. Great differences, the paper says, must be taken into account with regard to the:

1. Abundance of natural resources, by country;
2. Favourability of climate;
3. Population density, in cities, and outside;
4. Main types of land use;
5. Properties of the forests;
6. Level of education of the people;
7. Main source of fuel for cooking;
8. Ways of influencing the forest.

These differences add up to a rough classification, of regions characterized by:

1. Poor soils with little forest;
2. Potential for reforestation;
3. Overpopulation, little forest;
4. Economic poverty, but much forest;
5. Much forest, but high population pressure;
6. Economic wealth and much forest.

Each of them requires a different approach and planning.

The World Bank sees possible forestry projects in the following categories: (1) rural development, (2) management of natural forest and industrial plantations, (3) infrastructure improvement and logging, (4) industrial processing of wood, (5) improvement of institutions and staff. Each of these five categories is in itself large and complex.

But even though a welcome breakthrough, this *Sector Policy Paper* is still ambivalent. On the one hand, there is a desire to maintain the forest, and the World Bank is known for the severity of the environmental safety checks on its projects. It is also one of the rare agencies with an eye on the moderately far future: 50 years is now said to be its standard. On the other hand, there is an expressed willingness to finance ports, roads and railways "which are needed to open up new forest lands for exploitation".

Furthermore, the paper declares that all forests can (p. 10) regenerate themselves, (p. 15), without bothering to mention the fragility of tropical rain forest or the great differences between forest types. Biological realities are thus ignored, as others ignore them by regarding a forest as merely a quarry for timber (e.g. Lanly and Clement 1979).

Of course, it takes time to formulate an adequate vision, and to make it work. Even without bureaucratic obstruction, the processes of consultation, decision-making, financing, execution and evaluation are long. Most problems in development have turned out to be far more complicated than first assumed. By the time the first projects have actually alleviated exploitation pressure on the natural forests, the world's rain forest area may well again have been halved. In every year of delay, after all, some 150,000 km^2 is destroyed.

The Position of the Biologist: A Summary

The assumption that everything on earth exists for man keeps many people from speaking up for the rain forests and in general makes them behold the degradation of nature with an acquiescent eye. So deeply is the assumption embedded in man's thought and action that touching it evokes anxiety and defence reactions: if this certainty succumbs, what others will follow?

It is curious that man has created a system of morality which places him outside the context of living nature or creation: is it only man who has 'rights'? Insofar as rights are conferred on species of animals and plants, this is done incidentally. No general, fundamental, incontrovertible ethical or legal principle exists that could induce man to retreat in any conflict with nature. But at the same time, man's own existence presumes and depends on the rightless, continued existence of animals and plants. It is therefore most desirable to regulate the relations between man and nature, and one of the most impressive efforts to that effect is the *World Conservation Strategy* (IUCN 1980) even though it evokes no higher authority than that of human interest.

The plunder and destruction of tropical rain forest is 'justified' by a number of 18th century concepts, which in our times became common – although outdated – property. There is, in fact, no method or technique for exploitation of tropical rain forest which provides for the retention of all species ad infinitum within the context of a cash economy. Nor is there a clear vision of the future of rain forest countries which might serve as a basis for development. It is strange then, in absence of such a vision, to hear pleas for 'development as a necessity' all around. Into this vacuum, the rain forest falls – literally.

In addition, we must face the fact of species extinction. All damage inflicted on the rain forests can eventually be expressed in numbers of lost species. One can view this as the destruction of God's work, or as the effect of human stupidity, it amounts to the same decrease in future resources. What matters here is the position of the biologist, as a witness, a spokesman and an agent for change.

Biologists can make a real contribution to the course of world affairs. This contribution is threefold. Biologists:

1. **Know About the Biological Substratum** of man's existence, and the laws of nature that operate in it. They have access to all that is known about the variety of plants and animals on earth and can estimate the carrying capacity of ecosystems, predict the effect of damage and give prognoses for possible recovery.
2. **Approach Problems in an All-Inclusive Manner.** They have learned to look at all factors that influence the workings of the biological substratum. Unlike foresters, who deal with a handful of species only, biologists will never exclude any one species arbitrarily in discussing conservation priorities. In principle, all biological relationships concern him.
3. **Are Acutely Aware of the Time Factor.** Biologists have been trained to survey periods of time from very short to very long. They can think in greater detail than geologists can, but also more broadly and with more regard to the distant future than economists and foresters, or even historians. Their scale of time coincides most closely with that of geographers, but the latter cannot be supposed to have a working knowledge of the biological substratum, nor of evolution as a reality.

By virtue of these three qualifications, the biologists are professionally qualified to speak and act on behalf of the forest. Who else is in that position?

This imposes on a biologist the obligation to uphold under any circumstances, the criteria of his discipline which are the criteria of the forest itself. In observing them, he cannot overcome all eight disastrous factors, but he can ask critical questions, and give good advice.

But rarely or never does a biologist have the chance to decide the fate of a certain piece of land with its plants and animals. This is the task of 'authorities': persons invested with power derived from the highest power in the country. This highest power bears the ultimate responsibility for all decisions with regard to the use of land and other natural resources, and for their correct implementation. Others, later generations, will have to judge their wisdom or lack of it.

This separation of responsibilities is of essential importance, particularly in cases where a biologist may hesitate to say what he should say. Will he be obsequious for fear of being ridiculed? Or does he feel guilty, if he is a citizen of a rich country, towards the poor citizens of developing or undeveloped countries in the tropics, and therefore tells them what he thinks they want to hear? By such behaviour such a biologist would, in my opinion, put his discipline to shame. If because of his advice, certain species of plants and animals are sacrificed, and the biological substratum on which people depend therefore narrowed, who in a tropical country would continue to place trust in a biologist?

The truth is often hard to swallow. A biologist who specialized in tropical rain forest ecology has chosen a discipline whose facts and discoveries are not welcomed by everybody. This may be a reason to communicate them in a careful manner, with respect and conviction, but never to conceal them.

The Sacred Forest

In June 1860, shortly after publication of Darwin's *Origin of Species*, evolutionists and theologians met at Oxford in a public debate. This debate, as the evolutionists assure us, ended in a resounding victory for the evolutionists. So it became accepted doctrine that evolution is an autonomous process. Out of less organized forms, more organized forms are derived by purely natural forces, without the least involvement of a superior power. Throughout my own education I met with nothing but implicit denials of the existence of such a power. In no publication about the rain forest did I come across any considerations that went further than Corner's already quoted: "I measured my insignificance against the quiet majesty of the trees".

Yet I wonder whether in that way justice has been done to the real majesty of a tropical rain forest. Do we not experience that such a forest is *more* than a complicated ecosystem which can be described in strictly scientific terms? Isn't there a dimension of splendour, of indifference, of danger and intrigue that strikes us? Don't we see a rain forest as a reality in which both attraction and repulsion form a disquieting blend?

A tropical rain forest *has* something to convey to people, perhaps more than I originally thought. What it has to convey may not at all be clear at first. But in the course of time, when one has learned to look and has sorted out the multitude of impressions for oneself, a relationship with the forest is built up. 'Knowing' a forest is like 'knowing' a city. To grasp all that happens is impossible, yet a good deal of understanding and meaningful knowledge can be gathered as a basis for what we do, good or bad. In the process of getting acquainted, this relationship develops according to the interest, capacity, keenness, curiosity and fantasy of the person, and to such a relationship there is no end. New features constantly claim one's attention, while old ones assume a new significance.

Many people, of course, will remain indifferent to the rain forest, just as many stay indifferent to classical music. Such affinities or resistances are largely inborn. But he who has a musical disposition cannot help listening attentively. He memorizes the music and is aware that in a distinctive way it shapes his personality. He comes to recognize himself in the music, takes up playing, practices, makes progress, wants to know more about it, comes to recognize musicality in others; thus, music becomes part of his life. He then realizes that he takes part in something supra-personal, from which he receives more than he gives, an affair with terrible difficulties, to be sure, but which, once these are overcome, results in something sublime. To an unmusical person all this is meaningless. But that does not deny the nature of music.

Likewise you must have a sort of 'musicality' with regard to the tropical rain forest. He who hears and reads about it, studies pictures, feels puzzled and wishes to know how it is to be there is preparing himself for the great moment. He realizes that the rain forest can become something of meaning in his life. These experiences keep him busy in all sorts of places and moments; he is unable to shed them. Through them, he learns to see the world in a different light, and thereby feels himself growing in wisdom: capable of better acts. He has come to participate in a cause of higher order, worthy of serving.

Why do rain forests exist? This is a scientifically inadmissible question, but nevertheless it keeps people intrigued. If it is reasonable to be interested in our own origin and destiny, is it possible to leave that of the rain forests unconsidered? To me, it is not. Could the Great Architect of the Universe have nourished man with such loving attention and never bothered about the great forests?

Whatever the origins of the forests or of man, it seems an absurdity to presume that the rain forests evolved through aeons only to meet their demise at man's hands. He who is neither a hunter-gatherer nor a biologist often only defiles and impoverishes the forest. Everyone should try to understand the rain forests, and appreciate their products, if possible without damage. But we can only understand the forests if we accept them as they are. No one can *improve* an undisturbed rain forest. It is a culmination point of creation, as precious, worthy and sacred as any in the universe as we know it.

Last Look

This is the end of our journey to the tropical rain forests. It is the morning before departure. Take a last walk into the forest and sit down, anywhere, silent and curious, for a couple of hours, just to look around at what happens. It won't be much. But imagine every event in repetition, 10 times, a 100 times, a 1000 times, 10,000 times.

A large leaf comes down. It bears a cover of tiny mosses and lichens. The litter will be enriched through its decomposition with a tiny quantity of inorganic matter. Almost at the ground's surface roots wait for this nourishment. Ants are consuming a dead lizard; from this spot a steady column marches up the stem of a treelet to disappear into a hole near the tip of a slightly thickened twig. The plant takes advantage of the leftovers of the nourished ants. Very slowly, a shaft of sunlight moves, touching a few seedlings. They are already a bit higher than those in the shade.

A stinkhorn grows up; in half a morning it has unfurled a white veil under the dark green mass of spores. A carrion beetle lands on it and walks there for a time, then flies away. When, later in the morning, the forest sounds ebb away a bit, the heavy hum of the flying hornbills can be better heard. Their nest is somewhere in the hollow of a tree. High up in the canopy they feast on fruits whose remnants fall down like hailstones. But some seeds the birds will ingest and then drop them elsewhere in their faeces, thus another tree's progeny is spread. A band of monkeys whoops in the distance; the sound seems to be louder now than half an hour ago. And between the huge stems of the forest trees, a hand-sized butterfly moves, dancingly, alighting for a brief moment on the flowers of some dioecious plant.

This is about all that happens during those few hours. But each of those trifling movements in the great tranquil forest is an act that will be repeated ten thousand times, each with its effect on the processes we have witnessed, on the process . . . yes, of Creation itself. Because of the incredible gifts bestowed on us as human beings, we have been able to linger a moment in the corridors of time, but part of this Act of Creation lies ahead: far, far beyond our personal future.

References

Acosta-Solís, M. (1968) Division fitogeográfica y formaciones geobotánicas del Ecuador. Casa de la Cultura Equatoriana, Quito. 307 p

Akker, A., Groeneveld, W. (1984) Reconnaissance of Amazon rainforest architecture and insect habitats. LH Bosteelt, Wageningen, rapp. 84–19. 95 p

Alder, D., Brünig, E.F., Heuveldop, J., Smith, J. (1979) Struktur und Funktionen im Regenwald des internationalen Amazon-Oekosystemprojektes. Amazoniana 6(4): 423–444

Allen, P.H. (1956) The rain forests of Golfo Dulce. University of Florida Press, Gainesville. 417 p

Alvim, P. de T. (1977) The balance between conservation and utilization in the humid tropics with special reference to Amazonian Brazil. In: Prance & Elias (eds), Extinction is forever. New York Botanical Garden, New York. pp 347–352

Amadon, D. (1973) Birds of the Congo and Amazon forests, a comparison. In: Meggers (ed) Tropical forest ecosystems. Smithsonian Institute, Washington. pp. 267–277 2 fig

Anderson, J.A.R. (1963) The flora of the peat swamp forests of Sarawak and Brunei, including a catalogue of all recorded species of flowering plants, ferns and fern allies. Gard Bull 20: 131–228 1 map + 10 phot

Anderson, J.A.R. (1975) Illipe nuts (*Shorea* spp.) as potential agricultural crops. In: Williams et al. (eds) South East Asian plant genetic resources. Biotrop, Bogor. pp 217–230

Anonymous (1977) Peat and podzolic soils and their potential for agriculture in Indonesia. Soil Research Institute, Bogor. 198 p

Anonymous (1980) Verstandig met hout snÿdt hout. Brochure, Amsterdam/Renkum

Ashton, P.S. (1964) Ecological studies in the mixed dipterocarp forests of Brunei State. Clarendon Press, Oxford. 75 p + 37 phot + 20 tab + 70 fig

Ashton, P.S. (1969) Speciation among tropical forest trees: some deductions in the light of recent evidence. Biol J Linn Soc 1: 155–196, 14 fig

Ashton, P.S. (1973) The biological significance of complexity in lowland tropical rain forest. J Ind Bot Soc 50A: 530–537

Ashton, P.S. (1976a) An approach to the study of breeding systems, population structure and taxonomy of tropical trees. In: Burley & Styles (eds) Tropical trees. Academic Press, London. pp 35–42, 2 fig

Ashton, P.S. (1976b) Factors affecting the development and conservation of tree genetic resources in South-east Asia. In: Burley & Styles (eds) Tropical trees. Academic Press, London. pp. 189–198, 2 fig

Ashton, P.S. (1982) Dipterocarpaceae. Fl Mal I, 9: 237–552

Aublet, F. (1775) Histoire des plantes de la Guiane Françoise. Pierre-François Didot jeune, Londres, Paris. 4 vol

Aubréville, A. (1949) Climats, forêts et désertification de l'Afrique tropicale, Société d'éditions geografiques, maritimes et coloniales, Paris

Aubréville, A. (1950) Flore forestière Soudano-Guinéenne. Societé d'éditions géographique, maritimes et coloniales, Paris. 523 p

Aubréville, A. (1959) La flore forestière de la Côte d'Ivoire, 2me éd. CTFT, Nogent-sur-Marne. 3 vols

Aubreville, A. (1961) Etude écologique des principales formations végétales du Brésil. CTFT, Nogent-s-Marne. 265 p

Ayensu, E.S. (ed) (1980) Jungles. Cape, London. 200 p

Backer, C.A. & R.C. Bakhuizen van den Brink Jr (1963–1968) Flora of Java, 3 vol. Noordhoff Wolters, Groningen

Balgooy, M.M.J. van (1976) Phytogeography. In: Paijmans (ed) New Guinea vegetation. Anu Press, Canberra. pp. 1–22, 7 fig

Balgooy, M.M.J. van, I.G.M. Tantra. (1986) The vegetation in two areas in Sulawesi, Indonesia. Bul Penelitian Hutan (For Res Bull) special ed. 61 p

Balick, M.J. (1979) Amazonian palms of promise. Econ Bot 33: 11–28, 6 fig

Barney, G.O. (1980) The global 2000 report to the president. Government Printing Office, Washington. 3 vol

Barrera, A., A. Gomez-Pompa & C. Vazquez-Yanes. (1977) El manejo de las selvas por los Mayas: sus implicaciones silvicolas y agricolas. Biotica 2(2): 47–61

Bartels, M. & H. (1937) Uit het leven der neushoornvogels. Trop Natuur 26: 117–127, 140–147, 166–173, 15 phot

Bartlett, H.H. (1956) Fire, primitive agriculture, and grazing in the tropics. In: Thomas (ed) Man's role in changing the face of the earth. University of Chicago, London. pp 692–720

Baur, G.N. (1966) The ecological basis of rainforest management. The Forestry Commission of New South Wales, Sydney. xiv + 499 p

Beard, J.S. (1944) Climax vegetation in tropical America. Ecology 25: 127–158

Beccari, O. (1904) Wanderings in the great forests of Borneo. Travels and researches of a naturalist in Sarawak (Original title: Nelle foreste di Borneo, publ. 1902). Constable, London. xxiv + 424 p

Becker, G. (1976) Concerning termites and wood. Unasylva 111: 2–11, 14 illus

Berg, J.H. van den (1977) Gedane zaken. Twee omwentelingen in de westerse geestesgeschiedenis (Affairs concluded. Two revolutions in the history of western ideas) Callenbach, Nijkerk. 400 p, many illus

Biswas, K. (1942) Address of welcome on the 150th anniversary of the Royal Botanic Garden, Calcutta (1787, 1937) Anniversary vol, pp 1–12, illus

Bodegom, van (1981) (Author's reference: untraced)

Boerboom, J.H.A. & K.F. Wiersum (1983) Human impact on tropical moist forest. In: Holzner, W. et al. (eds) Man's impact on vegetation. Junk, The Hague. pp 83–106

Boom, B.M. & S.A. Mori. (1982) Falsification of two hypotheses on liana exclusion from tropical trees possessing buttresses and smooth bark. Bull Torrey Bot Club 109: 447–450

Botting, D. (1973) Humboldt and the cosmos. Sphere Books, London. 295 p

Bourlière, P. (1973) The comparative ecology of rain forest mammals in Africa and tropical America: some introductory remarks. In: Meggers (ed) Tropical forest ecosystems. Smithsonian Institute, Washington. pp 279–292, 4 fig

Braak, C. (1945) On the climate of and meteorological research in the Netherlands Indies. In: Honig & Verdoorn (eds) Science and Scientists in the Netherlands Indies. Board for the Netherland Indies, Surinam and Curacao, New York City. pp 15–22, fig 8–10

Brandis, D. (1906) Indian trees. An account of trees, shrubs, woody climbers, bamboos and palms, indigenous or commonly cultivated in the British Indian Empire. Archibald Constable, London. xxxiv + 767 p, 201 fig.

Brenan, J.P.M. (1963) The value of Floras to underdeveloped countries. Impact 13: 121–146

Brezhnev, D. (1970) Mobilization, conservation and utilization of plant resources at N.I. Vavilov All-Union Institute of Plant Industry, Leningrad. In: Frankel & Bennett (eds) Genetic resources in plants. Blackwell, Oxford Edinburgh. pp 533–538

Briscoe, C.B. (1983) Integrated forestry-agriculture-livestock land use at Jari Florestal e Agropecuária. In: Huxley, P.A. (ed) Plant research and Agroforestry. chap 5 ICRAF, Nairobi, Kenya

Brown, K.S. Jr. (1977) Centros de evolção, refúgios quaternários e conservação de patrimônios genéticos na regiaô neotropical: padrões de diferenciação em Ithomiinae (Lepidoptera: Nymphalidae). Acta Amazonica 7 (1): 75–137

Brown et al. (1976) Twenty-two dimensions of the population problem Worldwatch Paper, 5

Brown, W.H. (1919) Vegetation of Philippine mountains. The relation between the environment and physical types at different altitudes. Bureau of Science, Manila. 434 p, 30 fig + 41 pl

Browne, F.G. (1955) Forest trees of Sarawak and Brunei. Government Printing Office, Kuching, Sarawak. iv + 369 p + xviii

Brune, A. & G.H. Melchior (1976) Ecological and genetical factors affecting exploitation and conservation of forests in Brazil and Venezuela. In: Burley & Styles (eds) Tropical trees. pp 203–215, 4 fig

Brünig, E.F. (1977) The tropical rain forest – A wasted asset or an essential biosphere resource? Ambio 6: 187–191, 4 fig

Brünig, E.F. (1980) Leistungssteigerung, Nachhaltigkeit und Bestandesstruktur agroforstwirtschaftlicher Oekosysteme in den Feuchttropen. Mitt Bundesforschungsans Forst-Holzwirtsch, Hamburg 132: 27–58

Budowski, G. (1965) Distribution of tropical rainforest species in the light of successional process. Turrialba 15: 40–42

Budowski, G. (1976) Why save tropical rain forests? Some arguments for campaigning conservationists. Amazoniana 5: 529–538

Budowski, G. (1978) Agro-forestry: a bibliography. CATIE, Turrialba, Costa Rica

Burger, D. (1972) Seedlings of some tropical and shrubs mainly of South East Asia. Pudoc, Wageningen. 399 p, 155 fig

Burgess, P.F. (1971) The effect of logging on hill dipterocarp forests. Malay Nat J 24: 231–237, 1 fig + pl 65–70

Buringh, P. (1970) Introduction to the study of soils in tropical and subtropical regions, 2nd edn. Pudoc, Wageningen. 99 p, illus

Burkill, I.H. (1935) A dictionary of the economic products of the Malay Peninsula. Crown Agents for the Colonies, London. xi + 2402 p 2nd edn. Ministry of Agriculture, Kuala Lumpur 1966 xiv + 2444 p.

Burley, J. & B.T. Styles (eds) (1976) Tropical trees. Variation, breeding, conservation. Academic Press, London. 243 p

Burtt, B.L. (1977) Notes on rain-forest herbs. Gard Bull 29: 73–80

Cañadas Cruz, L. (1983) El mapa bioclimático y ecológico del Ecuador. Publicacion bajo auspicio del Banco Central del Equador, Quito. 210 p

Cañadas Cruz, L. (1965) Los bosques pantanosos en la zona de San Lorenzo. Turrialba 15(3): 225–230
Carte internationale du tapis végétal et des conditions écologiques à 1/1.000.000 (Institut français; Pondichéry 1961–1966), by H. Gaussen et al. Cape Comorin (1961), Ceylon (1964, notice de la feuille 1965), Mysore (1964, notice de la feuille 1966), Bombay (1966)
Centeno, J.C. (1984) El recurso forestal y el drama económico de la América latina. IFLA, Mérida. part 1: 31 p, part 2: 55 p
Chandrasekharan, C. (1977) A report on the forestry situation in Indonesia. FAO, Bangkok. 95 p
Chin, S.C. (1977) The limestone hill flora of Malaya. I. Gard Bull Singapore 30: 165–291
Chin, S.C. (1979) The limestone hill flora of Malaya. II. Gard Bull Singapore 32: 64–203
Chin, S.C. (1982) The limestone hill flora of Malaya. III. Gard Bull Singapore 35: 137–190
Chin, S.C. (1983) The limestone hill flora of Malaya. IV. Gard Bull Singapore 36: 31–91
Cohen, H. (1939) De economische betekenis der boschbijproducten van de buitengewesten (in Dutch, summary in English). The economic significance of minor forest products of the outlying areas. Tectona 23: 883–919
Cook, C.F. (1909) Vegetation affected by agriculture in Central America. Bureau of Plant industry Bull 145, USA
Corner, E.J.H. (1940) The wayside trees of Malaya. Government Printing Offices, Singapore. VII + 772 p, 269 fig + 228 phot
Corner. E.J.H. (1949) The durian theory or the origin of the modern tree. Ann Bot 13: 367–414, 36 fig
Corner, E.J.H. (1952) Wayside trees of Malaya. 2nd edn. Government Printing Office, Singapore. 2 vols
Corner, E.J.H. (1953) The durian theory extended – I. Phytomorphology 3: 465–476
Corner, E.J.H. (1954a) The durian theory extended – II. The arilate fruit and the compound leaf. Phytomorphology 4: 152–165
Corner, E.J.H. (1954b) The durian theory extended – III. Pachycauly and megaspermy – Conclusion. Phytomorphology 4: 263–274
Corner, E.J.H. (1963) Why *Ficus*, why Moraceae? Fl Males Bull (18) 1000–1004
Corner, E.J.H. (1964) The life of plants. Weidenfeld & Nicolson, London. xii + 315 pl, 103 fig + 41 pl
Corner, E.J.H. (1966) The natural history of palms. Weidenfeld & Nicolson, London. 393 p, 133 fig + 24 pl
Corner, E.J.H. (1970) *Ficus*. Identif Lists Males Spec (37) 537–648
Corner, E.J.H. (1976) The seeds of dicotyledons, 2 vol. Cambridge University Press, Cambridge
Corner, E.J.H. (1978) The freshwater swamp-forest of South Johore and Singapore. Botanic Gardens, Singapore. ix + 266 p, 6 tab, 18 fig + 40 phot
Cremers, G. (1973) Architecture de quelques lianes d'Afrique tropicale. Candollea 28: 249–280
Cremers, G. (1974) Architecture de quelques lianes d'Afrique tropicale. Candollea 29: 57–110
Curry-Lindahl, K. (1972) Conservation for Survival. An ecological strategy. Gollancz, London. xiv + 335 p
Curry-Lindahl, K. (1975) Is aid for developing countries destroying their environment? (Author's reference: untraced)
Curry-Lindahl, K. (1978) Background and development of international conservation organizations and their role in the future. Environm Conserv 5: 163–169
Dale, I.R. and Greenway, P.J. (1961) Kenya trees & shrubs. Buchanan's Kenya Estates, Nairobi, in association with Hatchards, London. xxvii + 654 p
Dammerman, K.W. (1935) The quinquagenary of the Foreigners' Laboratory at Buitenzorg, 1884–1934. Ann Jard Bot Buitenz 45: 1–60
Darwin, C. (1905) The movements and habits of climbing plants. Murray, London. ix + 208 p, 13 fig
Dasmann, R.F. et al. (1973) Ecological principles for economic development. Wiley, London. ix + 252 p, 31 fig
Davis, D.D. (1962) Mammals of the lowland rain-forest of North Borneo. Bull Nat Mus St Singapore 31: 1–129
Davis, T.A.W. & P.W. Richards. (1933) The vegetation of Moraballi Creek, British Guiana. Part I. J Ecol 21: 350–384
Davis, T.A.W. & P.W. Richards. (1934) The vegetation of Moraballi Creek, British Guiana. Part II. J Ecol 22: 106–155
Descourtilz, J. Th. (1960) Pageantry of Brazilian birds (reprint from 1855). Colibris ed. Ltda, Rio de Janeiro, Amsterdam. 60 paintings
Djajapertunda, S. (1978) The utilization of tropical hardwoods from Indonesia. Masyarakat Perkayuan Indonesia, Jakarta. iii + 63 + 5 p
Dobby, E.H.G. (1960) Southeast Asia. University of London Press. 415 p, 118 fig
Dransfield, J. (1974) A short guide to rattans. BIOTROP, Bogor. 69 p
Dunn, F.L. (1975) Rain forest collectors and traders. A study of resource utilization in modern and ancient Malaya. Monogr Mal Br Roy As Soc 5. Kuala Lumpur. xii + 151 p
Earl, D. (1975) A renewable resource of fuel. Unasylva 110: 21–26, phot
Eckholm, E.P. (1976) Losing Ground: Environmental stress and world food prospects. Norton, New York. 223 p
Eckholm, E.P. (1978) Disappearing species: the social challenge. Worldwatch Institute, Washington D.C. 38 p
Eckholm, E.P. (1979) Planting for the future: forestry for human needs. Worldwatch Institute, Washington. 64 p
Eddowes, P.J. (no date, actually 1978) Commercial timbers of Papua New Guinea. Their properties and uses, Department of Primary Industry, Port Moresby. 195 + xiv p

Eggeling, W.J. (undated, 1952) The indigenous trees of the Uganda Protectorate. The Government Printer, Entebbe. xxx + 490 p

Ellenberg, H. (1975) Vegetationsstufen in perhumiden bis perariden Bereichen der tropischen Anden. Phytocoenologia 2: 368–387

Endert, F.H. (1928) Geslachtstabellen voor Nederlandsch-Indische boomsoorten naar vegetatieve kenmerken (Genus tables of tree species from the Netherlands Indies according to vegetative characteristics). Veenman, Wageningen. ix + 242 p

FAO (1978) Forestry for local community development. FAO, Rome. 114 p

FAO (1979) Forestry for rural communities, FAO, Rome. 56 p, illus

FAO-Unesco (1979) Soil map of the world / 1: 5.000.000, vol. 9 Southeast Asia. UNESCO, Paris. xiv + 149 p + 7 b/w maps + col map

FAO (1981) Tropical forest resources assessment project. FAO, Rome. 4 vol

FAO (1982) (Author's reference: untraced)

Farvar, M.T. & J.P. Milton (eds) (1972) The careless technology. Natural History Press, Garden City, New York

Fiard, J.P. (1979) La forét martiniquaise. Parc National Regional, Ex-Caserne Bouillé, Fort de France. 67 p

Figuereido, J. & Cals, C. (1982) Fitogeografia Brasileira: classificacao fisionómica-ecolôgica de vegetacão neotropical. Bol Tec RADAMBRASIL, Sér. Vegetacão neotropical no 1. 85 p

Flenley, J.R. (1979) The equatorial rain forest: a geological history. Butterworths, London, Boston. viii + 162 p, illus

Fontanel, J. & A. Chantefort (1978) Bioclimats du monde Indonésien. Travaux de la section scientifique et technique, Inst fr de Pondichéry, Pondichéry. 102 pp

Fox, J.E.D. (1969) Climbers in the lowland dipterocarp forest of Sabah. Commonw For Rev 48: 196–198

Fox, J.E.D. (1976) Constraints on the natural regeneration of tropical moist forest. For Ecol Managem 1: 37–65, 4 phot

Francis, W.D. (1951) Australian rain-forest trees. 2nd edn. Forestry & Timber, Australia. xvi + 469 p

Furtado, J.I. (1979) The status and future of the tropical moist Forest in Southeast Asia. In: MacAndrews & Chia (eds) Developing economies and the environment/The Southeast Asian experience. McGraw-Hill, New York. pp 73–120

Gamble, J.S. (1902) A manual of Indian timbers. 2nd edn. Sampson Low, Marston, London. xxvi + 856 p (Reprinted 1922, London, 1972, Dehra Dun, India)

Gan, Y.Y. (1977) Genetic variation in the wild populations of rain forest trees. Nature 269: 323–324

Gentry, A.H. (1977) Endangered plant species and habitats of Ecuador and Amazonian Peru. In: Prance & Elias (eds) Extinction is forever. New York Botanical Garden, New York. pp 136–149, 3 fig

Gomez-Pompa, A. & C. Vazquez-Yanes (1976) Regeneración de selvas. Comp Ed Continental, México. 676 p

Good, R. (1974) The geography of the flowering plants, 4th edn. Longman, London. xvi + 557 p

Goodland, R.J.A. & H.S. Irwin (1975) Amazon jungle: from green hell to red desert?. An ecological discussion of the environmental impact of the highway construction program in the Amazon Basin. Elsevier, Amsterdam. ix + 155 p, 7 fig

Goodland, R.J.A. & H.S. Irwin (1977) Amazonian forest and cerrado: development and environmental considerations. In: Prance & Elias (eds) Extinction is forever. New York Botanical Garden, New York. pp 214–233, 4 fig

Gourou, P. (1958) The tropical world / Its social and economic conditions and its future status, 2nd edn. Longmans, Green & Co. London, New York, Toronto xii + 159 p, 16 fig + 39 phot

Graaf, N.R. de (1986) A silvicultural system for natural regeneration of tropical rain forest in Suriname. Thesis, Agric University, Wageningen. ?? + 250 p

Granville, J.J. de (1978) Recherches sur la flore et la végétation guyanaises. D. Sc. thesis, Montpellier University. 272 p

Gray, B. (1978) Pests and diseases in forests and plantations. In: UNESCO, Tropical forest ecosystems. UNESCO, Paris. pp 286–314

Grenand, F. (1982) Et l'homme devint jaguar. L'Harmattan, Paris. 427 p

Grubb, P.J. (1974) Factors controlling the distribution of forest-types on tropical mountains: new facts and a new perspective. In: Flenley J.R. (ed) Altitudinal zonation in Malesia. Dept of Geography, University of Hull, Misc ser 16. pp 13–46

Guillaumet, J.L. (1967) Recherches sur la végétation et la flore de la région du Bas-Cavally (Côte d'Ivoire). Mém ORSTOM 20. ORSTOM, Paris. x + 247 p

Haffer, J. (1969) Speciation in Amazonian forest birds. Science 165: 131–137

Hall, J.B. and Swaine, M.D. (1981) Distribution and ecology of vascular plants in a tropical rain forest. Dr W. Junk, The Hague. xiv + 382 p.

Hallé, F. & R.A.A. Oldeman (1970) Essai sur l'architecture et la dynamique de croissance des arbres tropicaux. Masson, Paris. English translation An essay on the architecture and dynamics of growth of tropical trees, by B.C. Stone, Penerbit Universiti Malaya, Kuala Lumpur (1975) xxv + 156 p, 75 fig + tab

Hallé, F., Oldeman R.A.A. & Tomlinson, P.B. (1978) Tropical trees and forests. An architectural analysis. Springer, Berlin Heidelberg New York. xvii + 441 p, 111 fig

Hamilton (1976) (Author's reference: untraced)
Hanbury-Tenison, R. (1980) Mulu. the rain forest. Weidenfeld & Nicolson, London. xi + 176 p, 2 maps + 45 phot
Harms, H. (1935) Rafflesiaceae. In: Engler & Prantl (eds) Die natürlichen Pflanzenfamilien, 2nd edn. Engelmann, Leipzig. 1 6b: 243–281, fig 124–147
Haverschmidt, F. (1968) Birds of Surinam. Oliver & Boyd, London. 445 p
Hedberg, I. & O., (eds) (1968) Conservation of vegetation in Africa South of the Sahara. Acta Phytogeogr. Suecica 54 xi + 320 p
Hedegart, T. (1976) Breeding systems, variation and genetic improvement of teak (Tectona grandis L.f.). In: Burley & Styles (eds) Tropical trees. Academic Press, London. pp 109–123, 3 fig
Heijnsbergen, P. van (1977) Declarations of the rights of animal and plant life. Environ Policy Law 3: 85–86
Henderson, M.R. (1954) Malayan Wild Flowers – Monocotyledons. Malay Nat Soc, Kuala Lumpur. 357 p
Hershkovitz, P. (1969) The evolution of mammals on southern continents, VI. The recent mammals of the neotropical region: a zoogeographic and ecological review. Quart Rev Biol 44: 1–70
Hesmer, H. (1975) Leben und Werk von Dietrich Brandis, 1824–1907. Westdeutscher Verlag, Opladen, BRD. xxiii + 476 p
Heyne, K. (1950) De nuttige planten van Indonesië, 3rd edn. van Hoeve, 's-Gravenhage Bandung. Slightly revised from the 1927 edn. 1660 + ccxli p
Hladik, A. (1974) Importance des lianes dans la production foliaire de la forêt équatoriale du Nord-Est du Gabon. CR Acad Sci Paris 278D: 2527–2530
Hladik, C.M. and Hladik, A. (1972) Disponibilités alimentaires et domaines vitaux des primates à Ceylan. Terre Vie (1972): 149–215
Hladik, A. & C.M. (1980) Utilisation d'un ballon captif pour l'étude du couvert végétal en forêt dense humide. Adansonia 19: 325–336, 4 fig
Holdridge, L.R. et al. (1971) Forest environments in tropical life zones, a pilot study. Pergamon Press, Oxford. xxxi + 747 p, 314 tab, 498 fig + 8 col phot + 3 maps
Holloway, H.L.O. (1977) Seed propagation of *Dioscoreophyllum cumminsii*, source of an intense natural sweetener. Econ Bot 31: 47–50, 3 fig
Holttum, R.E. (1953) A Revised Flora of Malaya, vol I. Orchids of Malaya. Government Printing Office, Singapore. v + 753 p
Holttum, R.E. (1954) Plant life in Malaya. Longman, London. viii + 254 p, 51 fig
Holttum, R.E. (1955) Growth-habits of monocotyledons – variations on a theme. Phytomorphology 5: 399–413, 18 fig
Hopkins, B. (1965) Forest and savanna. An introduction to tropical plant ecology with special reference to west-Africa. Heinemann, Ibadan, London. xii + 100 p, 31 fig + 5 pl.
Hueck, K. (1966) Die Wälder Südamerikas. Oekologie, Zusammensetzung und wirtschaftliche Bedeutung. Fischer, Stuttgart. xix + 422 p, 253 fig
Hueck, K. & P. Seibert (1972) Vegetationskarte von Südamerika/Mapa de la vegetación de America del Sur. Fischer, Stuttgart. 71 p, + map
Hutchinson, J. & J.M. Dalziel. Flora of West Tropical Africa. ed. 1: 1927–36, ed. 2: 1954–1972. London.
Irvine, F.R. (1961) Woody plants of Ghana. Oxford University Press, London. xcv + 868 p
IUCN (1975) The use of ecological guidelines for development in the American humid tropics, 249 p. IUCN Publications 31. Guidelines, by Duncan Poore, on pp 225–247
IUCN (1980) World Conservation Strategy. Living resource conservation for sustainable development IUCN, Gland. 4 items in folder, unpaged, incl. 5 maps
IUCN (1981) (Author's reference: untraced)
IUCN (1983) Ecological structures and problems of Amazonia. Comm Ecol Papers 5. 79 p
Jacobs, M. (1962) *Pometia* (Sapindaceae), a study in variability. Reinwardtia 6: 109–144
Jacobs, M. (1972) The plant world on Luzon's highest mountains. Rijksherbarium, Leiden. 32 p + 16 phot + 1fig + 1 map
Jacobs, M. (1974) Botanical Panorama of the Malesian Archipelago (vascular plants). In: UNESCO, Natural resources of humid tropical Asia. UNESCO, Paris. pp 263–294, 10 fig
Jacobs, M. (1976a) Het tropisch regenbos en de mens (Tropical Rain Forest and Man) Natura 73:2–20. Reprinted from Intermediair, 14 February 1975. 6 fig
Jacobs, M. (1976b) The study of lianas. Fl Males Bull (29) 2610–2618
Jacobs, M. (1978a) Bedreigde plantensoorten, universiteiten en botanische tuinen in ontwikkelingslanden (Endangered species, universities and botanical gardens in developing countries) Natura 75: 45–54, 3 fig
Jacobs, M. (1978b) Botanie en natuurbehoud in Indonesië (Botany and nature conservation in Indonesia) Panda 14: 137–142, illus
Jacobs, M. (1978c) Significance of the tropical rain forests on 12 points. Papers 8th World Forestry Congress, Jakarta. FQL/26-12, 19 p
Jacobs, M. (1979) A plea for S.E. Asia's forests. Habitat (Australia) 7 (4): 8–12, 31
Jacobs, M. (1982) The study of minor forest products. Fl Mal Bull 35: 3768–3782

Janzen, D.H. (1967) Synchronization of sexual reproduction of trees within the dry season in Central America. Evolution 21 (3): 620–637
Janzen, D.H. (1970) Herbivores and the number of tree species in tropical forests. Am Natur 104: 501–528
Janzen, D.H. (1973) Comments on host-specificity of tropical herbivores and its relevance to species richness. In: Heywood (ed) Taxonomy and ecology. The Systematics Association, London. pp 201–211
Janzen, D.H. (1974) Tropical blackwater rivers, animals, and mast fruiting by the Dipterocarpaceae. Biotropica 6: 69–103, 3 fig
Janzen, D.H. (1975) Ecology of plants in the tropics. Arnold, London. vi + 66 p, 4 pl
Janzen, D.H. (1976) Why tropical trees have rotten cores. Biotropica 8: 110
Jarrett, F.M. (1958, 1960) Studies in *Artocarpus* and allied genera III, IV. J Arn Arb 40: 115–368, 41: 73–140, fig 11–19
Jeník, J. (1978) Roots and root systems in tropical trees: morphologic and ecologic aspects. In: Tomlinson & Zimmermann (eds) Tropical trees as living systems. Cambridge University, Cambridge. pp 323–349, 5 fig
Jochems, S.C.J. (1930) De plantengroei van de omgeving der warmwaterbronnen van den boven-Petani op Sumatra's Oostkust (Vegetation around hot springs in the upper Petani region). Trop Natuur 19: 25–31
Johansson, D. (1974) Ecology of vascular epiphytes in west African rain forest. Acta phytogeogr Suec 59: 1–29, 119 fig
Kahn, F. (1983) Architecture comparée de forêts tropicales humides et dynamique de la rhizosphere. D. Sc. thesis, University of Montpellier. 426 p
Kartawinata, K. (1974) Report on the state of knowledge on tropical rain forest ecosystems in Indonesia. Herbarium Bogoriense, Bogor. 85 p
Kaur, A., et al. (1978) Apomixis may be widespread among trees of the climax rain forest. Nature (Land) 271: 440–442
Keay (1959) Vegetation map of Africa, South of the Sahara. UNESCO/AETFAT.
Keng, H. (1974) How many vascular plants are there in W. Malaysia and Singapore? Malay Nat J 28: 26–30
Kiew, R. (1978) Floristic components of the ground flora of a tropical lowland rain forest at Gunung Mulu National Park, Sarawak. Pertanika 1: 112–119, 2 fig
Kiew, R. (1982) Observations on leaf color, epiphyll cover, and damage on Malayan Iguanura wallichiana. Principes 26: 200–204
King, K.F.S. (1979) Concepts of Agroforestry. Int. Council for Research in Agroforestry, Nairobi.
Kira, T. (1978) Community architecture ... Pasoh Forest ... In: Tomlinson & Zimmermann (eds) Tropical trees as living systems. Cambridge University, Cambridge. pp 561–590, 13 fig
Kochummen, K.M. (3rd ed. by J. Wyatt-Smith) (1979) Pocket check list of timber trees. Forest Dept HQ, Kuala Lumpur. viii + 362 p
Koning, J. de (1983) La forêt du Banco. I. La Forêt. Med Landb hogesch Wageningen, 83 1. 156 p
Kostermans, A.J.G.H. (1973) A forgotten Ceylonese cinnamon-tree (*Cinnamomum capparu-coronde* Bl.) Ceylon J Sci Biol 10: 119–121
Kostermans, A.J.G.H. (1980) *Stemonoporus* Thw. (Dipterocarpaceae): a monograph. 2 (1981). Bull Mus nat Hist Nat Paris, sér 4, sect B, Adansonia 4 (3) 373–405
Kramer, F. (1933) De natuurlijke verjonging in het Goenoeng Gedeh complex (Natural rejuvenation in the Gunung Gedeh complex) Tectona 26: 155–185 + 3 fig
Küchler, A.W. (1967) Vegetation mapping. The Ronald Press Company, New York. vi + 472 p
Küchler, A.W. (ed) (1970) International bibliography of vegetation maps. 1: North America, 1965; 2: Europe, 1966; 3: USSR, Asia, Australia, 1968; 4: Africa, S. America, World maps, 1970. University of Kansas Libraries, Lawrence Kansas. 453; 584; 389; 561 p
Lam, H.J. (1927) Fragmenta Papuana III. Indrukken uit het Neder-Mamberamogebiet (Impressions from the Lower Mamberamo region) Natuurk Tijds Ned Ind. 87: 139–180
Lam, H.J. (1945) Fragmenta Papuana (Observations of a naturalist in Netherlands New Guinea). Sargentia V: 196 p
Lam, H.J. (1962) De nieuwe morfologie en haar invloed op onze inzichten in het systeem der kormofyten (The new morphology and her effect on our views concerning the system of cormophytes) Biologie in de XXe eeuw 6: 15–42
Lanly, J.P. & J. Clement (1979) Present and future forest and plantation areas in the tropics. FAO, Rome. Printed version in Unasylva 123: 12–20 (1979) 1 + 47 p
Lathrap, D.W. (1975) Ancient Ecuador, culture, clay and creativity 3000–300 BC. Field Mus Nat Hist, Chicago. 110 p
Laufs, R. & H. Steinke (1975) Vaccination of non-human primates against malignant lymphoma. Nature (Land) 253: 71–72
Lawrence, G.H.M. (1951) Taxonomy of vascular plants. MacMillan, New York. xiii + 823 p, 322 fig
Lee, D.W. (1977) On iridescent plants. Gard Bull Sing 30: 21–29
Leenhouts, P.W. (1958) Connaraceae. Fl Mal 5: 495–541, 15 fig
Leigh Jr., E.G., A.S. Rand & D. M. Windsor (eds) (1983) The ecology of a tropical forest: seasonal rhythms and long-term changes. Oxford University Press. 468 p
Leighton, M. & B. Thomas (1980) A canopy observation platform in East Kalimantan, Indonesia. Fl Males Bull 33: 3432–3434

Leslie, A. (1977) Where contradictory theory and practice co-exist. Unasylva 29: 2–17
Lindeman, J.C. and Mennega, A.M.W. (1963) Bomenboek voor Suriname (The trees of Surinam). Dienst 's lands bosbeheer Suriname, Paramaribo/Kemink, Utrecht. 312 p
Little, E.L. et al. (1964, 1974) Common trees of Puerto Rico and the Virgin Islands. U.S. Dept of Agric, Washington, 2 vols
Little, E.L. Dixon (1969) Arboles comúnes de la provincia de Esmeraldas. FAO (Complete reference untraced)
Löffler, E. (1977) Geomorphology of Papua New Guinea. ANU Press, Canberra. xvii + 195 p, 49 fig, 97 phot
Loh, C.L. (1975) Fruits in Peninsular Malaysia. In: Williams (ed) South East Asian plant genetic resources. Biotrop, Bogor. pp 47–52
Longman, K.A. & J. Jenık (1974) Tropical forest and its environment. Longman, London. x + 196 p, figs, 28 pl
Lowe-McConnell, R.H. (1969) Speciation in tropical freshwater fishes. Biol J Linn Soc 1: 51–75, 2 fig
Mabberley, D.J. (1977) The origin of the Afroalpine pachycaul flora and its implications. Gard Bull 29: 41–55, 1 fig
Martin, P.J. (1977) The altitudinal zonation of forests along the West Ridge of Gunong Mulu. Forestry Department, Kuching. iv + 77 p + 35 fig + 18 append + 23 tab
Martius, C.F.P. de, et al. (1840–1869) Tabulae physiognomicae. Flora Brasiliensis vol. 1 pars 1, cx p + 59 pl
Martius, C.F.P. von, A.G. Eichler, I. Urban (1906) Flora Brasiliensis. (Reprint 1964–1965). Stechert-Hafner, New York. 15 vol
Maxwell, N. (1975) Witch doctor's apprentice. (Reprint 1960). Macmillan Publ Cy, New York. 406 p
McClure, F.A. (1966) The bamboos. A fresh perspective. Harvard University Press, Cambridge Mass. xv + 347 p, 99 fig
McClure, H.E. (1966) Flowering, fruiting and animals in the canopy of a tropical rain forest. Malay For 29: 182–203, 28 fig
McCreagh, G. (1961) White waters and black. (repr from 1954). Anchor Books N5; Doubleday & Cy, Garden City, New York. 335 p
McIntyre, L. (1980) Jari: a massive technology transplant takes root in the Amazon jungle. Nat Geogr Mag 157: 686–711
Medway, Lord (1972) Phenology of a tropical rain forest in Malaya. Biol J Linn Soc 4: 117–146, 2 fig
Medway, Lord (1978) The wild mammals of Malaya (Peninsular Malaysia) and Singapore, 2nd edn. Oxford University Press. xxii + 128 p, 9 fig + 15 pl
Medway, Lord & D.R. Wells. (1971) Diversity and density of birds and mammals at Kuala Lompat, Pahang. Malay Nat J 24: 238-247
Meggers, B.J. et al. (ed) (1973) Tropical forest ecosystems in Africa and South America: A comparative review, Smithsonian Institute, Washington. viii + 350 p
Meijer, W. (1959) Plantsociological analysis of montane rainforest near Tjibodas, West Java. Acta Bot Neerl 8: 277–291, 3 fig
Meijer, W. (1974) Field guide to trees of West Malesia. Private edition available from University of Kentucky Book Store.
Meijer, W. (1975) Indonesian forests and land use planning. University of Kentucky, Lexington. 112 p, illus
Merrill, E.D. (1912) A Flora of Manila. Bureau of Printing, Manila. 430 p
Meyenfeldt, C.F.W.M. et al. (1978) Restoration of devastated inland forests in South Vietnam. 3 vol. Bosteelt, Agric University, Wageningen. Mimeo. c. 440 p
Mildbraed, J. (1922) Wissenschaftliche Ergebnisse der Zweiten deutschen Zentral-Afrika Expedition 1910–1911. Bd 2. Klinkhardt & Biermann, Leipzig. 202 p
Miracle, M.P. (1973) The Congo Basin as a habitat for man. In: Meggrs (ed) Tropical forest ecosystems. Smithsonian Institute, Washington. pp 335-344, 5 fig
Mitchell, A.W. (1981) Operation Drake-Voyage of discovery. Severn House, London, pp 68–69
Mitchell, A.W. (1982) Reaching the rain forest roof. A handbook on techniques of access and study in the canopy. Leeds Philosophical and Literary Society, Leeds. 25 p
Mohr. E.C.J. (1946) Bodem (Soils) In: C.J.J. van Hall & C. van de Koppel (eds) De landbouw in den Indische Archipel 1: 9–62 Van Hoeve, 's-Gravenhage
Mohr, E.C.J. & F.A. van Baren (1954) Tropical soils. A critical study of soil genesis as related to climate, rock and vegetation. Manteau; Bruxelles. 498 p, 103 fig + 4 pl
Molnar, P. & P. Tapponnier (1977) The collision between India and Eurasia. Sci Am 236: 30–41
Montgomery, G. G. & M.E. Sunquist, (1978) Habitat selection and use by two-toed and three-toed Sloths. In: Montgomery G.G. (ed) The ecology of arboreal folivores. Smithsonian Inst Press, Washington D.C. pp 329–359
Moore, H.E. (1973) Palms in the tropical forest ecosystems of Africa and South America. In: Meggers (ed) Tropical forest ecosystems. Smithsonian Institute, Washington. pp 63–88, 6 fig
Mörzer Bruyns, M.F. (1974) Natuurbeheer in de tropen (Nature management in the tropics) Landbouwhogeschool, Wageningen. 205 p
Most, B.H. et al. (1978) Thaumatococcus daniellii. Econ Bot 32: 321–335, 1 fig, 2 pl

Muller, J. (1970) Palynological evidence on early differentiation of angiosperms. Biol Rev 45: 417–450, 6 fig
Myers, N. (1979) The sinking ark. A new look at the problem of disappearing species. Pergamon Press, Oxford. xiii + 307 p
Myers, N. (1980) Conversion of tropical moist forests. Natl Acad Sci, Washington. ix + 205 p
Myers, N. (1984) The primary source, tropical forests and our future. Norton, New York London. xiv + 399 p
Myrdal, G. (1968) Asian drama, an enquiry into the poverty of nations. Pantheon, New York. 3 vol, 2284 p
Nix, H.A. & J.D. Kalma (1972) Climate as a dominant control in the biogeography of northern Australia and New Guinea. In: Walker, D. (ed) Bridge and barrier. ANU Press, Canberra. pp 61–91, 12 fig
NRDC, Natural Resources Defense Council (1978) The tropical moist forest bulletin 1. National Resources Defense Council, Washington
Ochse, J.J. & R.C. Bakhuizen van den Brink (1931) Vegetables of the Dutch East Indies. Archipel Drukkerij, Buitenzorg. 2nd edn. (1977) Asher, Amsterdam. xxxvi + 1005 p
Odum, H.T. (ed) (1970) A tropical rain forest/ A study of the irradiation and ecology at El Verde. U.S. Atomic Energy Commission, Puerto Rico. 3 vol
Oldeman, R.A.A. (1974) L'architecture de la forêt guyanaise. Mémoires ORSTOM 73. ORSTOM, Paris. 204 p, 113 fig
Oldeman, R.A.A. (1978) Architecture and energy exchange of dicotyledonous trees in the forest. In: Tomlinson & Zimmermann (eds) Tropical trees as living systems. Cambridge University, Cambridge. pp 535–560, 6 fig
Oldeman, R.A.A. (1979) Scale-drawing and architectural analysis of vegetations. Bosteelt, Wageningen. Syllabus, mimeographed, 42 p
Oldeman R.A.A. (ed) (1982) Tropical hardwood utilization: practice and prospects. Martinus Nijhoff Junk Publ, The Hague. 584 p
Oldeman, R.A.A. (1983) Tropical rain forest, architecture, silvigenesis and diversity. In: Sutton, S.L. et al. (eds) Tropical rain forest: ecology and management. Blackwell Science Publ, Oxford. pp 139–150
Packard, V. (1961) The waste makers. Longmans, London. xii + 340 p
Paijmans, K. (ed) (1976) New Guinea vegetation. ANU Press, Canberra. xvii + 213 p, some fig, 53 phot
Pape, R. (ed) (1973) New Horizons. Forestry in Papua New Guinea. Publ for the Minister for Forest by Jacaranda Press, Port Moresby. vi + 70 p
Pasoh (1978; this anonymous issue has thus been indicated for the sake of convenience) The Malaysian International Biological Program synthesis meetings: selected papers from the symposium at Kuala Lumpur (1974) IBP-PT. Malay Nat J 30(2): 119–447
Pennington, T.D. and Sarukhan, J. (1968) Arboles tropicales de México. Instituto Nacional de Investigaciones Forestales, México & FAO, Rome. vii + 413 p
Perdue, R.E. & J.L. Hartwell (ed) (1976) Plants and cancer. Cancer Treat Rep 60: 973–1215. See Fl Males Bull (31) 3079–3080 (1978)
Persson, R. (1974) World Forest Resources: Review of the world's forest resources in the early 1970s. Royal College of Forestry, Stockholm.
Philipson, W.R. (1978) Araliaceae: growth forms and shoot morphology. In: Tomlinson & Zimmermann (eds) Tropical trees as living systems. Cambridge University, Cambridge. pp 268–284, 3 fig
Pijl, L. van der (1960, 1961) Ecological aspects of flower evolution. Evolution 14: 403–416, 15: 44–59
Pijl, L. van der (1966) Ecological aspects of fruit evolution. Kon Ned Akad Wet Amst Proc C69; 597–640
Pijl, L. van der (1972) Principles of dispersal in higher plants, 2nd edn. Springer, Berlin Heidelberg New York. xi + 161 p, 26 fig
Plumwood, V., Routley, R. (1982) World rainforest destruction – the social factors. Ecologist 12: 4–22
Poore, M.E.D. (1968) Studies in Malaysian rain forest. I. The forest on Triassic sediments in Jengka Forest Reserve. J Ecol 56: 143–196, 19 fig + 5 pl
Poore, D. (1976a) The values of tropical moist forest ecosystems/ and the environmental consequences of their removal. Unasylva 112–113: 127–143, 145–146
Poore, D. (1976b) Ecological guidelines for development in tropical rain forests. IUCN, Morges. viii + 39 p, 6 illus
Powell, J.M. (1976) Ethnobotany. In: Paijmans (ed) New Guinea vegetation. Anu Press, Canberra. pp 106–199, 1 fig, pl 44–53
Prance, G.T. (1973) Phytogeographic support for the theory of Pleistocene forest refuges in the Amazon Basin, based on evidence from distribution patterns in Caryocaraceae, Chrysobalanaceae, Dichapetalaceae and Lecithydaceae. Acta Amazonia 3 (3): 5–28
Prance, G.T. (1977a) The phytogeographic subdivisions of Amazonia and their influence on the selection of biological reserves. In: Prance & Elias (eds) Extinction is forever. New York Botanical Garden, New York. pp 195–213, 13 fig
Prance, G.T. (1977b) Floristic Inventory of the tropics: where do we stand? In: Perspectives in Tropical Botany, proc of a symposium held Aug 1977, Michigan. Ann Mo Bot Garden 64: 659–684
Prance, G.T. & T.S. Elias (eds) (1977) Extinction is forever. Threatened and endangered species of plants in the Americas and their significance in ecosos today and in the future. New York Botanical Garden. vi + 437 p, illus

Purseglove, J.W. (1968) Tropical crops/ Dicotyledons. Longman; London. xv + 719 p, 102 fig
Purseglove, J.W. (1972) Tropical crops/ Monocotyledons. Longman London. x + 607 p, 32 fig
Putz, F.E. (1978) A survey of virgin jungle reserves in Peninsular Malaysia. FRI Research Pamphlet 73. Forestry Department; Kuala Lumpur. iv + 87 p, maps
Putz, F.E. (1980) Lianas vs. Trees. Biotropica 12: 224–225
Rappard, F.W. (1937) De damar van Benkoelen (in Dutch, summary in English) The damar of Benculu. Tectona 30: 897–916
Reinders, E. (ed) (1949) Leerboek der algemeene plantkunde (Handbook of general botany) 2de druk. Scheltema & Holkema, Amsterdam. deel 1, xvi + 597 p, 610 fig
Républica de Colombia. (1979a) La Amazonia Colombiana y sus recursos. Proyecto Radargramétrico del Amazonas, Bogota. 6 vol
Républica de Colombia. (1979b) Principales plantas utiles de la Amazonia Colombiana. Proyecto Radargramétrico del Amazonas, Bogotá. 263 p
Richards, P.W. (1952) The tropical rain forest. An ecological study. Cambridge University Press. xviii + 450 p, 43 fig + 15 pl
Richards, P.W. (1970) The life of the jungle. McGraw-Hill, New York. 232 p, many illus
Richards, P.W. (1973a) The tropical rain forest. Sci Am 229: 58–67 illus
Richards, P.W. (1973b) Africa, the 'odd man out'. In: Meggers (ed) Tropical forest ecosystems. Smithsonian Institute, Washington. pp 21–26
Rijksen, H.D. (1978) A field study on Sumatran orang utans (Pongo pygmaeus abelii Lesson 1827). Ecology, behaviour and conservation, Meded Landbouwhogesch, Wageningen 78–2. 421 p, 161 fig
Rizzini. C.T. (1978) The discovery of Sebastiào-de-Arruda, a fine Brazilian wood that was botanically unknown. Econ Bot 32: 51–58, 6 fig
Robbins, R.G. & J. Wyatt-Smith (1964) Dryland forest formations and forest types in the Malayan Peninsula. Malay For 27: 188–216, 12 fig
Rollet, B. (1978) Organization. In: UNESCO, Tropical forest ecosystems, pp 112–142
Rollet, B. (1980) Intérêt de l'étude des écorces dans la détermination des arbres tropicaux sur pied. Rev Bois Forêts Trop 194: 3–28
Routley, R. & V. (1980) Human chauvinism and environmental ethics. In: D. Mannison et al. (eds) Environmental philosophy. ANU Press, Canberra. pp. 96–189
Ruinen, J. (1953) Epiphytosis. A second view on epiphytism. Ann Bogor 1: 101–158 + pl 5–13
Ruinen, J. (1961) The phyllosphere, I. An ecologically neglected milieu. Plant Soil 15: 81–109, 3 fig + 9 pl
Ruinen, J. (1963) The phyllosphere, II. Yeasts from the phyllosphere of tropical foliage. Antonie Leeuwenhoek 29: 425–438
Ruinen, J. (1965) The phyllosphere, III. Nitrogen fixation in the phyllosphere. Plant Soil 22: 375–394, 2 fig
Salati, E., H.O.R. Shubart, W. Junk & A.E. de Oliveira (1983) Amazônia; desenvolvimiento, integracão, ecolôgia. São Paulo: Brasiliense; (Brasilia) CPNq. 327 p
Sandved, K.B. and Emsley, M. (1979) Rain forests and cloud forests. Abrams, New York. 264 p
Schenck, H. (1892, 1893) Beiträge zur Biologie und Anatomie der Lianen/ im Besonderen der in Brasilien einheimischen Arten, 2 vol. Fischer, Jena
Schimper, A.F.W. (1898) Pflanzengeographie auf physiologischer Grundlage. Fischer, Jena. xviii + 876 p, 502 fig + 4 maps
Schimper, A.F.W. (1964) Plant-geography on a physiological basis (Engl edn, Reprint 1903). J. Cramer, Wernheim. xxx + 839 p
Schmid, M. (1975) Végétation du Viet-Nam. Le massif sud-annamitique et les régions limitrophes. Mémoires ORSTOM 74. ORSTOM, Paris. 244 p, 15 fig + 16 pl
Schmithüsen, J. (1976) Atlas zur Biogeographie. Bibliographisches Institut Zürich. 80 p incl maps
Schnell, R. (1976) Introduction à la phytogéographie des pays tropicaux. Vol 3. La flore et la végétation de l'Afrique tropicale. Gauthier-Villars, Paris. x + 458 p
Schnell, R. (1977) Introduction à la phytogéographie des pays tropicaux. Vol 4. La flore et la végétation de l'Afrique tropicale. Gauthier-Villars, Paris 369 p
Schodde, R. (1973) General problems of fauna conservation in relation to the conservation of vegetation in New Guinea. In: Costin & Groves (eds) Nature conservation in the Pacific. ANU Press, Canberra. pp 123–144, fig 13
Schulz, J.P. (1960) Ecological studies on rain forest in northern Surinam. North Holland, Amsterdam. 267 p, 70 fig
Schumacher, E.F. (1973) Small is beautiful. Harper Torchbooks, New York (edn 1974), Abacus, London (edn 1975)
Shanahan, E.W. (1927) South America, an economic and regional geography with an historical chapter. Methuen, London. 318 p
Shaw, H.K. Airy (1975) The Euphorbiaceae of Borneo. Kew Bull addit ser 4, 245 p, 1 map
Sicco Smit, G. (1978) Shifting cultivation in tropical rainforest detected from aerial photographs. ITC Journal 1978–4: 603-632, 11 phot

Sleumer, H. (1966) Ericaceae. Fl Mal 6: 469–914. 126 fig
Smith, A. (1776) An inquiry into the wealth of nations. London
Soepadmo, E. & B.K. Eow (1977) The reproductive biology of *Durio zibethinus* Murr. Gard Bull 29: 25–33 + 6 pl
Sommer (1976) (Author's reference: untraced)
Sparks, B.W. (1972) Geomorphology, 2nd edn. Longman, London. xxi + 530 p, many illus
Special Environment Agency. (1977) Program of Ecological Stations. Ministry of the Interior, Brasilia, Brazil. 39 p
Spruce, R. (1908) Notes of a botanist on the Amazon & Andes, (ed) A.R. Wallace, 2 vol. McMillan, London
Stamp, L.D. (1925) The vegetation of Burma from an ecological standpoint. Thacker, Spink, Calcutta. vi + 65 p, 12 fig, map + 28 pl
Start, A.N. & A.G. Marshall (1976) Nectarivorous bats as pollinators of trees in West Malaysia. In: Burley & Styles (eds) Tropical trees. Academic Press, London. pp. 141–150
Stearn, W.T. (1958) Botanical exploration to the time of Linnaeus. Proc Linn Soc Lond 169: 173–196, 5 fig
Stebbins, G.L. (1950) Variation and evolution in plants, Columbia Biological Series. xx + 643 p, 55 fig
Steenis, C.G.G.J. van (1948) Introduction. Fl Mal i 4: i–xii, 1 fig
Steenis, C.G.G.J. van (1949) Notes on the genus *Wightia.* Bull Bot Gard Buit iii 18: 213–277, 2 fig
Steenis, C.G.G.J. van (1950) The delimitation of Malaysia and its main plant geographical subdivisions. Fl Mal i 1: lxx–lxxv, fig 20–26
Steenis, C.G.G.J. van (1954) Vegetatie en flora (Vegetation and flora) In: Klein, W.C. (ed) Nieuw Guinea, 2nd edn Staatsdrukkerij, 's-Gravenhage. 2: 218–275, 17 fig
Steenis, C.G.G.J. van (1957) Specific and infraspecific delimitation. Fl Mal i 5: clxvii–ccxxix, 7 fig
Steenis, C.G.G.J. van (1958) Introductory matter on ecology of mangroves. Fl Mal i 5: 431–444, illus
Steenis, C.G.G.J. van (1965) Brief survey of vegetation types. In: Backer & Bakhuizen van den Brink (eds) Flora of Java 2: 7–35, Noordhoff/Wolters, Groningen. 32 fig
Steenis, C.G.G.J. van (1969) Plant speciation in Malesia, with special reference to the theory of non-adaptive saltatory evolution. Biol J Linn Soc 1: 97–133, 1 fig
Steenis, C.G.G.J. van (1972) The mountain flora of Java. Brill, Leiden. x + 90 p, 26 fig + 20 pl + 57 col pl facing legend pages
Steenis, C.G.G.J. van (1976, actually 1977) Autonomous evolution in plants. Differences in plant and animal evolution. Gard Bull 29: 103–126, 3 fig
Steenis, C.G.G.J. van (1981) Rheophytes of the world. An account of the flood-resistant flowering plants and ferns and the theory of autonomous evolution. Sijthoff & Noordhoff, Alphen a/d Rijn. xv + 407 p, 47 fig + 23 phot
Steenis-Kruseman, M.J. van. (1950) Malaysian plant collectors and collections. Fl Mal i, 1: clii + 639 p Suppl: Fl Mal 1, 5: ccxxxv–cccxlii (1958); 1, 8: vii–cxv (1974)
Stone, B.C. (1966) *Pandanus Stickm.* in the Malayan peninsula, Singapore and Lower Thailand. Malay Nat J 19: 291–301
Summerfield, R.J. et al. (1977) *Dioscoreophyllum cumminsii.* Econ Bot 31: 331–339
Temple, S.A. (1977) Plant-animal mutualism: coevolution with Dodo leads to near extinction of plant. Science 197: 885–886
Thomas, W.L. (ed) (1956) Man's role in changing the face of the earth. University of Chicago Press. xxxviii + 1193 p, 180 fig
Tomlinson, P.B. & A.M. Gill (1973) Growth habits of tropical trees: some guiding principles. In: Meggers (ed) Tropical forest ecosystems. Smithsonian Institute, Washington. pp 129–143, 17 fig
Tomlinson, P.B. & M.H. Zimmermann (eds) (1978) Tropical trees as living systems. Cambridge University Press. xviii + 675 p, illus
Tosi, J.A., Jr. (1975) Some relationships of climate to economic development in the tropics. IUCN Publications 31: 41–58, 1 fig
Tran, V.N. (1974) Agrisilviculture: joint production of food and wood. Papers 8th World Forestry Congress, Jakarta 1978. FFF/7–0. 16 p
Troll, W. (1937, 1939, 1942) Vergleichende Morphologie der höheren Pflanzen. Borntraeger, Berlin. pp 815–859, 1920–1983, 2465–2500
Tropical Deforestation (1978) Proceedings of the U.S. Strategy Conference on tropical deforestation , June 12–14, 1978, Washington D.C. Department of State, Room 7820, Washington. 78 p, 4 fig
Troup, R.S. (1921) The silviculture of Indian trees, 3 vol. Clarendon Press, Oxford
UNEP (1980) Overview document/ Experts Meeting on tropical forests, Libreville, 25 February–1 March 1980. UNEP, Nairobi. Mimeographed, UNEP/WG.35/4. Compiled by Norman Myers. 70p
UNESCO (1974) Natural resources of humid tropical Asia. UNESCO, Paris. 456 p, some fig
UNESCO (1978) Tropical forest ecosystems. A state of knowledge report. 683 p
UNESCO (1981) Carte de la végêfation d'Amérique du Sud, Notice explicative. UNESCO, Paris. 189 p
USAID (1980) Forestry activities and deforestation problems in developing countries. Office of Science and technology, A.I.D., Washington D.C. 115 + 63 + 16 p
Vanzolini, P.E. (1970) Zoologia sistemática, geografia e a origem das espécies. Instituto de Geografica Universidad de Sao Paulo, Serie Teses e Monografias 3, Sao Paulo

Vanzolini, P.E. (1973) Paleoclimates, relief and species multiplication in equatorial forests. In: Meggers, Ayensu, Duckworth (eds) Tropical Forest ecosystems in Africa and South America: a comparative review. Smithsonian Institute, Washington

Vareschi, V. (1980) Vegetationsökologie der Tropen. Ulmer, Stuttgart. 293 p

Veevers-Carter, W. (1978). Nature Conservation in Indonesia. Intermasa, Jakarta. 86 p, 11 maps, 19 fig

Veevers-Carter, W. (1984). Riches of the Rain Forest. An introduction to the trees and fruits of the Indonesian and Malaysian rain forests. Oxford University Press. Singapore Oxford New York. 103, XI p, 34 fig

Verdoorn, F. (ed) (1945) Plants and plant science in Latin America. Chronica Botanica, Waltham. xxxvii + 384 p, 45 fig + 38 pl

Verstappen, H. Th. (1956) The physiographic basis of pioneer settlement in southern Sumatra. (Indonesian/English). Djawatan Topografi Angkatan Darat/Balai Geografi, publ 6. 25 p

Vogel, E.F. de (1980) Seedlings of dicotyledons. Structure, development types. Descriptions of 150 woody Malesian taxa, Pudoc, Wageningen. 465 p, 178 fig + 16 pl

Voorhoeve, A.G. (1965) Liberian high forest trees. Agric Res Rep, Wageningen 652. 416 p

Wassink, J.T. (1977) Agroforestry. Agriculture and forestry teaming up for the sake of humanity and its natural environment. Koninklijk Instituut voor de Tropen, Amsterdam. 19 p., illus

Wassink, J.T. (1979) Choice patterns for tropical timber, imported in 7 European countries. Koninklijk Instituut voor de Tropen, Amsterdam. Pamphlet. 33 p

Wealth of India (1948–1976) The Wealth of India. A dictionary of Indian raw materials and industrial products, 11 vol. Publications and Information Directorate, New Delhi

Webb, L.J., J.G. Tracey & W.T. Williams. (1976) The value of structural features in tropical forest typology. Austr J Ecol 1: 3–28

Went, F.W. (1940) Soziologie der Epiphyten eines tropischen Urwaldes. Ann Jard Bot Buit 50: 1–98 + 17 pl + 1 tab

White, F. (1962) Forest flora of Northern Rhodesia. Oxford University Press, London. xxvi + 455 p

White, F. (1977) The underground flora of Africa: a preliminary review. Gard Bull Sing 29: 57–71

White, F. (1983) The vegetation of Africa. A descriptive memoir to accompany the UNESCO/AETFAT/UNSO vegetation map of Africa. UNESCO, Paris. 356 p, 3 maps, legends

Whitmore, T.C. (1975) Tropical rain forests of the Far East. Clarendon Press, Oxford. xiii + 282 p, many illus

Whitmore, T.C. (1977) A first look at *Agathis*. Forestry, Oxford. ix + 54 p. + 12 pl

Whitmore, T.C. & C.P. Burnham (1969) The altitudinal sequence of forests and soils on granite near Kuala Lumpur. Malay Nat J 22: 99–118, 1 fig + pl 18–22

Whitmore, T.C. et al. (eds) (1972) Tree Flora of Malaya. Longman, London. 3 vols, 4th in prep

Wiebes, J.T. (1976) A short history of fig wasp research. Gard Bull 29: 207–232, 8 fig

Williams, J.T., Lamoureux, C.H., Wulijarni-Soetjipto N. (eds) (1975) South East Asian Plant Genetic Resources. Proceedings Symp. South East Asian Plant Genetic Resources Bogor, 20–22 March 1975. Biotrop, Bogor. 272 p

Wit, H.C.D. de (1963) De Wereld der Planten. Gaade, Delft

Wolterson, J.F. (1977) De vernieuwbare grondstof. Bosbouw, houtgebruik en verbruik toen, nu en straks (The renewable raw material. Forestry, use and consumption of wood-then, now and in the future) Staatsuitgeverij, Den Haag. 88 p, illus

Woodwell, G.M. et al. (1978) The biota and the world carbon budget. Science 199: 141–146, 3 fig

World Bank (1978) Forestry. A sector policy paper. World Bank, Washington. 65 p

World Resources Institute, World Bank & U.N. Development Programme (1985) Tropical Forests: a call for action. 3 parts (World Resources Institute)

Wyatt-Smith, J. (1965) Manual of Malayan silviculture of inland forests, parts i–iii. Malay For Rec 23

Zimmermann, W. (1930) Die Phylogenie der Pflanzen. Fischer, Jena. xii + 452 p

Zimmermann, W. (1959) Die Phylogenie der Pflanzen. Ein Ueberblick über Tatsachen und Probleme, 2nd edn. Fischer, Stuttgart. xxiv + 777 p, 331 fig

Name and Subject Index

Page numbers in *italics* refer to illustrations